A Structural Account of Mathematics

A Structural Account of Mathematics

CHARLES S. CHIHARA

CLARENDON PRESS · OXFORD

OXFORD

UNIVERSITY PRESS

Great Clarendon Street, Oxford OX2 6DP

Oxford University Press is a department of the University of Oxford.
It furthers the University's objective of excellence in research, scholarship,
and education by publishing worldwide in

Oxford New York

Auckland Bangkok Buenos Aires Cape Town Chennai
Dar es Salaam Delhi Hong Kong Istanbul Karachi Kolkata
Kuala Lumpur Madrid Melbourne Mexico City Mumbai Nairobi
São Paulo Shanghai Taipei Tokyo Toronto

Oxford is a registered trade mark of Oxford University Press
in the UK and in certain other countries

Published in the United States
by Oxford University Press Inc., New York

British Library Cataloguing in Publication Data
Data available

Library of Congress Cataloging in Publication Data
Data available

ISBN 0–19–926753–7

1 3 5 7 9 10 8 6 4 2

Typeset by Newgen Imaging Systems (P) Ltd., Chennai, India
Printed in Great Britain
on acid-free paper by
T. J. International Ltd., Padstow, Cornwall

Dedicated to

**My Brother
Paul**

whose music
soars around the world

with a logic all its own

Preface

This work develops and defends a structural view of the nature of mathematics, which is used to explain a number of striking features of mathematics that have puzzled philosophers for centuries. It rejects the most widely held philosophical view of mathematics (Platonism), according to which mathematics is a science dealing with mathematical objects such as sets and numbers—objects which are believed not to exist in the physical world. Instead, it makes use of the constructibility theory of my earlier work, *Constructibility and Mathematical Existence* (Oxford University Press, 1990), to develop a view of mathematics that is distinct from Structuralism and yet makes use of some key ideas of Structuralism.

The structural view is used to show that, in order to understand how mathematical systems are applied in science and everyday life, it is not necessary to assume that its theorems either presuppose mathematical objects or are even true. My previous work had a different aim: its goal was to present and develop a new system of mathematics that did not make reference to, or presuppose, mathematical objects. Both works support a nominalistic point of view. However, whereas the earlier book was aimed at creating a new nominalistic system of mathematics, the present work analyzes mathematical systems currently used by scientists to show how such systems are compatible with a nominalistic outlook. The present work also advances several new ways of undermining the heavily discussed indispensability argument for the existence of mathematical objects made famous by Willard Quine and Hilary Putnam.

I also endeavor, in this book, to present a rationale for the nominalistic outlook that is quite different from those generally put forward by nominalists. I do this, to a great extent, because I believe that serious misunderstandings of the nominalistic outlook have been fostered by the type of rationale for nominalism that is typically discussed in the recent philosophical literature.

A number of criticisms that have been leveled at my constructibility theory by the Structuralists. Since these criticisms have for many years been largely unanswered, they may appear to students and non-specialists to be unanswerable. In this work, the criticisms will be rebutted.

This work grew out of some graduate seminars in the philosophy of mathematics I gave at Berkeley in the late 1990s. I am grateful to the students and faculty members who participated in these seminars and were instrumental in getting this project off the ground. In addition some of the ideas of this work were tried out in lectures I gave at the following conferences: The Seventh Asian Logic Conference, held in Hsi-Tou, Taiwan, in June 1999 (a revised version of the paper delivered, "Five Puzzles About Mathematics in Search of Solutions", is to be published in the proceedings of the conference); "One Hundred Years of Russell's Paradox", the International Conference in Logic and Philosophy, held at the University of Munich, June 2001 (where I discussed Shapiro's objections to my earlier work); the symposium in the philosophy of mathematics, Pacific Division, American Philosophical Association, held in Seattle in March 2002 (where Mark Wilson responded to my paper on the van Inwagen puzzle); and the Hawaii International Conference on Arts and Humanities, held in Honolulu in January 2003 (where I discussed nominalism). Ideas from the present work found their way into philosophy lectures I gave at the following institutions: Massachusetts Institute of Technology, Cambridge, November 1999; Institut für Philosophie, Logik und Wissenschaftstheorie at the University of Munich in November 2000; University of Saarlandes in November 2000; and the Logic Colloquium of the University of California, Berkeley, October 2001. I am grateful to those who raised interesting objections or made helpful suggestions at these lectures (some of whom are mentioned later in footnotes).

A number of philosophers have aided the writing of this book. Some were kind enough to read and comment on parts of preliminary versions of this work; others have responded to my queries or requests for prepublications or references. I am especially grateful to Susan Vineberg for her careful study of Chapters 5 and 11 and for providing me with many very helpful objections to early versions of these chapters; Paul Teller and Guido Bacciagaluppi for their helpful comments on my discussion of the mathematics of quantum mechanics; Alan Code for his many useful insights and references pertaining to Greek mathematics and philosophy; Paolo Mancosu for his help in improving my discussions of the history of geometry; John MacFarlane for his careful reading and criticisms of an early version of the chapter on structuralism; Ellery Eells and Elliot Sober for their many helpful comments on the sections dealing with the holistic version of the indispensability argument; Geoffrey Hellman for his detailed comments on my early objections to his modal structuralism; Stewart Shapiro for his many responses to my queries about his version of structuralism; Richard Zach for his helpful replies to my queries about Hilbert; and Penelope Maddy for looking over the sections dealing with her criticisms of the indispensability argument.

Two mathematicians should also be thanked for their assistance: Theodore Chihara for his helpful comments on the section dealing with Fermat's Last Theorem and James T. Smith for providing me with a useful list of references pertaining to the Hilbert–Frege dispute. I also wish to thank two budding philosophers, Jonathan Kastin and Jukka Keränen, for allowing me to read prepublications of their papers on Shapiro's structuralism. Many thanks also to two unnamed readers for OUP for their many genuinely useful suggestions.

My Ph.D. student William Goodwin has ably served as my research assistant for this work, reading the whole of the manuscript, making useful suggestions and corrections, and constructing the index. For this, I am most grateful. Thanks also to Angela Blackburn, for her excellent editorial assistance.

As always, my beloved wife Carol has aided me in a variety of ways throughout the writing of this work, but she has been especially helpful by serving as my in-house specialist dealing with the many problems that arose involving the computer and also by serving as my consultant on all matters pertaining to biology and genetics.

Finally, I would also like to express my deep thanks to Drs Lolly Schiffmann and Paul Li of Kaiser Permanente for extending the time I have left for the kind of productive research needed to complete this work.

NOTATIONAL CONVENTIONS

The notational conventions I use in this work are those of my earlier works, Chihara, 1990 and Chihara, 1998. Briefly, double quotation marks are used for direct quotation and as scare quotes. Single quotation marks are used to refer to linguistic items such as words and symbols. Greek letters are used as meta-variables. The primitive symbols of an object-language discussed are frequently used autonymously. For additional explanations, with examples, of the conventions I use, see Chihara, 1998: pp. x–xi. Throughout this work, I use 'iff' as an abbreviation for 'if and only if'.

C. S. C.

Berkeley, California
June 2003

Contents

Introduction

I begin this work with a personal view of philosophy: a view that is set forth not as something to be argued for and defended with abundant quotations from philosophical journals and the writings of great philosophers, but as a kind of orientation piece, aimed at setting out much of the motivation for the philosophical developments to follow.

1. A NOMINALIST'S VIEW OF PHILOSOPHY: THE BIG PICTURE

The field of philosophy is divided into a number of specialties. Among these, there is philosophy of language, philosophy of mind, philosophy of science, philosophy of logic, philosophy of art, philosophy of history, philosophy of religion, and so on. For practically any area X of intellectual study, there is a philosophy of X. As a general rule, one can say that the philosophy of X is aimed at achieving a kind of understanding of X that is unique to philosophy. One might call this sort of understanding "Big Picture understanding". What one seeks in philosophy is the really "Big Picture": what, in general and in broad outlines, is the universe like? What, in general and in broad outlines, is our (i.e. humanity's) place in the universe? How, in general and in broad outlines, do we (humans) gain an understanding of the universe? And, more specifically, how, in general and in broad outlines, does X fit into this Big Picture?

Of course, this goal of producing such a Big Picture should not mislead one into thinking that subtle distinctions, careful and detailed examination of conceptual matters, and lengthy and intricate reasoning about minute points should not matter. Philosophers are concerned with very fundamental concepts upon which much rests, so that their analyses, even about apparently small matters, have very far-reaching consequences for the Big Picture being constructed.

In this search for the Big Picture, *coherence* is an essential ingredient. We seek an understanding of X that is consistent with, and holds together with, the other views we accept about the universe and about us. Take the philosophy of language, for example. Here, we seek an understanding of the nature of language and our mastery of language that is consistent with our general scientific, epistemological, and metaphysical views, both about the universe we inhabit and also about us as organisms with the features attributed to us by science. An account of the nature of language that made our ability to learn a language into a complete mystery would be considered by most philosophers of language to be in serious trouble. We seek a coherent and comprehensive Big Picture, where all the different Xs fit together. In general, one would not expect a contemporary philosopher's account of language to contradict any of our prevailing views of science and scientific knowledge without very compelling reasons.

Revolutions in science

This conception of philosophy suggests an explanation of a striking feature of the history of philosophy. Following the development and acceptance of a *revolutionary scientific theory*—a theory that undermines fundamental and central beliefs of the well-educated elite—there tends to appear a heightened amount of philosophical theorizing. Think of the important (and radical) philosophical writings that appeared following the seventeenth-century scientific discoveries that undermined much of the medieval conception of the universe. Or consider the philosophical activity that arose from the publication of Darwin's work on evolution.[1] Other examples are the enormous number of philosophical works dealing with Freud's writings on mental illness and childhood sexuality, and the lively discussions in present-day philosophy of science dealing with the remarkable implications of relativity theory and quantum mechanics that conflict with so many fundamental beliefs underlying Newtonian physics.

The above-noted activity of philosophers is fitting, given the conception of philosophy I have been describing. When science undermines fundamental and central beliefs—fractures our Big Picture of the universe—then philosophers understandably feel a pressing need to put the pieces together again, to develop a new and coherent Big Picture of the universe.

The importance of paradoxes

Another characteristic of philosophy is its great attention to, and serious interest in, uncovering and solving paradoxes or antinomies. A paradox is an argument that starts with premises that seem to be incontestable, that proceeds according to rules of inference that are apparently incontrovertible,

---

[1] A work that emphasized the philosophical responses to these two cases, the seventeenth-century scientific discoveries and the Darwinian theory of evolution, is Girvetz et al., 1966.

but that ends in a conclusion that appears to be obviously false. In many cases, a paradox ends in a conclusion that is downright self-contradictory. From the earliest beginnings of philosophy in Classical Greece (think of Parmenides, Heraclitus, and Zeno), paradoxes have played an important role in motivating and developing philosophical theories.

Consider what is undoubtedly one of the most striking cases of philosophical fervor brought about by the discovery of paradoxes: the discovery of the various paradoxes of mathematics and set theory in the late nineteenth and early twentieth centuries. These paradoxes stimulated much research in the foundations of logic and mathematics.[2] They led Frege eventually to abandon his logicism.[3] They also stimulated Poincaré to come up with his vicious-circle principle.[4] The paradoxes figured in Zermelo's defense of his axiomatization of set theory.[5] Russell, who discovered the paradox that bears his name, was led to develop his Theory of Types and his "no-class" theory during the many years he spent searching for a solution to the paradoxes.[6] Hilbert motivated certain aspects of his formalist philosophy of mathematics by the need to make certain that such paradoxes would never again be produced in mathematics.[7]

Reasons for the importance philosophers attach to paradoxes are not hard to find, given the above view of philosophy. An antinomy starts from premises that appear to be obviously true and proceeds according to principles of inference that seem to be clearly valid. These premises and principles may be fundamental to our belief system (some may belong to a body of beliefs classified as "folklore"). An antinomy may show that some such beliefs and/or principles clash with recent developments in science, mathematics, or logic or are simply inconsistent.

Alfred Tarski once wrote: "The appearance of an antinomy is for me a symptom of disease" (Tarski, 1969: 66). What is diseased, evidently, is a body of beliefs or system of principles that had been largely unquestioned or taken

---

[2] Cf. Raymond Wilder's assessment that "symbolic logic itself had its beginnings long before the discovery of the contradictions ... There can be little doubt, however, of the great impetus given to its development by the logical contradictions" (Wilder, 1952: 56).

[3] Near the end of his life, Frege completely abandoned his logicism and came to the conclusion that the source of our arithmetical knowledge is what he called "the Geometrical Source of Knowledge". See Frege, 1979a.

[4] See Chihara, 1973: ch. 1, sect. 1 for a discussion of Poincaré's reasoning.

[5] See Bach, 1998: 21–2 for supporting arguments.

[6] Chihara, 1973: ch. 1, contains a detailed discussion of Russell's attempt to solve the paradoxes and his development of the Theory of Types, as well as of his "no-class" theory.

[7] In Hilbert, 1983: 190–1.

for granted. In those cases, philosophers take it upon themselves to try to heal the body or system. In the case of the mathematical and set-theoretical paradoxes, it was thought that the basic principles of mathematics or logic were shown to be diseased.[8] It is no wonder that the discoveries of these paradoxes brought about so much unease and disquiet.[9] One can see why foundationalists—mathematicians, logicians, and philosophers—put so much effort into an attempt to put the mathematical house in order. Since a paradox is a symptom that our body of beliefs and principles is not coherent, the above account of one of the principal goals of philosophy allows us to see why philosophers feel the need to attempt to refashion our beliefs and principles into a more coherent Big Picture in which the paradoxes can no longer be constructed.

Philosophy of mathematics

Philosophy of mathematics may seem, at first sight, not to fit the overall scheme I have sketched above. Some philosophers of mathematics frequently seem not at all to be interested in fitting mathematics into the sort of Big Picture I have been discussing. Many researchers classified as philosophers of mathematics seem to be primarily mathematical logicians intent upon proving theorems and only vaguely aware of questions of epistemology or metaphysics. Big Picture questions seem to be too far removed from their expertise to be of any interest to them. Indeed, some philosophers of mathematics make no attempt to understand our actual mathematical practices at all, but rather set about constructing a new kind of mathematics. A striking case in point is the philosophy of mathematics known as "intuitionism" that was advanced by L. E. J. Brouwer, Herman Weyl, and Arend Heyting. Heyting tells us that he simply does not understand classical mathematics and that, for him, statements of classical mathematics have no clear sense.[10] So he, with his other intuitionist colleagues, set themselves the task of developing a new kind of mathematics: "intuitionistic mathematics". Intuitionists have thus appeared to some mathematicians and philosophers to be revolutionaries, seeking to overthrow the established

---

[8] Cf. Russell's attitude toward the paradoxes. He was convinced that logic itself needed to be reformed. See Chihara, 1973: 1.

[9] Frege was dismayed because the very foundations of his system of arithmetic were shaken by Russell's paradox. Russell, at first, thought that some relatively trivial error was responsible for the paradoxes. It was only later that he came to think some radical changes in logic were necessary to resolve the paradoxes. See in this regard Chihara, 1973: ch. 1.

[10] See Chihara, 1990: 21–2 for references and more details.

institutions of mathematics in order to replace them with their own brand of mathematics.[11] This will appear strange when compared to what philosophers of science, religion, music, and language do. We do not, for example, find many philosophers of science claiming not to understand actual science and setting out to develop a new kind of science. Philosophers of religion, in general, do not complain of their inability to understand the religions now being practiced and set out to form a new kind of religion. How many philosophers of language have decided that they do not understand any of the languages now being spoken and, as a result, have set out to develop a new kind of language that they can understand?

Of course, there is nothing in my view that precludes there being individual scholars who are classified as philosophers of mathematics even though they do not seem to be pursuing research that is directed at contributing to the kind of Big Picture described earlier. An individual philosopher of mathematics, especially one whose background is primarily mathematical, might very well undertake research aimed at achieving the kind of results judged successful by the criteria appropriate to the field of mathematics (for example, *credit for theorems*).[12] Furthermore, despite appearances to the contrary, I believe that one can understand the researches of even those philosophers of mathematics who are primarily interested in theorem-proving, as contributing towards the long-range goal of developing the kind of Big Picture I have been discussing. Such a strategy is apparent in the technical work of the "reconstructive nominalist" to be discussed in the next section.

2. NOMINALISTIC RECONSTRUCTIONS

In the latter part of the twentieth century, there have been many attempts to develop either an alternative kind of mathematics or an alternative kind of mathematical physics that is classified as "nominalistic". (A number of such attempts will be discussed in great detail throughout this work.) Since no one advocates replacing our mathematical and physical theories with such

[11] Hilbert is quoted as saying: "I believe that as little as Kronecker was able to abolish the irrational numbers ... just as little will Weyl and Brouwer today be able to succeed. Brouwer is not, as Weyl believes him to be, the Revolution—only the repetition of an attempted *Putsch*, in its day more sharply undertaken yet failing utterly, and now, with the State armed and strengthened ..., doomed from the start!" (Reid, 1970: 157). Frank Ramsey wrote of preserving mathematics "from the Bolshevik menace of Brouwer and Weyl" (Ramsey, 1931: 56).

[12] William Thurston, commenting on Jaffe and Quinn, 1993, writes: "Jaffe and Quinn analyze the motivation to do mathematics in terms of common currency that many mathematicians believe in: credit for theorems."

alternatives, some philosophers have questioned just what point there is to such "reconstructions".[13] What, it may well be asked, do these nominalistic versions of mathematics and/or physics contribute to our understanding of the mathematics and/or physics that mathematicians and scientists actually use? Later, I shall provide detailed explanations of what some of these reconstructive nominalists were attempting to accomplish and also how the nominalistic reconstructions they were devising contribute to the task of fitting our actual mathematical theories into the Big Picture discussed above. But at this point, the reader needs to have some idea of the basic philosophical orientation of the kind of nominalist I have in mind.

As I view philosophy of mathematics, philosophers of mathematics hope to achieve for mathematics what philosophers of language hope to achieve for language: they seek to produce a coherent overall general account of the nature of mathematics (where by 'mathematics' I mean the actual mathematics practiced and developed by current mathematicians)—one that is consistent not only with our present-day theoretical and scientific views about the world and also our place in the world as organisms with sense organs of the sort characterized by our best scientific theories, but also with what we know about how our mastery of mathematics is acquired and tested. Given this description of the aims of philosophy, all of the technical work published by the reconstructive nominalists can be judged by its contribution to the goal of producing such an account.

What is a *nominalistic* reconstruction of mathematics? To answer this question, one needs to know what a nominalist is. The kind of nominalist I have in mind is an anti-realist (or anti-Platonist). Thus, in order to understand nominalism in the philosophy of mathematics, one needs to understand the realist's (or Platonist's) view of mathematics that the nominalist opposes.

What is realism (Platonism)?

In the philosophy of mathematics, the realist maintains that mathematical objects exist; the nominalist takes the opposing position that there are no such things (or that we have no compelling reasons to believe such things exist). Thus, realists base much of their view of mathematics on the hypothesis that such things as numbers, sets, functions, vectors, matrices,

---

[13] The pun in the title of the Burgess–Rosen book, *A Subject with No Object*, suggests that there is no object (goal) to the nominalistic reconstructions. I shall examine the Burgess–Rosen criticisms of nominalism in Chapter 6.

spaces, and so on truly exist: they generally assert, for example, that the theorems of set theory are true statements that tell us what sets in fact exist and how these mathematical objects are related to one another by the membership relationship. Now mathematical entities are not supposed to be things that can be seen, touched, heard, smelled, tasted, or even detected by our most advanced scientific instruments. So a problem for the realist is to explain how mathematicians have been able to gain knowledge of such things. Realists, however, have backed their belief in the existence of such mathematical entities with a variety of philosophical arguments (to be discussed in great detail in Chapter 5). Nominalists have found these arguments far from convincing, and remain skeptical of the realists' philosophical accounts of mathematics. Thus, nominalists have attempted to develop an account of mathematics that does not require the existence of mathematical objects which the mathematician is able somehow to discover. The "nominalistic reconstructions" of mathematics are all supposed to contribute to such an account.

Puzzles about the nature of mathematics

The focus of the following chapter will be on a number of puzzles about the nature of mathematics. I look upon these puzzles much as the early twentieth-century foundationalists looked upon the set-theoretical paradoxes: as furnishing us with valuable clues and guides as to whether or not one's view of mathematics is at all accurate. Thus, I am convinced that no philosopher attempting to account for the nature of mathematics can afford to ignore such puzzles.

Five Puzzles in Search of an Explanation

I shall describe five puzzles about mathematics, each in need of investigation and solution. In succeeding chapters, I shall offer analyses and explanations of these puzzles, in the course of which my own views on the nature of mathematics will emerge.

1. A PUZZLE ABOUT GEOMETRY

Consider the first three postulates of Euclid's version of plane geometry:

Postulate 1: A straight line can be constructed from any point to any point.

Postulate 2: A straight line can be extended indefinitely in a straight line.

Postulate 3: A circle can be constructed with its center at any point and with any radius.[1]

---

[1] The first four postulates of Euclid's geometry are translated by Thomas Heath (in Heath, 1956: 154) as follows:

Let the following be postulated:

1. To draw a straight line from any point to any point.
2. To produce a finite straight line continuously in a straight line.
3. To describe a circle with any centre and distance.
4. That all right angles are equal to one another.

But he understands Postulate 1 to be "asserting the possibility of drawing a straight line from one point to another", Postulate 2 to be "maintaining the possibility of producing a finite straight line continuously in a straight line" (196), and in the case of Postulate 3, he tells us that "Euclid's text has the passive of the verb: 'a circle can be drawn'" (199). Commenting on Postulate 4, Heath notes that according to Proclus, "Geminus held that this Postulate should not be classed as a postulate but as an axiom, since it does not, like the first three Postulates, assert the possibility of some *construction* but expresses an essential property of right angles" (200). E. G. Kogbetliantz (in Kogbetliantz, 1969: 554) gives the straightforwardly modal translation of Euclid's postulates as follows: Postulate 1: A straight line can be drawn from any point to any point. Postulate 2: A straight line may be produced, that is, extended indefinitely in a straight line. Postulate 3: A circle can be drawn with its center at any point and with any radius. Peter Wolff explains Euclid's first three postulates as follows: "The root meaning of the word 'postulate' is to 'demand'; in fact, Euclid demands of us that we agree that the following things can be done: that any two points can be joined by a straight line; that any straight line may be extended in either direction indefinitely; that given a point and a distance, a circle can be drawn with that point as center and that distance as radius" (Wolff, 1963: 47–8).

Now compare those postulates with the first three axioms of Hilbert's version published in his *Foundations of Geometry*.[2]

Axiom 1: For every two points A, B there exists a line L that contains each of the two points A, B.

Axiom 2: For every two points A, B there exists no more than one line that contains each of the points A, B.

Axiom 3: There exist at least two points on a line. There exist at least three points that do not lie on a line.

Notice that Hilbert's axioms are existential in character: they assert the existence of certain geometric objects, that is, points and lines. Euclid's postulates, on the other hand, do not assert the existence of anything. Rather, what is asserted is the *possibility* of making some sort of geometric construction.[3] This constructive aspect of Euclid's geometry is fundamental, for as Ernest Adams suggests:

Euclid was concerned with *applications* in a way that modern pure geometry tends not to be. Many of the propositions of *The Elements* resemble *recipes* for doing various things, e.g., for constructing equilateral triangles. Of course there is much more than that to *The Elements*, since many if not most of its propositions are theorems that state *facts*, such as the Pythagorean Theorem ... but persons seeking to understand the applications of Geometry are not ill-advised to look to its older and more technologically oriented formulations. (Adams, 2001: 38)

---

[2] Hilbert, 1971: 3–4.

[3] Cf. Mueller's comments (in Mueller, 1981: 14): "Hilbert asserts the existence of a straight line for any two points, as part of the characterization of the system of points and straight lines he is treating. Euclid demands the possibility of drawing the straight-line segment connecting the two points when the points are given. This difference is essential. For Hilbert geometric axioms characterize an existent system of points, straight lines, etc. At no time in the *Grundlagen* is an object bought into existence, constructed. Rather its existence is inferred from the axioms. In general Euclid produces, or imagines produced, the objects he needs for a proof ... It seems fair to say that in the geometry of the *Elements* there is no underlying system of points, straight lines, etc. which Euclid attempts to characterize." My colleague Paolo Mancosu has informed me that his researches indicate that some Greek mathematicians read the modal Euclidean postulates as having an existential commitment to eternal objects and that, even at the time of Euclid, there were disputes about the "ontological commitments" of the postulates of geometry. This is not surprising, if Reviel Netz is right in claiming that, in cultural settings such as that of the Greeks, "polemic is the rule, and consensus is the exception" (Netz, 1999: 2). In any case, it would seem that the historical facts are more complicated than I have indicated above, but my basic point is that there is a way of understanding the postulates of Euclid's geometry which is such that no commitment to mathematical objects is presupposed by them and that many mathematicians and philosophers understood them in that way.

In contrast to Hilbert's existential geometry, then, Euclid's geometry is modal in character, asserting what it is possible to construct.[4] For over two thousand years, geometry was understood and developed by many mathematicians as a modal theory, but for some reason, by the twentieth century, geometry had became straightforwardly existential.

Hilbert was by no means the first mathematician to think and reason about geometrical objects (such as points and lines) in terms of existence rather than constructibility. Mathematicians had begun to shift to the existential mode of expressing geometrical theorems hundreds of years before Hilbert had written on the topic. This shift in geometry from making constructibility assertions to asserting existence raises some fundamental questions:

(a) No one seems to have made a fuss about the change that took place or even to have taken note of it. No one believes that an ordinary existential statement such as "There are buildings with over three hundred stories" is equivalent to a modal statement of the form "It is possible to construct buildings with over three hundred stories". Why weren't there serious debates among mathematicians and philosophers over the validity of making such an apparently radical ontological change in the primitives of one of the central theories of mathematics?

(b) The applications to which geometry was put evidently did not change as a result of the described shift. How is it that, in our reasoning about areas, volumes, distances, and so on, it seems to make no difference whether the geometry we use asserts what it is *possible to construct* or whether it asserts the *existence* of mathematical entities?

A completely adequate answer to these questions would require a detailed investigation into the history of mathematics spanning many hundreds of years—something that is well beyond the scope of this work. So, although I shall give some indications in later chapters how I would answer the above puzzles, I shall concentrate my investigation on the following (restricted but closely related) puzzle about Hilbert's view of geometry:

In the introduction to his *Foundations of Geometry*, Hilbert writes: "The establishment of the axioms of geometry and the investigation of their relationships is a problem ... equivalent to the logical analysis of our perception of space" (Hilbert, 1971: 2). And he goes on to say that his axioms

---

[4] What does it mean to say that an assertion or theory is "modal"? For those unfamiliar with the notion of modality, an explanation will be given below in Chapter 5, sect. 3.

express "facts basic to our intuition" (3). But does our perception of space tell us that there exist the infinity of points and lines postulated by his geometrical axioms? Does our intuition inform us that such points and lines truly exist? How could Hilbert have felt justified in postulating axioms which express such existential "facts", when Euclidean plane geometry had been developed and applied for a multitude of centuries without any such commitment to an apparent ontology of imperceptible objects?

I now turn to the second of my puzzles, which brings out a striking difference between the attitudes of set theorists regarding the mathematical theories they work on and those of physical scientists regarding the empirical theories they work on.

2. DIFFERENT ATTITUDES OF PRACTICING MATHEMATICIANS REGARDING THE ONTOLOGY OF MATHEMATICS

Some outstanding mathematicians have believed that the set theorist is reasoning about some non-physical entities that truly exist (I have in mind such Platonic or realist researchers in set theory as Kurt Gödel, Robert Solovay, and John Steel), whereas other outstanding mathematicians, such as Alfred Tarski, Paul Cohen, and Abraham Robinson, have maintained that set theorists are not reasoning about things that truly exist at all, with Cohen opining that "probably most of the famous mathematicians who have expressed themselves on the question have in one form or another rejected the Realist position".[5] What is striking about the latter group is that, despite such skeptical beliefs about the ontology of mathematics, many of them have produced, and continue to produce, important and fruitful mathematical results.

---

[5] Cohen, 1971: 13. Robinson has expressed his anti-realist views in Robinson, 1965; Cohen in Cohen, 1971. For discussion of the formalist views of Robinson and Cohen, see Pollard, 1990: ch. 5, sect. 3. I personally heard Tarski express his nominalistic views on several occasions in Berkeley. Other outstanding mathematicians at Berkeley who reject the realist position are Jack Silver and John Addison. Solomon Feferman has expressed his anti-realist views quite forcefully: "Briefly, according to the platonist philosophy, the objects of mathematics such as numbers, sets, functions, and spaces are supposed to exist independently of human thoughts and constructions, and statements concerning these abstract entities are supposed to have a truth value independent of our ability to determine them. Though this accords with the mental practice of the working mathematician, I find the viewpoint philosophically preposterous" (Feferman, 1998c: p. ix).

This is clearly very different from what is typical in the other sciences. A chemist who does not believe in phlogiston does not theorize in the phlogiston theory, perform experiments based upon the theory, and explain phenomena with the theory. Few, if any, chemists who were skeptical about the existence of phlogiston continued to produce fruitful developments in phlogiston theory.[6] Similarly, there do not exist large groups of outstanding geneticists who deny that there are such things as genes. How is it, then, that there are so many mathematicians who are thoroughly skeptical of the existence of mathematical objects, and yet continue to work fruitfully in such fields as set theory? Evidently, there is something about the nature of mathematics in general, and of set theory in particular, that allows mathematicians to behave in such an apparently strange manner (when compared with the behavior of empirical scientists). So the puzzle is: How does the science of mathematics differ from the other sciences that accounts for this striking difference?

The next of my puzzles concerns a strange feature of mathematical objects that sets them apart from the physical objects with which we are familiar.

3. THE INERTNESS OF MATHEMATICAL OBJECTS

As Jody Azzouni has emphasized, mathematical practice supports a view implying that mathematical objects are very different in kind from the objects the empirical scientists study:

A crucial part of the practice of empirical science is constructing means of access to (many of) the objects that constitute the subject matter of that science. Certainly this is true of theoretical objects such as subatomic particles, black holes, genes, and so on. ... Empirical scientists attempt to interact with most of the theoretical objects they deal with, and it is almost never a trivial matter to do so. Scientific theory and engineering know-how are invariably engaged in such attempts, which are often ambitious and expensive. Nothing like this seems to be involved in mathematics.[7]

Mathematicians do not give seminars or conference presentations on the newest means of detecting sets, numbers, or functions. Detecting or

---

[6] There may be some areas of physics, such as quantum mechanics, in which significant numbers of researchers are skeptical about the existence of the theoretical entities that the principal theories of the area postulate. But such areas, I contend, would not be typical.

[7] Azzouni, 1994: 5. He goes on to note that the closest the mathematician comes to doing anything that might be called "engaging" mathematical objects is to perform some kind of act of introspection.

interacting with mathematical objects does not seem to constitute any part of our mathematical practice. It would seem that the objects of mathematics do not interact with us or with anything in our world. This inertness feature of mathematical objects, indicated by the contrast between our mathematical and scientific practices, gives rise to three closely related philosophical questions, which I shall take up and discuss in the following sections.

How do we know of the existence and properties of mathematical objects?

The first is a question of epistemology. Given that mathematical objects are causally inert and that no human interacts with any of these entities, how then are mathematicians able to discover mathematical truths? The reason this is a puzzle can be found in three *apparent* facts about mathematics.

(a) Mathematicians talk about such things as numbers, sets, functions, and spaces.
(b) The theorems of our mathematical theories—theories such as number theory and set theory—are true assertions about these things talked about by mathematicians.
(c) Among the theorems of our mathematical theories are existential assertions that affirm the existence of these things talked about by mathematicians—such as numbers and sets.

Evidently, our mathematical theories tell us that such things as numbers and sets truly exist. But the inertness feature of mathematical objects gives rise to the question of how mathematicians were able to discover such existential truths in the first place. How are human beings able to gain knowledge of the existence of these inert mathematical objects? How are humans able to learn what properties they have and what relations they are in?

Brown's answer

These naturally arising questions have engendered a considerable literature. For example, James Brown, in his recent book in the philosophy of mathematics, answers the above questions in the following way: "We have mathematical knowledge and we need to explain it; the best explanation is that there are mathematical objects and that we can 'see' them" (Brown, 1999: 45).

Now one objection that has been raised to such a Platonic view of mathematical knowledge is based upon a doctrine known as the "causal theory of knowledge".[8] Brown characterizes this doctrine with the words: *"To know anything at all, there must be some sort of causal connection between the object known and the knower"* (Brown, 1999: 16, italics his). From this doctrine and the fact that there can be no causal connection between a human and any mathematical object (because of the inertness feature of mathematical objects), the objector arrives at the conclusion that 'seeing' any mathematical object to obtain mathematical knowledge in the way postulated by the Platonist is impossible and that the Platonist's explanation of mathematical knowledge collapses.

Brown's response to this objection is to describe an admittedly "bizarre situation" which can arise according to one standard interpretation of quantum mechanics. In what is described as an Einstein–Poldolsky–Rosen type set-up, an experiment is described in which a pair of photons arrive at opposite ends of the set-up each with the property *spin-up* or *spin-down*. According to quantum theory, we supposedly have the following situation:

Suppose that I am at one wing of the measuring apparatus and get the result: spin-up. Then I can immediately infer that you at the other wing have the result: spin-down. I know the distant outcome *without being causally connected* to the remote wing. ... So the causal theory is simply refuted by example. *We can have knowledge even without a causal connection.* (Brown, 1999: 17)

Let us accept (for the sake of argument) Brown's "refutation" of the causal theory. What then can we infer about his Platonic "explanation" of how we are able to obtain knowledge of the "inert" mathematical entities? Here is what Brown concludes:

Once the causal theory is rejected, there is no objection to our knowing about abstract entities without being causally related to them. The problem of access is a pseudo-problem; resistance to Platonism is motivated by misplaced scruples. (Brown, 1999: 18)

To conclude that the problem of access is a pseudo-problem merely on the basis of a refutation of one version of the causal theory of knowledge seems to me to be both hasty and illogical. Reconsider Brown's quantum mechanical example. We may not be in any sort of causal contact with the remote wing at the particular time in question and we may not be able to causally interact with the photon's particular spin state or with the photon in question at the

---

[8] This sort of objection, frequently attributed to Paul Benacerraf, will be discussed again in Chapter 6.

particular time at which it makes contact at the remote wing, but we could have causally interacted with the photon at other times (the photon is, after all, detectable by our instruments). Furthermore, we are in all sorts of causal contact with a multitude of other photons and other particles on the basis of which we have been able to empirically verify many of the implications of quantum physics. And we are able to make the sort of inference Brown describes because we have this heavily studied and remarkably verified theory, which forms the basis for the unusual inference that Brown describes.

Could we have causal contact with mathematical entities, at least some of the time, as we do with photons? And do we have a theory about mathematical entities that has been comparably empirically verified and whose predictive accuracy is close to that of quantum physic? What heavily studied and remarkably verified theory of mathematical entities serves as the basis for our inference to the properties and relations of the mathematical entities that we, supposedly, are unable to "see"? To infer that the problem of access is a pseudo-problem merely from the fact that one version of the causal theory of knowledge is refutable is like concluding that the problem of global warming is a pseudo-problem merely on the grounds that one specific way of calculating carbon dioxide in the atmosphere is refutable.

It is not as if Brown is able to answer such questions as: How did mathematicians discover that the empty set exists? Have mathematicians literally seen the empty set? Of all the Platonists I know and know of (which happens to be quite a few), not a single one has ever claimed, so far as I know, to be able to "see" (in any relevant sense) the empty set. If we can't "see" it, then how do we come to know of its existence? What allows us to assert its existence? Perhaps Brown believes that mathematicians "see" with the "mind's eye" that such an entity exists. But that just gives rise to further questions: What sort of "seeing" is this seeing with the mind's eye? What does such an "explanation" come to (if one eliminates the metaphorical element)? Is it a scientifically respectable explanation? I suspect that few philosophers or scientists would regard Brown's defense of the Platonist's epistemological claims as a reasonable way of defending the belief in the empty set.

The problem of referential access to mathematical objects

Closely related to the above epistemological question is the question of reference: how are we able to refer to these inert mathematical objects? Unlike ordinary physical objects, with which we are in causal contact, the mathematical objects described in our mathematical theories do not appear to be things that even exist in our (physical) world. So by what mechanism, physical or mental, are we able to refer to these esoteric objects? This puzzle of reference to inert objects was stressed by Jonathan Lear (in Lear, 1977), taken up by Penelope Maddy (in Maddy, 1980, as well as in Maddy, 1990: 37–41), and I have discussed it myself (in Chihara, 1990: ch. 10, sect. 2). However, the philosopher who has written most extensively about this problem is Jody Azzouni.

Azzouni takes up the problem in the following way:

We have to tell a story of how the terms we use refer to what they refer to, and this story is not supposed to be nonnatural. That is, whatever the story is, it must be consistent with our current scientific picture of the sort of creature we are. (Azzouni, 1994: 7).

He then briefly sketches some attempts to answer this problem about how we refer in general. One popular class of answers he considers are what are called "causal theories of reference".[9] But such theories seem doomed to failure for the case of reference to mathematical entities, since such entities are evidently causally inert. In this way, we arrive at what Azzouni calls "the puzzle of referential access to mathematical objects" (Azzouni, 1994: 7).

I shall not go into the many analyses Azzouni provides of the various solutions that have been offered to the puzzle of referential access. Suffice it to say that he examines a significant number of different "solutions" all of which he finds to be wanting. As a result, we seem to be faced with a serious philosophical puzzle:

[S]omething must be fixing the references of our mathematical terms. If it isn't definitions and axioms, it isn't practices, and it isn't intellectual intuition, then *what* could be doing it? (Azzouni, 1994: 31)

This takes me to the third of the puzzles that the inertness feature of mathematical objects engenders.

Why is knowledge of these causally inert objects needed by empirical scientists?

This is the question of why knowledge of these esoteric objects is so very important for the empirical scientist. The scientist deals with various physical objects with which we are related, if only indirectly, by a multitude of causal relationships. Yet, our scientific theorizing requires the use of the mathematical theories discussed above that evidently make reference to

[9] The reader can find references to some works defending such theories in Azzouni, 1994: 7 n. 10.

a wide variety of mathematical entities. But why is it crucial for scientists to refer to and to discover relationships among the inert mathematical objects discussed and referred to in our mathematical theories in order to discover facts about the physical entities discussed in their scientific theories? What epistemic role are these mathematical entities playing in our scientific theorizing?[10]

Although there have been many attempts to answer the three daunting questions described above, none of the solutions proposed to date has satisfied the majority of philosophers of mathematics. So from my point of view, we are left with a group of fundamental puzzles about the nature of mathematics, in need of analyses and explanations.

I turn now to the next of my puzzles—a puzzle about the very nature of mathematical existence.

4. CONSISTENCY AND MATHEMATICAL EXISTENCE

If one looks back at the history of mathematics, one finds a string of brilliant mathematicians making what seems to be the same basic point, namely, that *mathematical existence is a matter of consistency*. Consider what Henri Poincaré wrote in *Science and Hypothesis*:

The word 'existence' has not the same meaning when it refers to a mathematical entity as when it refers to a material object. A mathematical entity exists provided there is no contradiction implied in its definition, either in itself, or with the proposition previously admitted. (Poincaré, 1952: 44)

The above point is amplified in *Science and Method*, where he wrote:

If ... we have a system of postulates, and if we can demonstrate that these postulates involve no contradiction, we shall have the right to consider them as representing the definition of one of the notions found among them. (Poincaré, 1953: 152)

The following quotation from the writings of David Hilbert expresses a thought that is close to what Poincaré expressed in the previous quotations:

If contradictory attributes are assigned to a concept, I say, that mathematically the concept does not exist. So, for example, a real number whose square is -1 does not

---

[10] This question is closely related to what Azzouni calls "the epistemic role puzzle": "Given standard mathematical practice, there seems to be *no* epistemic role for mathematical objects" (Azzouni, 1994: 55). It will be seen that this puzzle is also closely related to a puzzle about the applications of mathematics in science which Mark Steiner calls the "metaphysical problem of application" (to be discussed in Chapter 9).

exist mathematically. But if it can be proved that the attributes assigned to the concept can never lead to a contradiction by the application of a finite number of logical inferences, I say that the mathematical existence of the concept (for example, of a number or a function which satisfies certain conditions) is thereby proved. In the case before us, where we are concerned with the axioms of real numbers in arithmetic, the proof of the consistency of the axioms is at the same time the proof of the mathematical existence of the complete system of real numbers or of the continuum.[11]

A third outstanding mathematician seems to have maintained a view that is very similar to the above. According to the historian of mathematics Joseph Dauben,

Logical consistency was the touchstone which Cantor applied to any new theory before declaring it existent and a legitimate part of mathematics. ... Since he took his transfinite numbers to be consistently defined, ... there were no grounds to deny his new theory. This kind of formalism, stressing the internal conceptual consistency of his new numbers, was all mathematicians needed to consider before accepting the validity of the transfinite numbers. (Dauben, 1990: 129)

The above three are by no means the only mathematicians who have expressed such thoughts. Indeed, such an idea is considered a commonplace by some. For example, Paul Bernays has written: "It is a familiar thesis in the philosophy of mathematics that existence, in the mathematical sense, means nothing but consistency."[12] For my purposes, these examples should suffice.

What is there about mathematical practice and mathematical theorizing that accounts for the striking attractiveness (to these outstanding mathematicians) of the doctrine that mathematical existence amounts to freedom from contradiction? One cannot plausibly maintain that this doctrine appeals to these mathematicians because they are simply ignorant of mathematics. Nor can we reasonably claim that the doctrine appears tempting to them primarily because of their peculiar philosophical prejudices. After all, in their time, these researchers were leading figures in mathematics and can hardly be considered ignorant of the subject. Similarly, to attribute their views about mathematical existence to some extreme philosophical attitude they had towards mathematics itself does not appear to be reasonable, since the

---

[11] This quotation is from an excerpt of Hilbert's famous address to the International Congress of Mathematics of 1900, in which he put forward his list of 23 important unsolved problems, published in Calinger, 1982: 662.

[12] From a translation of Bernays, 1950 by Richard Zach (unpublished ms.).

philosophical views they expressed at various times in their careers about the nature of mathematics seem to be so diverse.[13]

The puzzle is: *what must mathematics be like if the mere consistency and coherence of the definition of a totality of mathematical objects seems to be sufficient to brilliant practitioners of the science to yield the acceptability of the reality of such objects?*

I now take up the last of my puzzles—one that is quite different from the others and that has been raised and pondered by metaphysicians.

5. THE VAN INWAGEN PUZZLE

I call this puzzle the "van Inwagen puzzle" because it was put forward by the metaphysician Peter van Inwagen. Actually, van Inwagen did not put forward this bit of reasoning *as a paradox or puzzle*, but rather as part of his ad hominem defense of Alvin Plantinga's modal realism against an objection found in David Lewis's *On the Plurality of Worlds*.[14]

Here's how the puzzle gets started. We are to imagine that a philosopher has advanced the following theory.

The typosynthesis theory

The theory of typosynthesis makes the following assertions:

(1) There are exactly ten cherubim.
(2) Each human being bears a certain relation, typosynthesis, to some but not all cherubim.
(3) The only things in the domain of this relation are human beings.
(4) The only things in the converse domain of this relation are cherubim.

This is all the typosynthesis theory says.[15] The theory is concerned solely with cherubim and the relation of typosynthesis that humans bear to cherubim, and all it says about the cherubim is given by the four assertions above.

A classification of types of properties and relations

To understand van Inwagen's puzzle, one needs to have some understanding of the basic concepts he used to pose his puzzle. In particular, one needs to

---

[13] Poincaré was a predicativist; Hilbert, in his later years, founded formalism; and Cantor was a mathematical Platonist.

[14] For a discussion of Lewis's objection and van Inwagen's defense, see Chihara, 1998: ch. 3, sect. 9, where one can find references to the relevant works.   [15] Van Inwagen, 1986: 206.

grasp the following classification of types of properties and relations used by many metaphysicians.

a. Intrinsic vs. extrinsic relations   We need, first of all, to distinguish *intrinsic* from *extrinsic relations*. This is done by noting that there are two types of intrinsic relations which I shall characterize in what follows. An extrinsic relation is a relation that is not intrinsic.

The two kinds of intrinsic relations are called 'internal' and 'external' relations. However, in order to characterize these two types of intrinsic relations, we need to understand what an "intrinsic property" is.

Definition: The *intrinsic properties* of a thing are those the thing has just in virtue of the way it is—not in virtue of any relation it is in with other things.

For example, you have a heart, a kidney, you are covered with skin, you have hair, and so on. Thus, you can be said to have the intrinsic properties of having a heart, of having a kidney, of being covered with skin, of having hair, and so on. The property of being married, on the other hand, is not one of your intrinsic properties, since you have that property not in virtue of the way you are, but in virtue of your being in a certain relation to another person (your spouse).

b. Extrinsic properties   The property of being married is classified by philosophers as one of your *extrinsic properties*—properties that a thing has in virtue of its relation or lack of relation to other things.

c. Internal and external relations   Now here is how we distinguish the two kinds of intrinsic relations:

Definition: *Internal relations* are relations that "supervene" on the intrinsic properties of the relata.

What does this mean? To assert supervenience is to deny independent variation. As David Lewis describes it: "To say that so-and-so supervenes on such-and-such is to say that there can be no difference in respect of so-and-so without difference in respect of such-and-such" (Lewis, 1983: 358). Thus, if X_1 and Y_1 are in the internal relation R, but X_2 and Y_2 are not, then necessarily, either the Xs have different intrinsic properties or the Ys do (or both). Thus, suppose that there were two possible people (in some other possible world) who were exact duplicates of you and your spouse (insofar as one had all the intrinsic properties that you have and the other had all the intrinsic properties that your spouse has).[16] Then (assuming that you are, in fact, taller than your spouse), since necessarily your duplicate would be taller than the duplicate of your spouse, we can infer that the relation of being taller than is an internal relation.

> Definition: A relation is *external* if its holding depends on the intrinsic properties of the composite of the relata, but does not depend on just the relata themselves.

To use an example Lewis gives, the relationships of distance holding between the electron orbiting the proton of a classical hydrogen atom are not internal, since these relations do not depend upon just the intrinsic properties of the electron and the proton taken separately. But if we take the composite—the hydrogen atom—then the relations' holding does depend upon the intrinsic properties of the composite. So we have an example of an external (intrinsic) relation.

As I noted earlier:

Definition: A relation that is not intrinsic is *extrinsic*.

An extrinsic relation, then, is one that is neither internal nor external. An example of an extrinsic relation is that of being an owner of something: a person is related to a piece of property by this relation. The relation is not internal because its holding does not depend solely on the intrinsic properties of the person and the piece of property. It is not external because its holding does not depend purely on the intrinsic properties of the composite consisting of the person and the piece of property.

The unsatisfactory nature of the typosynthesis theory

A question now arises: is it possible for us to place the relation of typosynthesis within this classification? In other words, can we classify typosynthesis as internal, external, or extrinsic? Let us see.

---

[16] One might wonder how anything can have a duplicate, since it would be reasonable to suppose that one intrinsic property any thing X would have is the property of being identical to X. So it would be more accurate to define a duplicate in terms of intrinsic qualitative properties—where a qualitative property is what Adams calls a "suchness" (as opposed to a "thisness"). See, in this regard, Adams, 1979.

Suppose that typosynthesis could be classified as intrinsic. Then the relation must hold of a particular human being and some cherub because of the intrinsic properties of that human and that cherub (since the relation must supervene on the intrinsic properties of the relata). In other words, there must be something about the intrinsic properties of that human and that cherub in virtue of which the relation obtains. But we have no idea of what intrinsic properties a cherub has. Do they have wings? Are they in physical space and time? Do they have thoughts? Are they changeable? The theory doesn't say. Clearly, we are in no position to say what intrinsic properties a cherub has in virtue of which a particular cherub is related by typosynthesis to a particular human being. In short, we are in no position to classify the relation as internal.

Suppose that typosynthesis could be classified as external. In that case, we can bring into consideration the intrinsic properties of the composite of that human and that cherub. But what is it about the intrinsic properties of the composite in virtue of which the relation holds? Knowing nothing about the intrinsic properties of cherubim, we have no idea what the composite consisting of that human and that cherub would be. Adding a composite to the situation does not help in the least. Again, we are in no position to classify the relation as external.

Suppose that typosynthesis is extrinsic. In that case, there would have to be some other object or objects in the universe in virtue of which that particular human was in the typosynthesis relation to that cherub. Do we have any idea of what this thing or things could be? Not at all. Again, we are completely in the dark and in no position to classify the relation as extrinsic.

The conclusion seems to be that we cannot classify what sort of relation typosynthesis is. But in attempting to classify the relation, it has become apparent that we have no clear idea either of what this relation of typo-synthesis is or of what sort of thing cherubim are. Thus, we have a theory purporting to be about a specific relation (typosynthesis) and a specific type of thing (cherubim) but failing utterly to tell us what the nature of this relation is or to tell us what sort of thing the theory is about. In that case, says van Inwagen, the philosophical theory must be rejected as philosophically unsatisfactory.

Are the above distinctions too vague?

Here, it is useful to consider an objection. Some philosophers have objected to the above reasoning on the grounds that the philosophical distinctions being used are too vague and "metaphysical" to allow us to draw any conclusions about the theory of typosynthesis with any confidence.[17] It is not important to my analyses, however, that these distinctions be defended and retained. The above classification is only being used to facilitate seeing that we have no real understanding of the nature of the relation of typosynthesis or of the kind of thing cherubim are. The main idea of the above objection to the typosynthesis theory can be discerned, even in the absence of the above distinctions.

How might one arrive at the unsatisfactory nature of the typosynthesis theory without bringing in such a metaphysical system of classification? Consider some typical relations, say the relations *taller than* and *weighs more than*. We know, in general, what features or properties of John and Mary must be taken into account to determine if John is taller than Mary or whether John weighs more than Mary. But what properties of Hillary Clinton and some cherub must be taken into account to determine if Hillary Clinton is in the relation typosynthesis to the cherub? Who knows? The above theory does not tell us. In virtue of what properties of Hillary Clinton and what features of some cherub is Hillary related by typosynthesis to that cherub? The answer is: we haven't the slightest idea.

Perhaps typosynthesis is not that sort of relation. Perhaps typosynthesis is like the relation of being married. Perhaps it is in virtue of some things Hillary Clinton, the cherub, and some third being have done that brings it about that they are in the relation in question. Again, we haven't the vaguest notion of what actions, if any, are required for the relation to obtain. Perhaps typosynthesis holds of Hillary Clinton and the cherub in virtue of the relative spatial or temporal relationships they have to one another. But if so, we have no idea of what these relationships could be, since we don't even know if cherubim are in physical space or time. Thus, it is hard to see how we can have anything like a true understanding of this relation of typosynthesis. And so it would seem that the theory of typosynthesis can hardly be a satisfactory one.

Clearly, we know nothing about the "intrinsic natures" of cherubim, that is, we know nothing about the properties or qualities that cherubim possess that are not purely relational properties (such as the property of having some human related to them by typosynthesis). We don't know if they are intelligent, sentient, space occupying, visible, physical, or whatever. It seems clear, then, that we have no genuine understanding of what cherubim are or what this relation of typosynthesis is.

---

[17] This objection was raised, in particular, by Kate Elgin at my 12 Nov. 1999 MIT Philosophy Colloquium lecture.

A similar problem with set theory

What relevance has this theory of typosynthesis to the philosophy of mathematics? Why do I even consider such a strange theory here? Because essentially the objections to the typosynthesis theory raised above can also be raised to set theory. As in the typosynthesis case, the set theorist cannot tell us anything about the true nature of the relationship of membership. Thus, consider the enormous totality of unit sets that are supposed to exist. Only one of these unit sets is the one that has, as its only member, Bill Clinton. Then, what properties of Bill Clinton and this singleton determine that it is Bill Clinton and nothing else that is in the membership relation to this unit set? Who knows? Set theory does not tell us. Perhaps membership is not that sort of relation. Perhaps membership is like the relation of being married. Perhaps it is in virtue of some things Bill Clinton, the singleton, and some third being have done that brings it about that Clinton is in the membership relation to the set in question. Again, we haven't the vaguest notion of what actions, if any, are required for the relation to obtain. Perhaps Bill Clinton is related by the membership relation to the singleton in virtue of the relative spatial or temporal relationships they have to one another. But if so, we have no idea of what these relationships could be. It is hard to see how we can have anything like a true understanding of this relation of membership. And so it would seem that set theory can hardly be a satisfactory one.

What do we know about sets, given what set theory tells us? Everything we know about the empty set is relational. We do know that nothing is in the membership relation to it. And we know that the empty set is a member of, say the set whose only member is the empty set. But what do we know about the empty set that is not just relational? In other words, what do we know about its "intrinsic properties"? Recall that these are the properties it has because of the way it is and not because it is in some relation. We don't know if it occupies space, is visible, has mass, has any detectable features. Like the cherubim, it is a mysterious something.

For those who feel they have a grasp of the metaphysical distinctions with which van Inwagen reasons, the puzzle can be taken to proceed from the argument that, as in the typosynthesis case, the set theorist has such a poor grasp of the membership relation that she cannot classify the relationship that an object has to its singleton as intrinsic or extrinsic, internal or external. Thus, consider again the question as to whether the relation that Clinton bears to his singleton is an internal relation? If so, there must be something about the intrinsic properties of Clinton and his singleton in virtue of which the membership relation holds. Well, what is it about the intrinsic properties of just that one unit set in virtue of which Clinton is related to just it and not to any of the other unit sets in the set-theoretical universe? We haven't the vaguest idea. We know no more about the intrinsic properties of sets than we know about the intrinsic properties of cherubim. Set theory gives us no information about the intrinsic properties of sets. Furthermore, since we are not in any sort of causal relationship with sets, it does not seem possible for us to learn something about the intrinsic properties of sets by empirical means. So it is a complete mystery how any one could have understood this relation.

The conclusion seems to be that set theory is an unsatisfactory theory that should be rejected, just as we concluded above for the case of the typo-synthesis theory. But van Inwagen is reluctant to draw that conclusion. He cannot bring himself to conclude that set theory is an unsatisfactory theory, because he feels that would imply that mathematics is an unsatisfactory theory that should be rejected—for him, a complete absurdity. So *he concludes instead that there must be something wrong with his reasoning.* But he has no idea what it is.[18]

Lewis, upon whose reasoning van Inwagen modeled his typosynthesis reasoning, agrees that the above considerations do seem to lead to the conclusion that we do not understand the primitive membership relation of set theory:

It's a nasty predicament to claim that you somehow understand a primitive notion, although you have no idea how you could possibly understand it. That's the predicament I'm in. (Lewis, 1991: 36)

But Lewis goes on to say (in apparent agreement with van Inwagen's sentiments) that he cannot accept the conclusion that he does not grasp the membership relation of set theory, for that implies, he believes, that he should reject "present-day set-theoretical mathematics" (1991: 36). He concludes:

If there are no classes, then our mathematical textbooks are works of fiction, full of false 'theorems'. Renouncing classes means rejecting mathematics. That will not do. (1991: 58)

"Hot though it is in the frying pan, that fire is worse," he says (1991: 36). He tells us that he is moved to laughter "at the thought of how *presumptuous* it

---

[18] Michael Jubien puts forward a solution to van Inwagen's quandary in Jubien, 1991. I give reasons for questioning Jubien's solution in Chihara, 1998: 108–9.

would be to reject mathematics for philosophical reasons" (1991: 59). "How would *you* like the job of telling the mathematicians," he continues, "that they must change their ways, and abjure countless errors, now that *philosophy* has discovered that there are no classes?" "Not me," says Lewis: so he continues to maintain his belief that he somehow grasps the fundamental primitive of set theory, even though he has no idea how he could understand it.

In summary, van Inwagen concludes that there must be something wrong with the argument, but he doesn't know what, whereas Lewis concludes that he must grasp the membership relation, but he doesn't know how. In either case, we are left with a real puzzle.

In various places in what follows, I shall take up each of these five puzzles, starting in the next chapter with the ones about geometry and mathematical existence.

2

Geometry and Mathematical Existence

This chapter is concerned with the first and fourth puzzles described in the previous chapter. Both of these puzzles concern, to a great extent, Hilbert's views of mathematics. Hence, before I attempt to tackle these puzzles and, in particular, before even discussing the puzzle concerning the shift from Euclid's modal version of geometry to the contemporary Hilbertian existential version, I will first undertake an examination of just how *logically* different Hilbert's geometry is from Euclid's.

1. The Frege–Hilbert Dispute Concerning the Axioms of Geometry

An insight into these logical differences can be obtained from an investigation into the dispute that began near the end of the nineteenth century between Frege and Hilbert concerning the axioms of Hilbert's geometry. This dispute was carried on by letter between 1899 and 1903, and it was triggered by the pioneering research Hilbert was conducting on geometry, which culminated in the publication of his groundbreaking *Foundations of Geometry* in 1899.[1] It is a dispute that has received a great deal of attention from historians of philosophy, logic, and mathematics. On the whole, what I've read suggests that scholars have tended to side with Hilbert. For example, the editors of Frege's correspondence (Frege, 1980), in their introduction to the Frege–Hilbert exchanges, give a brief survey of the literature on this dispute in which they quote the mathematician H. Scholz as expressing the "dominant view" with the words: "Frege's critical remarks, though very acute in themselves and still worth reading today, must nevertheless be regarded as essentially

---

[1] "The axiomatic approach adopted by Hilbert in this book was to have an enormous influence on the development of twentieth century mathematics and on the way mathematicians looked at their science" (Corry, 1999: 145). This is not to suggest that the axiomatic view promulgated by Hilbert was a completely new approach. As Jeremy Gray notes, some of the elements of Hilbert's view are to be found in the work of Moritz Pasch on projective geometry and also in the writings of the "Italian Formalists". See Gray, 1999: 63–5.

besides the point".[2] Despite this "dominant view", I believe, there still are aspects of this dispute that have not been well understood. In particular, I am inclined to believe that Frege's side of the dispute has, on the whole, not been adequately appreciated.

In studying the correspondence at issue, I was struck by both Hilbert's lack of responsiveness to what seemed to me to be reasonable objections raised by Frege and also by Frege's inability to get Hilbert to see what he found so objectionable in Hilbert's positions. So I looked for something that might have obscured the successful communication of ideas between these powerful thinkers.

I shall not attempt here to go over all the various criticisms and objections that Frege raised to Hilbert's presentation of his version of geometry. Instead, I shall concentrate on the following three targets of Frege's criticisms: (1) Hilbert's claim that the axioms of his geometry are definitions; (2) Hilbert's method of proving the consistency and independence of his set of geometric axioms; and (3) Hilbert's doctrine that if a set of axioms is consistent, then the axioms are "true" and the things defined by the axioms exist. It is this third target that is most obviously relevant to the fourth of my five puzzles. (Here, the reader may wish to review that puzzle.)

How can axioms be definitions?

Let us first consider Hilbert's claim that the axioms of his geometry are definitions. What is wrong with that claim? Here, we have to ponder more than one kind of characterization that Hilbert gave of his axioms. In the introduction to his Festschrift on geometry, Hilbert had written: "Geometry requires ... for its consequential construction only a few simple facts. These basic facts are called axioms of geometry."[3] Notice that in this quotation, Hilbert is claiming that the axioms of his geometry express simple, basic facts.

[2] Frege, 1980: 31. See also Shapiro, 1997: ch. 5, sect. 3.3, and the references given there. A philosopher who bucks this trend is Michael Resnik, who exhibits a very sympathetic understanding of Frege's position in the dispute in his 1980: 106–19. Where my account of the dispute differs from Resnik's is in its emphasis on the contrast between Hilbert's structural view of the axioms and Frege's traditional view. Resnik's account occurs in a chapter entitled 'Deductivism' and, not surprisingly, he emphasizes the deductivistic elements in Hilbert's view of geometry. (What deductivism is and how this view of mathematics is related to my own will be discussed in detail in Chapter 10 below.)

[3] This is quoted in Frege, 1971a: 25 and then criticized. Frege also writes: "Mr. Hilbert's Festschrift concerning the foundations of geometry prompted me to write to the author, setting forth my own divergent views; and out of this grew an exchange of letters which unfortunately was soon terminated" (Frege, 1971a: 22).

In his *Foundations of Geometry*, he tells his readers that the axioms express "certain related facts basic to our intuition" (Hilbert, 1971: 3). From Frege's perspective, these characterizations fit rather closely his own views about the axioms of geometry. "Traditionally," he writes, "what is called an axiom is a thought whose truth is certain without, however, being provable by a chain of logical inferences" (Frege, 1971: 23). In another place, he writes:

In Euclidean geometry certain truths have traditionally been accorded the status of axioms. No thought that is held to be false can be accepted as an axiom, for an axiom is a truth. Furthermore, it is part of the concept of an axiom that it can be recognized as true independently of other truths.[4]

Again, he writes:

Axioms *do not* contradict one another, since they are true; this does not stand in need of proof. Definitions *must not* contradict one another. ... The usage of the words "axiom" and "definition" as presented in this paper is, I think, the traditional and also the most expedient one. (Frege, 1971a: 25, italics mine)

Had Hilbert rested content with the above characterizations of his axioms as simple facts, basic facts, or as expressing facts basic to our intuition, Frege probably would not have been so incensed by what is said in the *Foundations of Geometry* about the axioms.[5] However, in section 3, Hilbert asserts that the axioms there "define the concept 'between'", and he goes on to say, in section 6, that the "axioms of this group define the concept of congruence or of motion". These characterizations prompt the following response from Frege: "How can axioms [that express facts basic to our intuition] define something?" (Frege, 1971a: 25). Furthermore, if the axioms do define something, how then

---

[4] Frege, 1979b: 168. Notice that this quotation sheds light on what Frege meant by 'a priori'. According to Frege, an a priori proposition is one that is provable from general laws "which themselves neither need nor admit of proof" (Frege, 1959: 4). Now what did Frege mean by characterizing the general laws, from which a priori propositions are provable, as laws that do not need proof? Evidently, that these laws do not need a proof to be known. Since the axioms of Euclidean geometry are held by Frege to be propositions "whose truth is certain without, however, being provable by a chain of logical inferences" and which "can be recognized as true independently of other truths", it would seem that these axioms are just such "general laws". It is not surprising, then, that Frege would agree with Kant that truths of Euclidean geometry are a priori, writing: "In calling the truths of geometry synthetic and a priori, [Kant] revealed their true nature" (Frege, 1959: 102).

[5] The mathematician Giovanni Vailati wrote, in a letter to Frege: "I believe that if only Mr Hilbert could make up his mind to renounce his opinion that the axioms represent the 'fundamental facts of intuition' ... all the rest of his exposition could be given an irreproachable form" (Frege, 1980: 173–4).

can they also express facts basic to our intuition, as Hilbert claimed in the beginning?

If they do [express facts], then they assert something. But then, every expression that occurs in them must already be fully understood. However, if axioms are components of definitions, then they will contain expressions such as "point" and "straight line" whose references are not yet settled but are still to be established. (Frege, 1971a: 26)

On this point of dispute, Frege's objection is surely plausible. The problem for Hilbert is that he wanted to have it both ways: the axioms are truths that are basic to our intuition (presumably of space), and also they are definitions of such concepts as congruence and motion.[6] For Frege, that is nonsense. From a logical point of view, the axioms can't be both. If the axioms do express facts basic to our intuition, then they are assertions and cannot be definitions (since definitions for Frege are stipulated or postulated). And if they are definitions, then they cannot express facts basic to our intuition. That Hilbert gave these apparently conflicting characterizations of his axioms implied to Frege that Hilbert hadn't yet arrived at a clear understanding of his own approach to geometry.

Alberto Coffa has analyzed Frege's objection as resting upon the Fregean theory of concepts.[7] The suggestion is that, since Frege's criticisms of Hilbert rest upon an idiosyncratic view of concepts, those who do not share such a conception can safely ignore Frege's objections. But it seems to me that the above objection is quite independent of any such theory. Hilbert allows that definitions are laid down or postulated. On the other hand, propositions that are simple facts, basic facts, or that express facts basic to our intuition are truths, since propositions that express facts are truths. But such truths are truths, *whether or not we lay them down or take them to be truths*. So how can they be definitions which we lay down? Thus, it is reasonable for Frege to ask: how can an axiom be both? Frege's objection that the axioms of Hilbert's geometry cannot be both does not rest upon any appeal to his theory of

---

[6] Perhaps Hilbert could have responded that he was using the term 'axiom' in two different ways; according to the first way, the axioms are taken to be uninterpreted sentences, whereas according to the second, these same axioms are taken to be interpreted in such a way that they express facts basic to our intuition. But such a response presupposes a much more perspicuous grasp of what he was claiming than he, Hilbert, evidently had at this point (judging from the historical evidence).

[7] See Coffa, 1986: 33, where he writes: "It is hardly necessary to emphasize either the extent to which Frege's criticisms depended on his theory of the concept or, indeed, the centrality of that doctrine within Frege's philosophy."

concepts. It is an objection that I, who do not accept many aspects of Frege's theory of concepts, find quite compelling.

To obtain a more perspicuous grasp of Hilbert's point of view regarding his axioms, let us imagine that Hilbert's geometrical theory had been developed as a formalized first-order theory. Hilbert's geometric axioms would then appear as axioms in a deductive theory formalized in the first-order predicate calculus, where the undefined terms 'point', 'line', and 'plane' of Hilbert's book are given as non-logical constants in the vocabulary of this deductive theory. Within this formal setting, we can see why Hilbert might regard his axioms as "definitions": a number of commentators have observed that Hilbert's axioms can be regarded as defining a class of models.[8] Furthermore, since any first-order structure satisfying the axioms would have to be such that the non-logical constants refer to things in the domain of the structure, the things to which these constants refer would have to be related to one another in a definite way, and hence the axioms could be regarded as, in some sense, "implicitly defining" the non-logical constants.[9] One can see why Hilbert would claim, in a letter to Frege, that "to try to give a definition of a point in three lines is to my mind an impossibility, for the whole structure of axioms yields a complete definition" (Hilbert, 1980: 40). It is, of course, the whole set of axioms—and not just a single axiom—that determines what a model of the theory is.

Although the notion of a first-order model had not been explicitly laid out when Hilbert was carrying on his dispute with Frege, there are passages in Hilbert's letters to Frege that suggest that Hilbert was thinking in terms of models of the axioms. He wrote, for example:

[I]t is surely obvious that every theory is only a scaffolding or schema of concepts together with their necessary relations to one another, and that the basic elements can be thought of in any way one likes. If in speaking of my points I think of some system of things, e.g. the system: love, law, chimney sweep ... and then assume all my axioms as [specifying] relations between these things, then my propositions, e.g. Pythagoras' theorem, are also valid for these things. In other words: any theory can always be applied to infinitely many systems of basic elements. (Hilbert, 1980: 40)

[8] See Demopoulos, 1994: 219 n. 20, for references.

[9] Here's how Erik Stenius describes this situation:

[T]he axioms of Euclidean geometry are definitary—not implicitly but rather explicitly—of something that may be called a 'Euclidean system', which is a system having a definite structure like a [sic] algebraic group. And now we may take 'point', 'between' as terms referring to structural entities relatively to such a system, like the different names of chessmen, which refer to structural entities in the game of chess. (Stenius, 1974: 188 n. 12)

Paul Bernays, writing in the *Encyclopedia of Philosophy*, describes Hilbert's axiom system "not as a system of statements about a subject matter but as a system of conditions for what might be called a relational structure" (Bernays, 1967: 497). Similarly, Ian Mueller describes the content of Hilbert's geometrical axioms as "structural" and characterizes the Hilbertian geometry as "the study of structure".[10]

It should be noted that, even within this framework, we can grant that Frege's objection against characterizing the axioms as expressing facts basic to our intuitions is reasonable: if we take the axioms as sentences of a first-order theory in the above manner, then we should also allow that these sentences are not true—not true in the straightforward sense which Frege had in mind. Such first-order sentences could be taken to be true only in the technical sense of being true *in a structure* or true *under an interpretation*. Thus, Hans Freudenthal describes this revolutionary aspect of Hilbert's geometry with the words: "[T]he bond with reality is cut. Geometry has become pure mathematics. ... Axioms are not evident truths. They are not truths at all in the usual sense" (Freudenthal, 1962: 618).

Hilbert never gave an adequate reply to the above objection of Frege's and he continued to provide confusing and conflicting characterizations of his axioms. No doubt, he felt that Frege's objections were mere quibbles and that, mathematically, he was on firm ground in claiming that his axioms were definitions. When Frege found that Hilbert had not altered his

---

[10] Mueller, 1981: 9. Alessandro Padoa was, in some respects, clearer about the foundations of his "deductive theories", as can be seen from what he said in a paper he delivered at the Third International Congress of Philosophy, held in Paris in August 1900:

[D]uring the period of *elaboration* of any deductive theory we choose the ideas to be represented by the undefined symbols and the *facts* to be stated by the unproved propositions; but, when we begin to *formulate* the theory, we can imagine that the undefined symbols are *completely devoid of meaning* and that the unproved propositions (instead of stating *facts*, that is, *relations* between the *ideas* represented by the undefined symbols) are simply *conditions* imposed upon the undefined symbols.

Then, the *system* of *ideas* that we have initially chosen is simply *one interpretation* of the *system of undefined symbols*; but from the deductive point of view this interpretation can be ignored by the reader, who is free to replace it in his mind by *another interpretation* that satisfies the conditions stated by the *unproved propositions*. ...

Logical questions thus become completely independent of *empirical* or *psychological* questions (and, in particular, of the *problem of knowledge*), and every question concerning the *simplicity of ideas* and the *obviousness of facts* disappears. (Padoa, 1967: 120–1)

That same year, Padoa gave a lecture course at the University of Rome on algebra and geometry as "deductive theories". In his lectures, he explicitly took "a model-theoretic view" of his system of logic and set theory (Grattan-Guinness, 2000: 260).

characterizations of the axioms in the second edition of the book, he wrote: "Evidently Mr. Hilbert himself does not know what he means by the word 'axiom'; and consequently it also becomes quite doubtful whether he knows what thoughts he connects with his propositions" (Frege, 1971b: 51).

Hilbert's proofs of the independence and consistency of his axioms

I turn now to Frege's objection to Hilbert's method of proving the consistency and independence of his set of geometric axioms. From the contemporary point of view, Hilbert's proofs can be regarded as straightforwardly model-theoretic.[11] Within the framework of first-order logic, a set of sentences is by definition "consistent" iff there is an interpretation (or structure) in which the set of sentences is true, and to prove the consistency of a set of axioms, one need only show that there is a model or structure in which the axioms are all true. Thus, Hilbert proved the consistency of his set of geometric axioms by constructing a model of the axioms from the real number system (Hilbert, 1971: 28–9).

In order to distinguish the above sense of consistency from the intuitive notion Frege had in mind, the expression 'model-theoretic consistency' will be used with this contemporary sense. ('Model-theoretic consequence' will be used in an analogous way.) For Frege, on the other hand, a set of sentences is consistent if (roughly) it is not possible to deduce a contradiction from it using logic and the definitions of the terms in the sentences. Similarly, a sentence ϕ is independent of a set of sentences Γ if (roughly) it is not possible to deduce either ϕ or the negation of ϕ from Γ using logic and the definitions of the terms in ϕ and the members of Γ. In Frege's sense of the term 'consistent', then, to say that a set of sentences is consistent is to attribute senses to the sentences in the set in such a way that the truth of all these sentences is logically compatible with the meanings they have.[12] In what follows, I shall

---

[11] Cf. "It is one thing to build up geometry on sure foundations, another to inquire into the logical structure of the edifice thus erected. If I am not mistaken, Hilbert is the first who moves freely on this higher 'metageometric' level: systematically he studies the mutual independence of his axioms and settles the question of independence from certain limited groups of axioms for some of the most fundamental geometric theorems. His method is the *construction of models*" (Weyl, 1970: 265).

[12] Cf. "If we take the words 'point' and 'straight line' in Hilbert's so-called Axiom II.1 in the proper Euclidean sense, and similarly the words 'lie' and 'between', then we obtain a proposition that has a sense, and we can acknowledge the thought expressed therein as a real axiom. ... Now if one has acknowledged [II.1] as true, one has grasped the sense of the words 'point,' 'straight line,' 'lie,' 'between'; and from this the truth of [II.2] immediately follows, so that one will be unable to avoid acknowledging the latter as well. Thus one could call [II.1] dependent upon [II.1]" (Frege, 1971b: 103).

use the term *'propositional consistency'* and *'propositional independence'* to refer to the kind of consistency and independence Frege had in mind.

Taking Hilbert's axioms to be expressions of facts basic to our intuition, as Frege did, we can see why specifying a set-theoretical model, using the real numbers, of a set of geometric axioms in no way shows that these axioms have propositional consistency. Indeed, it is easy to see why Frege would maintain: "If Euclidean geometry is true, then non-Euclidean geometry is false, and if non-Euclidean geometry is true, then Euclidean geometry is false" (Frege, 1979b: 169). As Frege was viewing the situation, if the terms 'point', 'straight line', and 'parallel' mean the same thing in both geometries, the axioms of both geometries cannot be basic truths about physical space, since they contradict one another. Furthermore, if the axioms are characterized as basic truths, then there is simply no need for a proof of consistency, for as Frege notes: "Axioms do not contradict one another, since they are true" (Frege, 1971a: 25).

On the other hand, if we regard the axioms as definitions as Hilbert did, then it takes the totality of the axioms to do the defining. In that case, writes Frege:

Those axioms that belong to the same definition are therefore dependent on each other and do not contradict one another; for if they did, the definition would have been postulated unjustifiably. However, neither can one investigate before they are postulated, whether these axioms contradict one another, since they acquire a sense only through the definition. There cannot be any question of contradiction in the case of senseless propositions. (Frege, 1971a: 28)

Of course, Frege has in mind here "propositional contradiction"—not model-theoretic inconsistency. The reason there cannot be a question of contradiction (i.e. "propositional contradiction") if the axioms are senseless (or uninterpreted) is because uninterpreted sentences do not express propositions and hence cannot be propositional contradictories. On the other hand, there is no problem specifying a pair of senseless (uninterpreted) axioms that contradict one another, if one is talking about model-theoretic contradictories.

Let us attempt to gain some appreciation of Frege's classical perspective on this question, by viewing Hilbert's geometric axioms as truths. We can still regard Hilbert's theory as a theory formalized in the first-order predicate calculus, if we make use of what I call "natural language interpretations" (or "NL interpretations" for short).[13] These are the "interpretations" of first-order

---

[13] See Chihara, 1998: 186–7.

languages that *philosophically trained* logicians are apt to consider when "translations" of the logical language into some natural language are seriously contemplated. These "interpretations" do more than what mathematical structures do: they not only assign the relevant sort of sets and objects to the parameters of the logical language in question, they also supply meanings or senses to the parameters. "Interpretations" of this sort assign to each individual constant the sense of some English name or definite description; and they provide each predicate of the language with the sense or meaning of an *English predicate*, where English predicates are obtained from English declarative sentences by replacing occurrences of names or definite descriptions with occurrences of circled numerals.[14] When a logical language is given an NL interpretation, the sentences of the language can be regarded as expressing statements that are true or false (in the straightforward sense which Frege had in mind). For example, if the interpretation $\mathcal{T}$ assigns to 'R²' the sense of the English predicate '① is taller than ②', and to 'a' and 'b' the senses of 'The Governor of California' and 'Barbara Boxer' respectively, then the sentence 'R²ba' expresses the statement 'Barbara Boxer is taller than the Governor of California'. I shall say that a logical language equipped with the meanings or senses provided by an NL interpretation is an *NL interpreted language*.[15]

Notice that the sentences of an NL interpreted language can be true (and not merely true in a structure). Thus, by giving Hilbert's axioms this sort of natural language interpretation, we can have axioms that would satisfy the Fregean requirement that the axioms of geometry express genuine truths, while at the same time allowing the axioms to be syntactic objects that can be said to be satisfied by a set-theoretical structure (or model). Supposing that Hilbert's axioms are so understood, does the production of a set-theoretical structure (or "interpretation") under which the axioms (regarded as syntactic

---

[14] Actually, Benson Mates requires that the occurrences be "direct occurrences", in order to avoid having the circled numerals occur within the scope of terms of psychological attitude or modal operators, but it is not necessary, for our purposes, to go into these complications. See Mates, 1972: 77 for his introduction to English predicates, which he tells us he borrowed from Quine. See for example Quine, 1959: 131–4. However, it should be noted that Quine used the expression 'predicate' to refer to what I am here calling an 'English predicate'. I follow Mates both in calling these linguistic items 'English predicates' in order to distinguish them from the predicates of the formal language, and in treating English predicates as devices used in giving a kind of natural language interpretation to formal languages. See Mates, 1972: 77–86, for a discussion which more accurately gives the point of view of this work.

[15] See Chihara, 1998: chs. 5, 6, and 7 for more detailed discussion of NL interpretations and NL interpreted languages.

strings) are true amount to a proof of the propositional consistency of these axioms?

Consider the following simple example. Let the theory T be a deductive theory formalized in an NL interpreted language whose interpretation has been given as follows:

D: the set of living adult human beings

B: ① is a bachelor

M: ① is married

and let the axioms of T consist of the following:

$(\exists x)Bx$

$(x)Mx$

Obviously, one can prove the model-theoretic consistency of this set of axioms in the standard way (that is, by constructing a set-theoretical model of the set). But would such a proof establish the propositional consistency of this theory? Clearly not. One can see why Frege was skeptical of Hilbert's method of proving consistency and independence.[16]

Hilbert's criterion of truth and existence

Let us now consider Frege's objection to Hilbert's doctrine that, if a set of axioms is consistent, then the axioms are "true" and the things defined by the axioms exist. Frege submitted to Hilbert the following example of a set of axioms:

(A1) *A* is an intelligent being

(A2) *A* is omnipresent

(A3) *A* is omnipotent,

suggesting that if this set is consistent, then it should follow by Hilbert's doctrine that the axioms are true and that there exists a thing that is intelligent, omnipresent, and omnipotent (Frege, 1980: 47). Frege clearly thought any such inference would be absurd, but he could not believe that Hilbert actually maintained any such implausible doctrine. So he asked Hilbert to clarify what he was espousing.

---

[16] It can be seen that I disagree with Shapiro's evaluation of the Frege–Hilbert dispute, expressed with the words, "Frege and Hilbert did manage to understand each other, for the most part. Nevertheless, they were at cross-purposes in that neither of them saw much value in the other's point of view" (Shapiro, 1997: 165).

After Hilbert broke off the correspondence, Frege published a paper sharply criticizing Hilbert's views about the foundations of geometry, repeating many of the objections in his letters. In the essay, the gloves came off and Frege expressed his true attitude toward the above Hilbertian doctrine. This time, he set forth the following set of axioms:

EXPLANATION: We conceive of objects which we call gods.

AXIOM 1. Every god is omnipotent.
AXIOM 2. There is at least one god.

He then wrote: "If this were admissible, then the ontological proof for the existence of God would be brilliantly vindicated" (Frege, 1971a: 32).

This objection again illustrates Frege's misunderstanding of Hilbert's views. Hilbert's axioms of geometry are not assertions about the real world. The terms occurring in Hilbert's axioms, such as 'point' and 'line', are parameters, unlike the terms 'intelligent being', 'omnipresent', and 'omnipotent' occurring in Frege's examples. One would think that Hilbert could have pointed out such differences without much trouble, and in this way advanced the discussion considerably. But he didn't.

The gulf between Hilbert's views and those of Frege

It can be seen that the gulf between Hilbert's new conception of geometry and Frege's classical Euclidean conception was enormous. Frege regarded geometry as a theory of physical space: its theorems were true assertions about space.[17] According to Hilbert's new view, geometry is a branch of pure mathematics, to be developed from axioms that would characterize a kind of structure. Since the sentences of Hilbert's new geometry are uninterpreted sentences, the theorems of the geometry turn out to be not even true statements. What seems clear to most contemporary scholars studying this episode is that the dispute involved a great deal of misunderstanding and arguing at cross purposes, and that these two eminent and brilliant minds were defending quite different conceptions of geometry. Since Frege was obviously approaching Hilbert's pronouncements about geometry from the long-standing traditional perspective, and since Hilbert was developing

---

[17] Such a view of geometry was not idiosyncratic: it was widely held by mathematicians from the classical Greeks to the nineteenth century, and even such a logically acute and geometrically knowledgeable nineteenth-century mathematician as Moritz Pasch held such a view. For Pasch, "geometry was still the science of physical space" (Artmann, 1999: 50).

geometry from a radically new perspective,[18] the burden was surely on Hilbert to explain his approach to Frege. It is evident from the correspondence, however, that Hilbert was either unable or unwilling to provide Frege with a clear and accurate explanation of this new approach. So the question arises: why? Why did Hilbert not grasp the source of Frege's objection and clearly set out his new position on the nature of mathematical axioms? Why did Hilbert fail to see any validity to the objections Frege was raising?

Part of the explanation for the unsatisfactory nature of Hilbert's responses to Frege's objections may be that Hilbert's own views on the nature of geometry were only partially formed at this time, and he may not have adequately sorted out the conflicting tendencies in his own beliefs about the axioms. During the period in which he was developing his ideas on geometry, Hilbert expressed ideas about the axioms of geometry that were strikingly similar to Frege's traditional views about the topic, as is indicated by the following quotation taken from the introduction to a course he gave on geometry in 1891:

Geometry is the science dealing with the properties of space. It differs from pure mathematical domains such as the theory of numbers, algebra, or the theory of functions. The results of the latter are obtained by pure thinking ... The situation is completely different in the case of geometry. I can never penetrate the properties of space by pure reflection, much the same as I can never recognize the basic laws of mechanics, the law of gravitation or any other physical law in this way. Space is not a product of my reflections. Rather, it is given to me through the senses.[19]

Even as late as 1898, Hilbert continued to characterize geometry as a sort of empirical science, asserting:

[G]eometry [like mechanics] emerges from the observation of nature, from experience. To this extent, it is an *experimental science* ... But its experimental foundations are so irrefutably and so *generally acknowledged*, they have been confirmed to such a degree, that no further proof of them is deemed necessary. Moreover, all that is needed is to derive these foundations from a minimal set of

---

[18] I do not wish to suggest that there were no other mathematicians developing geometry in the axiomatic deductive way in which Hilbert proceeded. Herman Weyl opines: "In all this [development of geometry as a 'deductive science'], though the execution shows the hand of a master, Hilbert is not unique. An outstanding figure among his predecessors is M. Pasch, who had indeed travelled a long way from Euclid when he brought to light the hidden axioms of order and with methodical clarity carried out the deductive program for projective geometry" (Weyl, 1970: 265).

[19] This passage is quoted in Corry, 1999: 151.

independent axioms and thus to construct the whole building of geometry by *purely logical means.*[20]

When Hilbert says that geometry is "the science dealing with the properties of space" and refers to the axioms of geometry as "experimental foundations", it is evident that he is not regarding geometry as an uninterpreted formal theory, nor is he taking the axioms of his geometry to be implicit definitions of structures. It seems likely that he is, in effect, presupposing an "intended interpretation" or "standard interpretation" of the language of his geometry. Thus, the traditional view, according to which the axioms of geometry are genuine truths about physical space, is still to be found lurking among his beliefs about geometry, even as he was developing his new view of geometry. In other words, Hilbert's unsatisfactory part in the dispute may have been partly due to the fact that his ideas about his new account of geometry had not yet crystallized and were infused with an earlier and incompatible view.

Still, one wonders why Hilbert failed to appreciate the logical problems inherent in his own account, when Frege had so acutely pointed them out to him. It should be apparent to all fair-minded scholars studying the letters that passed between these two creative thinkers that the amount of effort Frege put into understanding Hilbert's position and argumentation was far greater than the amount of effort Hilbert put into understanding Frege's. At times, Hilbert did little more, in his response to Frege's objections, than to restate his original position. Why, one wants to ask, did he not concern himself more with the problems that Frege was raising? After all, to contemporary philosophers, Frege is a giant. It is hard for us today not to take any objection he raised very seriously. But to Hilbert, who evidently had not the slightest inkling that Frege was a great philosopher, Frege was a relatively unknown mathematics professor, whose standing in the field of mathematics was insignificant compared to his own.[21] The historian of science Leo Corry

---

[20] This quotation is taken from the introduction to a course on mechanics Hilbert taught in the winter semester of 1889. See Corry, 1999: 152 (italics in the text). In claiming that Hilbert's characterization of geometry was similar to Frege's, I do not wish to suggest that Frege held that the properties of space are given to us through the senses, that geometry is an experimental science, or that the axioms of geometry are a posteriori. Frege maintained that the axioms of Euclidean geometry are synthetic a priori propositions: above, n. 4. See also, in this regard, Frege, 1959: 20e–21e, and the quotations from Frege's writings given earlier in this chapter regarding the axioms of geometry.

[21] Cf. the following description, given by Feferman, of Hilbert's standing in the field of mathematics in 1900: "Hilbert was at that point a rapidly rising star, if not superstar, in mathematics, and before long he was to be ranked with Henri Poincaré as one of the two greatest and influential mathematicians of the era" (Feferman, 1998*a*: 3).

has noted that "Hilbert was undoubtedly among the most influential mathematicians at the beginning of this century, if not indeed the most influential one" (Corry, 1999: 149). If, as Corry claims, Hilbert "measured the quality of a mathematician's work by the number of earlier investigations it rendered obsolete",[22] then Frege's comparatively meager output of important mathematical research must have seemed tiny to Hilbert, especially given that mathematicians had paid almost no attention to Frege's logical writings. Hilbert evidently regarded the task of responding to Frege's philosophical scruples as little more than a formal duty, hardly worthy of a great deal of his time and effort.[23]

Frege wished to publish the correspondence; Hilbert did not. Frege wished to continue the dispute; Hilbert broke off the correspondence. It's a pity that Hilbert did not have a higher regard for Frege's abilities—the clouds of confusion surrounding the early development of axiomatic thought would no doubt have been dispersed more rapidly had Hilbert taken Frege's objection more seriously.

Before ending my discussion of the Hilbert–Frege dispute, I would like to note that the Hilbert–Frege correspondence concerning the nature of geometry took place before 1904. The Hilbert quotation in Chapter 1 about consistency and mathematical existence is from his address to the International Congress of Mathematics of 1900. I wish to make clear that I am not claiming that any of the views I attribute to Hilbert in this chapter were retained in his later "philosophical" period. So far as I know, Hilbert nowhere claimed that he viewed all of mathematics in the way he viewed geometry in his early period. There are no indications that, during his later period, Hilbert regarded the totality of axioms of his formalized versions of arithmetic to be mere implicit "definitions" or that he believed that these axioms characterized a kind of structure.

2. SOME SUGGESTIONS REGARDING THE NATURE OF MATHEMATICS

The above discussion of the Frege–Hilbert dispute does suggest an intriguing view of mathematics that needs to be explored: might we not view practically all axiomatized mathematical theories in the way Hilbert regarded geometry?

[22] See Corry, 1999: 153, where it is reported that David Rowe argued that Hilbert used such a criterion of quality.

[23] Grattan-Guinness describes Frege's "intense disappointment" in not being able to leave his "second-ranking university [Jena] for his entire career" (Grattan-Guinness, 2000: 178).

Suppose we take the axioms of the theory as, in effect, characterizing (or implicitly defining) a type of structure. We could then regard the theorems of the theory as being about this type of structure without worrying about what the assertions really say about mathematical objects or about "reality". Using the previous notion of a deductive theory formalized in an NL interpreted first-order language, we can regard the assertions of such a theory as characterizing a type of structure (the first-order models)—ignoring for this purpose the NL interpretation of the language. One advantage of understanding mathematical theories in this way is that we can avoid having to justify any analysis of what the assertions of the mathematical theories truly mean. Trying to understand what mathematical assertions mean usually comes down to trying to figure out what actual individual mathematicians, scientists, engineers, and everyday ordinary working people have in mind when they utter mathematical sentences—a none too easy task.

That mathematicians reason about structures is not controversial. That they even see themselves as reasoning about structures is evident. For example, William Thurston describes mathematical progress thus: "As our thinking becomes more sophisticated, we generate new mathematical concepts and new mathematical structures: the subject matter of mathematics changes to reflect how we think" (Thurston, 1994: 162). Arthur Jaffe and Frank Quinn describe the stages of mathematical discovery in the following way:

Typically, information about mathematical structures is achieved in two stages. First, intuitive insights are developed, conjectures are made, and speculative outlines of justifications are suggested. Then the conjectures and speculations are corrected; they are made reliable by proving them. (Jaffe and Quinn, 1993: 1)

Describing the relationships that mathematicians have to theoretical physicists, they write: "It is now mathematicians who provide [physicists] with reliable new information about the structures they study" (Jaffe and Quinn, 1993: 3). These are ideas that the philosopher of mathematics would do well to ponder. I shall explore them in detail in succeeding chapters. Let us now take up the puzzles.

3. THE FIRST PUZZLE

The structural conception of geometry did not arise in Hilbert's mind in complete isolation from the outside influences of the ideas and pronouncements of other mathematicians. Hans Freudenthal has described the hotbed of ideas in nineteenth-century mathematics, from which Hilbert's conception of a new foundations for geometry arose (Freudenthal, 1962). However, the seeds of the structural view of geometry were sewn hundreds of years earlier. At some period, and I won't speculate exactly when, but surely by the seventeenth-century birth and development of analytic geometry, especially starting with the work of Fermat, it became commonplace to regard the lines and curves discussed in geometry to be existing in space, independently of our constructions. Giorgio Israel has recently amplified this point:

Fermat avance, de manière plutôt explicite, le principe de la correspondance biunivoque entre algèbre et géométrie, en admettant que c'est à partir d'une equation algébrique qu'il est possible de donner un lieu géométrique. La centralité des constructions géométriques est éliminée d'un seul coup. Pour être admissible, il n'est plus necessaire que la courbe soit constructible ... La courbe existe uniquement parce que son équation est donnée. Elle est définie *non pas au moyen d'une construction*, mais comme *le lieu des points qui satisfont à l'équation.*[24]

Underlying this way of conceiving of geometric curves is a way of representing points in space by means of coordinate systems. In the case of a one-dimensional coordinate system, any simple scale makes use of a correlation between positions on a rigid object, such as a ruler or a thermometer, and a quantity, magnitude, or length. In the case of the frequently used *rectangular Cartesian system of coordinates*, a unit distance is arbitrarily selected and each point on a plane is regarded as corresponding to an ordered pair of real

---

[24] Israel, 1998: 202. Here is a loose translation of the above passage: "Fermat put forward, rather explicitly, the principle of the one-one correspondence between algebra and geometry, in allowing that, by means of an algebraic equation, it is possible to give a geometric locus. The centrality of geometric constructions is eliminated in a single shot. To be admissible, it is no longer necessary that the curve be constructible. The curve exists uniquely because its equation is given. It is defined *not by means of a construction*, but as *the locus of points which satisfy the equation.*" Thus, although René Descartes is widely thought to be the originator of analytic geometry, there are good grounds for crediting Fermat with making the decisive move that gave rise to this new branch of mathematics. Here's how Carl Boyer describes the crucial ideas on analytic geometry Fermat put forward in his short treatise entitled *Ad Locos Planos et Solidos Isagoge*: "[Fermat] states in clear and precise language the fundamental principle of analytic geometry:

Whenever in a final equation two unknown quantities are found, we have a locus, the extremity of one of these describing a line, straight or curved.

"This brief sentence represents one of the most significant statements in the history of mathematics. It introduces not only analytic geometry, but also the immensely useful idea of an algebraic variable. ... *Beginning* with an algebraic equation, he showed how this equation could be regarded as defining a locus of points—a curve—with respect to a given coordinate system" (Boyer, 1956: 75). For more on this aspect of Fermat's contributions to mathematics, see Boyer, 1985: 380–2.

numbers, the first member of which gives the distance in units from the origin along the x-axis, and the second member of which gives the distance in units from the origin along the y-axis, the two axes being regarded as perpendicular lines intersecting at a point called the "origin". By means of such correlations, points in space can be regarded as numbers or quantities.

In this way, mathematicians and scientists began to represent lines and curves in space by means of algebraic equations, representing physical space itself as having a mathematical structure which we would now describe as a structure isomorphic to the set of all ordered triples of real numbers, ordered in the familiar way. Under this representation, a position in this space corresponds to an ordered triple of real numbers, a straight line in space corresponds to the set of all such ordered triples that satisfy some linear equation, and a curve in space corresponds to the set of all ordered triples of real numbers that satisfy some algebraic equation. Thus, physical space itself came to be seen as corresponding to the totality of all such ordered triples of real numbers. In this way, physical space—the space about which geometers had been theorizing—was given a mathematically definite structural characterization that could eventually lead to an axiomatization of such a structure. It should be mentioned that this structural representation of space is linked conceptually to an array of methods of measurement and comparison of length of physical lines, as well as procedures for carrying out geometric constructions in physical space, such as drawing straight lines and arcs on flat surfaces. (I shall return to a fuller discussion of these connections in Chapter 9.)

Needless to say, this way of conceiving of space has been extremely theoretically fruitful and practically successful. Indeed, Hilary Putnam once argued that this very success counts as a sort of "quasi-empirical" justification of the principle that there is such a correspondence between the points in space and the ordered triples of real numbers. (These ideas of Putnam will be discussed in more detail in Chapter 5.)

Whether or not, and to what extent, scientists are justified in accepting such a correspondence, one can see why, once this new structural perspective took hold, the Euclidian constructive, modal way of viewing geometry began to lose favor among practicing mathematicians. It is one thing to conceive of geometers as literally drawing (and in this way constructing) points, lines, and curves on a plane; quite another to view them as literally constructing sets of points, or (what comes to the same thing under the above described correspondence) sets of ordered pairs of real numbers. Sets exist; they are not constructed. An indication of this changing mode of regarding the geometric objects is to be found in Georg Mohr's work of 1672, *Euclides Danicus*, in which it is shown that any pointwise construction that can be carried out by compasses and straightedge can be carried out by compasses alone.[25] The historian of mathematics Carl Boyer has noted that although "one cannot draw a straight line with compasses ... if one regards the line as known whenever two distinct points on it are known, then the use of a straightedge in Euclidean geometry is superfluous" (Boyer, 1985: 406). In the above fashion, one can take Mohr to be thinking of geometrical objects, such as lines and curves, as objects to be found, determined, or "known" rather than constructed.

Now at some point in time before Hilbert's foundational work on geometry was published, the vast majority of mathematicians began to conceive of the points and lines of Euclidean geometry in the existential manner. So it is not surprising that Hilbert would be inclined to express the axioms of his version of Euclidean plane geometry in such terms. Furthermore, one of his accomplishments was to construct a model of his geometric axioms from the algebra of real numbers and to show that this model was unique up to isomorphism[26] — something that cries out for an existential version of geometry.

Let us now focus on the first puzzle. In particular, let us take up the question:

> How could Hilbert have felt justified in postulating axioms which assert that there exist an infinity of points and lines, when Euclidean plane geometry was developed and applied for a multitude of centuries without any such commitment to an apparent ontology of imperceptible objects?

We can now see that Hilbert's axioms, when properly understood, do not, in fact, make any such existential assertions. Indeed, we have seen that his geometrical axioms are not assertions at all. They are like the uninterpreted sentences of a first-order theory that, in effect, characterize a kind of structure. Hence, in claiming that his axioms express "facts basic to our intuition", Hilbert was not maintaining that his axioms are assertions about the contents of physical space—that they assert that geometrical points and lines in fact exist.

---

[25] Boyer, 1985: 405. Boyer comments: "So little attention did mathematicians of the time pay to this amazing discovery that geometry uses compasses only, without straightedge, [that the principle] bears the name not of Mohr but of Mascheroni, who rediscovered the principle 125 years later" (406). For a discussion of Mascheroni's proof of the theorem, see Dorrie, 1965: 160–4.

[26] See Harry Goheen's introduction to Hilbert, 1971: 1.

Still, Hilbert's geometry can be applied, much as Euclid's geometry was, to draw conclusions about figures drawn on sheets of paper. One may well wonder how geometry, regarded in the structural way described above, can be reasonably applied in practical everyday situations and in science and engineering, even though the theorems are not, strictly speaking, true statements.

I shall not attempt to explain here in any detail how geometry is applied, but a few generalities may be helpful. Let us begin with the fact that the axioms of this geometry characterize a type of structure. Since the rules of inference by which theorems are derived yield sentences that must hold in any structure that is characterized by the axioms, and since physical space itself can be represented as having a mathematical structure of the sort that is characterized by the axioms—Hilbert proves that his geometry "is identical to the ordinary 'Cartesian' geometry" (Hilbert, 1971: 28–32)—it follows that the theorems proved in the geometry (when given an appropriate interpretation) must hold of the represented structure. Thus, one is justified in drawing conclusions about the lines and points constructed in physical space from the theorems of Hilbert's geometry.

The above point can be illustrated perspicuously by considering the matter in terms of a first-order version of Hilbert's geometry. In that case, none of the theorems of the theory can be regarded as straightforwardly true. Still, we know that the rules of derivation of first-order logic preserve truth in a model, so we can conclude that the theorems will hold in any structure that is a model of the axioms. Since the mathematical structure that physical space is represented as having constitutes, in effect, a model of the axioms, it follows that these theorems must hold of the represented structure, which justifies using these theorems to draw conclusions about the points and lines constructed in physical space. Of course, the points and lines we actually construct do not (and cannot) have all features that geometrical points and lines are characterized as possessing. So the above discussion of applications of the geometry is incomplete. I shall defer for a later chapter a more detailed examination of this topic, when more of my structural view has been developed. For now, I should like to quote what Hilbert wrote about this topic in a letter to Frege:

[T]he application of a theory to the world of appearances always requires a certain measure of good will and tactfulness; e.g., that we substitute the smallest possible bodies for points and the longest possible ones, e.g., light rays, for lines. We also must not be too exact in testing the propositions, for these are only theoretical propositions. (Hilbert, 1980: 41)

4. THE FOURTH PUZZLE

The fourth of the puzzles to be explained in this work looks back at the history of mathematics and finds a string of brilliant mathematicians making roughly the same basic point, namely, that mathematical existence is a matter of consistency. Consider again the quotation from Hilbert. What the great mathematician had in mind will be clearer, I believe, if we replace the word 'concept' that appears in the quotation with the phrase 'type of object'. Then we have him writing:

If contradictory attributes are assigned to a type of object, I say that mathematically the type of object does not exist. So, for example, a real number whose square is − 1 does not exist mathematically. But if it can be proved that the attributes assigned to the type of object can never lead to a contradiction by the application of a finite number of logical inferences, I say that the mathematical existence of the type of object (for example, of a number or a function which satisfies certain conditions) is thereby proved. In the case before us, where we are concerned with the axioms of real numbers in arithmetic, the proof of the consistency of the axioms is at the same time the proof of the mathematical existence of the complete system of real numbers or of the continuum.

Some light is shed on this quotation by the following response he gave to an objection of Frege's during their famous dispute about the nature of axioms (something I discussed earlier):

[I]f the arbitrarily given axioms do not contradict one another with all their consequences, then they are true and the things defined by the axioms exist. This is for me the criterion of truth and existence. The proposition 'Every equation has a root' is true, and the existence of a root is proven, as soon as the axiom 'Every equation has a root' can be added to the other arithmetical axioms, without raising the possibility of contradiction, no matter what conclusions are drawn. This conception is indeed the key to an understanding ... of the lecture I recently delivered in Munich on the axioms of arithmetic, where I prove or at least indicate how one can prove that the system of all real numbers *exists*. (Hilbert, 1980: 39–40)

Now why should a proof of the consistency of the axioms of the real number system be tantamount to a proof of the mathematical existence of the system of all real numbers? It can be seen, from the previous discussion of the Frege–Hilbert correspondence, that Hilbert saw the axioms of his geometrical system as defining a kind of structure. So let us investigate the question of what Hilbert had in mind when he identified mathematical existence with consistency within the framework of the model theory of first-order logic. Let us suppose, in other words, that the axioms Hilbert is writing about were formulated in first-order logic. We know, from a theorem of first-order logic, that a proof of the syntactic consistency of a set of axioms (which is the kind Hilbert seems to have in mind in these quotations)[27] implies that the set of axioms is semantically consistent and hence that there could be a structure of the kind specified by the axioms. In other words, the consistency proof establishes that the mathematical theory of real numbers concerns a genuinely possible kind of structure. This shows us that the theorems of such a system will result in the kind of truths that are of interest to mathematicians, namely, *truths that tell us what would have to be the case in structures of the type in question.* Thus, the consistency proof would give us a sort of guarantee of the real number system's mathematical legitimacy, and this allows us to take the existential statements about the real numbers to be mathematically significant. In this way, we can make good sense of the quotes from Hilbert.[28]

Similarly, one can understand Poincaré's views on the same basic model, especially taking into account his dispute with Bertrand Russell on the nature of geometric axioms.[29] Take is quote from *Science and Method*:

If ... we have a system of postulates, and if we can demonstrate that these postulates involve no contradiction, we shall have the right to consider them as representing the definition of one of the notions found among them. (Poincaré, 1953: 152)

We can understand Poincaré to be asserting that if we have defined some new kind of mathematical object by giving a system of postulates that supposedly characterize the entities in question, and if these postulates involve no contradiction (that is, the postulates are consistent), then these postulates define an acceptable mathematical notion and the entities in question can be considered to "exist" in the special mathematical sense. We can make sense of Poincaré's thoughts on the above structural model, for if the postulates are consistent, then the postulates can be regarded as characterizing a genuinely possible kind of structure so that, in this way, the postulates can be regarded as defining an acceptable kind of mathematical entity.

---

[27] Note, especially, Hilbert's words: "if it can be proved that the attributes assigned to the type of object can never lead to a contradiction by the application of a finite number of logical inferences".

[28] Of course, such a result linking syntactic consistency with model-theoretic consistency cannot be obtained in second-order logic, but we cannot expect Hilbert to have known this, given that the very distinction between first- and second-order logic had not been made when Hilbert was writing on these matters. It should also be noted that Michael Resnik gives an explanation of "Hilbert's puzzling insight that in mathematics consistency is sufficient for existence" that is similar in certain respects to the one I give above. See Resnik, 1988a: 419 n. 20.

[29] See Shapiro's detailed discussion of the Poincaré–Russell dispute in his 1997: 153–7.

Obviously, the thoughts being attributed to Cantor by Dauben (the third of the examples I gave in posing puzzle 4) fit this structural conception nicely. If it is agreed that Cantor's transfinite numbers have been consistently defined, then the above considerations will explain why he would have thought there would be no grounds to reject his new theory. For the consistency of his characterization of his new numbers guarantees that he has characterized a genuinely possible kind of structure—which is all that mathematicians should require in order to accept the validity of these transfinite numbers.

5. SOME CONCLUDING THOUGHTS

Before leaving this chapter, I should like to compare again the structural attitude exhibited by Hilbert towards the axioms of his geometry with the Fregean view that the axioms of geometry are true assertions about the real world. Revert again to the example of the NL interpreted theory formalized in the first-order predicate calculus. Hilbert can be regarded as taking the sentences of the theory as formal (uninterpreted) sentences that are either true or false *in a structure*. Frege can be regarded as taking the sentences of the theory as NL interpreted sentences which are either true or false. From Hilbert's perspective, the axioms of the theory determine a class of structures—the models of the theory—and this is so independently of the meaning or truth of the sentences insofar as they constitute a consistent set. From Frege's perspective, what is important is the truth or falsity of the sentences, and consequently the objective facts of physical space to which the sentences are answerable. Much of the confusion and misunderstanding that infused the Frege–Hilbert controversy arose because these two radically different perspectives were not clearly discerned and appreciated. The confounding of these two perspectives continues to take place even today. This contrast and opposition between the perspectives of these philosophical mathematicians is absolutely fundamental to the investigations of this work, and as we shall see, it will come up again and again in the following chapters.

3

The van Inwagen Puzzle

In this chapter, I shall focus on the fifth of my puzzles, the one concerning van Inwagen's reasoning about typosynthesis and set theory. In order to explain my response to this puzzle, I need to give a preliminary explanation of my use of the term 'structure', which differs slightly from how some of the structuralists use the term.

1. STRUCTURES

What is a structure? Let's start with what is more or less the standard explanation of the concept of structure by saying that *a structure is a domain of objects, with one or more relations on that domain*. What needs to be emphasized from the very beginning is that mathematicians and logicians frequently give what are called "specifications of structures"—what I prefer to call "structural descriptions"—that are abstract and general, leaving out of their specifications any mention of the particular objects in the domain of the structure. Thus, the practice is such that:

(1) One does not have to say what the things are that are in the domain of the structure. They can be regarded as simply "points" or "places" in the structure, to be filled or replaced by actual or possible things to yield an "instance" of the structure.

(2) One does not have to say *how* the things in the relations are related to one another: more specifically, one does not have to explain what properties things must have or what the things must do, or what spatial or temporal relationships the things must bear to one another, to be in these relations. Thus, from the perspective of set theory, one can specify the (binary) relations of a structure by simply giving the relations as sets of ordered pairs.

Resnik's notion of structure

For Michael Resnik, structures are what he calls "patterns". He explains what patterns are thus: "I take a pattern to consist of one or more objects, which

I call positions that stand in various relationships", where these objects ("positions") in the pattern are held to have "no distinguishing features other than those it has in virtue of being the particular position it is in the pattern to which it belongs" (Resnik, 1997: 203–4). What Resnik means to exclude by this last condition can best be explained by way of an example. Consider the particular pattern made by football teams which is called the "single-wing formation". Now a particular tailback of a team that happens to be in this formation will have many distinguishing features other than those he has in virtue of being the tailback in the formation: he will be of a certain age, have a particular weight and height, be enrolled in particular classes, etc. Obviously, then, no particular physical objects can be the positions in a structure, as Resnik uses the term, since every such object will have distinguishing features other than those it has in virtue of being the particular position it is in the pattern to which it belongs.

Compare Resnik's characterization of structures with the mathematician Marc Barbut's characterization: "An algebraic structure" he tells us, "is a whole made up of any *laws of combination* or, synonymously, *operations*" (Barbut, 1970: 374). Notice that there is no suggestion in Barbut's characterization that the objects in the field of the operations must lack any identity or distinguishing features outside the whole that is the structure. I think Resnik's extra condition is something he came up with for philosophical reasons and that it is not essential to the mathematician's conception of structure. I suspect that the following sorts of questions, which philosophers are apt to ask, have not been addressed by most mathematicians:

> Does any object in the natural number structure have the property of having no distinguishing feature "outside the structure"? And if so, is that very property a distinguishing feature "outside the structure"? On the other hand, if that property is not a distinguishing feature, is that because every such natural number has some distinguishing feature "outside the structure"?

> Do the objects in the natural number structure actually exist in the physical world? If so, is that very feature (of existing in the physical world) a feature "outside the structure"? If not, then do not these objects have the feature of not existing in the physical world? And is not that feature a feature "outside the structure"?

For the above reasons, in describing the mathematician's conception of structure, I shall drop the last clause of Resnik's characterization. Mathematicians may not attribute to the objects in a structure any such distinguishing features, and they may have no idea what distinguishing features "outside the structure" such objects could have, but that is far from denying that such objects could have any such features.

In any case, I wish to use the term 'structure' in a way that will permit a structure to have physical objects as positions in the structure. Thus, my use will differ from Resnik's insofar as I allow the possibility of structures with specific physical objects in the domain of the structure. As for the practice of mathematicians and logicians of specifying structures without specifying what the things are that are in the domain of the structure or without saying *how* the things in the relations are related to one another, one can understand such scholars to be specifying a *type of structure* which specific structures may exemplify or instantiate. In summation, my use of the term will conform to one standard use of the term according to which a structure is any domain of objects, with one or more relations on that domain (where by 'relation' I mean to include both monadic relations or "properties" on the one hand, and operations on the other).[1]

Although I shall say much more about structures in later chapters, it may be useful to some readers to examine a simple example. The following theory of linear ordering, expressed in first-order logic, has only one non-logical constant: 'L^2'. The assertions of this theory are all the sentences of the theory that are derivable from the following axioms:

1. $(x)–L^2xx$
2. $(x)(y)(z)((L^2xy \ \& \ L^2yz) \rightarrow L^2xz)$
3. $(x)(y)(L^2xy \ v \ L^2yx \ v \ x=y)$

This theory is not supposed to describe a single structure; it is only meant to characterize a type of first-order structure.

A model M of the above theory is a structure which is such that all the assertions of the theory are true in M. In other words, M is a model of the theory if all the assertions of the above theory are true when the domain of the quantifiers of the theory is regarded as the domain of M, and the non-logical constant 'L^2' is interpreted as standing for the one relation of M.[2]

---

[1] "What is meant by a structure is usually a domain of objects together with certain functions and relations on the domain, satisfying certain given conditions" (Parsons, 1996: 274). I don't mention functions (and operations of any sort) in the above characterization because I take those things to be special sorts of relations.

[2] For a precise definition of 'model' when we are concerned with first-order theories, see Enderton, 1972: 79–83.

Here is a specification of a particular model of the above theory:

The domain of this structure is simply the set: {Al Gore, Tipper Gore, G. W. Bush, Laura Bush}; the one relation of this structure is the *taller than* relation on this domain. I leave it to the reader to verify that all the axioms can be seen to be true when the domain of the quantifiers is taken to be the domain of this structure and when the constant 'L^2' is taken to be the *taller than* relation on this domain.

With these preliminaries out of the way, I turn to my analysis of the van Inwagen's puzzle.

2. My Resolution of the Puzzle

I shall begin my analysis by investigating van Inwagen's belief that, since we cannot classify the membership relation as internal, external, or extrinsic, we can infer that we do not grasp the membership relation and hence do not understand mathematics. Since van Inwagen could not accept such an inference, he concluded that there must be something wrong with his reasoning, even though he knew not what. But why should we be strongly tempted to draw such an inference?

Let us ask: what is it to understand mathematics? Here, we should restrict our considerations to just some area of mathematics, say set theory, and ask: how would it be determined whether some student, who claims to understand set theory, really does understand the subject matter of the theory?

By the usual criteria that we use to determine if someone understands a substantial amount of set theory (imagine a Ph.D. qualifying examination in the subject), we see whether the student can explain the principal concepts of the theory, by giving the relevant definitions and by applying these definitions within the theory. We test the student to see if she knows the axioms or fundamental assumptions of the area, and can also cite, prove, explain, and apply (in a variety of set-theoretical contexts) the principal theorems of the area (both basic and advanced).

It is important to note that the student is not required to know anything about the nature of sets or about the relation of membership apart from knowing what sets exist and how they are related by membership to other sets. In particular, students are not required to know anything about the "intrinsic properties" of sets (that is, the properties a particular set may have other than the relational properties it has—for example, those it has in virtue of being related by the membership relation to other sets). All students of set theory know, for example, that the empty set has no members and that no other set has no members, but such students are not taught, in their mathematics courses, anything about the intrinsic properties of the empty set.

Now nothing in the van Inwagen argument shows that mathematicians do not understand mathematics *according to the above criteria*. Thus, I agree with van Inwagen's reasoning to this extent: If his reasoning truly shows that no one understands mathematics according to the above criteria, then we really would have an absurdity. But of course, it doesn't.

So what is going on? Let us reexamine the typosynthesis theory. Suppose that we regard the theory not as a theory about particular things called 'cherubim' and a particular relation called 'typosynthesis', but rather as simply giving us (or characterizing) a type of structure. The type of structure in question will consist of all the living humans (henceforth labeled 'H') plus ten other objects, which we can label 'C' (for cherubim). Then there is a relation R that is such that:

(a) For every x that is an H, there is both a y that is a C such that xRy and also a z that is a C such that $-xRz$.

(b) For every x and y, if xRy, then x is an H and y is a C.

Now if one thinks of the axioms of typosynthesis theory as just describing a type of structure,[3] then one will feel no need to answer such questions as "Is the relation R an internal or external relation?" Not being able to answer such questions does not indicate a lack of understanding of the theory, since according to this way of regarding the theory, the letter 'R' is not supposed to stand for a specific relation. The letter can be regarded as functioning as a parameter, which stands for different relations in different specific structures.

Set theory reconsidered

Reconsider now the case of set theory. The above considerations are upsetting to Lewis and van Inwagen because, as they understand set theory, it is a theory about real objects, in the way biology is a theory about real things. They have a view of set theory that is similar to the view put forward by Gödel implying that the objects of set theory are things that "exist independently of our constructions"; and that the general mathematical concepts we employ in set theory are "sufficiently clear for us to be able to recognize their

---

[3] Readers who are unfamiliar with the notion of structure will find a detailed discussion of the notion in Chapter 4 below.

soundness and the truth of the axioms concerning" these objects (Gödel 1964b: 262). For Lewis and van Inwagen, the membership relation is like the relation of being taller than: whether some actual living person is taller than another living person is a determinate question of fact, to be determined by going out into the world and measuring. In other words, Lewis and van Inwagen are treating the axioms of set theory in much the way Euclid regarded the axioms of his geometry: as assertions about the real world. So for these philosophers, the questions as to whether the membership relation is intrinsic or extrinsic, internal or external, are genuine questions about a relation that actually obtains among real objects. Similarly, the question as to whether the membership relation obtains because of the properties of the things that are in the relation, or because of what these things may have done, or because of the spatial or temporal relationships that these things have to one another is a genuine question about a relation that obtains among real objects. The fact that we can't answer it strongly suggests, to Lewis and van Inwagen, that our grasp of this relation is fundamentally flawed.

But think of set theory in the way we regarded the typosynthesis theory: think of the axioms of set theory along the lines Hilbert regarded the axioms of his geometry, as specifications or characterizations of a type of structure. According to this conception, any model of the axioms will have some domain of objects to be called 'sets', and a relation among these objects to be called 'the membership relationship'. The axioms tell us what (so-called) 'sets' there must be in the domain of a model and also how the things in the domain must be related by this relation in order that we have a structure of the type in question.[4] Under this way of interpreting the symbols of the language, existential quantification signifies *existence in a structure*, that is, 'There is' tells us that there is something in the domain of the model such that . . . This is like taking the theorem of Euclidean geometry asserting "Any triangle is such as to have angles that sum to 180 degrees" as asserting that any triangle *in a Euclidean structure* will be such as to have angles that sum to 180 degrees.

Return to the criteria of understanding on page 52 above. You will recall that, in ordinary life, to determine if someone understands set theory,

---

[4] Cf. Ian Mueller's comments in Mueller, 1981: 10: "One specifies the structure under consideration by specifying the conditions which it fulfills, i.e., by giving the axioms which determine it. In some cases, the axioms are the only characterization of the structure. For example, in algebra, a group is defined to be any system of objects satisfying certain axioms. In other cases the specification of axioms is an attempt to characterize precisely a roughly grasped structure. Peano's axiomatization of arithmetic and Hilbert's *Grundlagen* are examples."

we see if the person can explain the principal concepts of the theory, by giving the relevant definitions and by applying these definitions within the theory. We test the person to see if she knows the axioms or fundamental assumptions of the area, and also can cite, prove, explain, and apply (in a variety of set-theoretical contexts) the principal theorems of the area (both basic and advanced).

These criteria are sensible and appropriate, if we regard set theory in the way I have been suggesting. These criteria test the candidate's understanding of the type of structure being characterized and his/her grasp of the detailed information that has been developed about this type of structure. Furthermore, if we regard set theory in the way being suggested above, there is no reason why we should expect anyone with the sort of structural understanding of set theory tested by the above criteria to be able to answer the sort of questions about the relation of membership that van Inwagen and Lewis pose. In short, if we accept this way of regarding set theory, the criteria we in fact use to determine if someone understands a substantial amount of set theory are appropriate and the criteria implicit in the van Inwagen–Lewis puzzle are not appropriate, so the puzzle can, in this way, be resolved.[5]

My response to the van Inwagen puzzle, then, does not take the form of attempting to show that 'typosynthesis', 'cherub', 'membership', and 'set' are perfectly well-defined terms. It seems to me that these terms are not adequately defined if the assertions of the respective theories are taken to be genuine assertions about the real world. Similarly, set theory is an unsatisfactory theory if it is taken to be a theory of the sort that genetics or Newtonian physics is. So we can understand the temptation to conclude that we do not understand what membership is: we don't know what particular relation it is. But it would be wrong to conclude that we do not understand set theory and that therefore set theory is an unsatisfactory theory. We can understand the axioms to be merely characterizing a type of structure, in which case 'membership' can be regarded as a parameter—a constant that can stand for different relations in different structures. Then we can see that the sort of criteria we use to determine whether someone understands a substantial amount of set theory is perfectly appropriate. So we don't have to go to mathematicians and say: "You are all wrong and you have to abandon set theory".

[5] My attempt to frame an account of the nature of set theory that is consistent with the criteria we use to determine if someone understands set theory is part of my overall project of attempting to develop an account of mathematics that will fit into the Big Picture discussed in the Introduction.

There is nothing wrong with set theory when seen, in this way, as a characterization of a kind of structure.[6]

Implications of the above solution

Let us survey the positions to which the above reasoning has taken us. It was concluded that, if we take the axioms of set theory to be merely characterizing a type of structure, then the criteria we in fact use to determine if a student understands set theory can be seen to be reasonable and perfectly appropriate. We can, in this way, avoid the absurdity of being forced to conclude that mathematicians do not understand set theory. In other words, if we take set theory to be the mathematical study and development of certain types of structures, we can deliver a nice solution to van Inwagen's puzzle. Since no other equally plausible solution has thus far appeared, we have some grounds for adopting this structural way of viewing set theory.

On the other hand, if we adopt the Gödelian position that the axioms of set theory are genuine assertions about what exists in the real world, than it is hard to see how we can avoid the paradoxical position that mathematicians (even experts in set theory) do not truly understand set theory. All of this throws some doubt on the Gödelian view if it is taken to give the whole story, or even the main story, about the semantics of set theory.

A clarifying point about my position needs to be made. Despite appearances, I am not giving an analysis of the meaning of set-theoretical sentences. Nor am I claiming what individual set theorists are actually asserting when they affirm some theorem of set theory. I am instead staking out a position about *the practice of mathematics*: I am claiming that we can make good sense of this practice if we view it, not as the development of theories about abstract mathematical objects that do not exist in the physical world, but rather as the development of theories about structures.

3. THE GENETICS OBJECTION

Some light can be shed on the view I am putting forward here by considering an interesting objection that was raised to my analysis of the present puzzle.[7]

---

[6] Erik Stenius suggested a view of this kind in Stenius, 1974: 188 n. 12.

[7] The objection was raised on 12 November 1999 by Richard Cartwright at the MIT. Philosophy Colloquium at which I presented an earlier version of this chapter. During the discussion, the objection was taken up and developed by Kate Elgin, with this one difference: whereas Cartwright raised the objection in terms of Spinoza's metaphysical theory, Elgin discussed it in terms of genetics. Here I have chosen to discuss the objection in terms of genetics.

The objection is that the criteria of understanding for set theory described earlier would be sensible and appropriate even for genetics (a scientific theory which cannot be taken to be the kind of structural theory that the above considerations suggest set theory and the typosynthesis theory are). If this objection were correct, then my analysis of this puzzle would seem to provide no support at all for the idea that set theory can be understood to be a mere characterization of a kind of structure, so some response to this objection is called for.

A response to the genetics objection

Let us suppose that a graduate student's understanding of genetics were being assessed in a qualifying examination. Would the student be expected to be able to cite and explain the axioms of genetics and then *prove* the principal "theorems" (both basic and advanced) of the theory? First of all, it is doubtful that there is any *definite set* that contains the fundamental assumptions or axioms of genetics which the student would be expected to formulate or explain in any precise way. Secondly, I doubt very much that any student of genetics would be expected to prove theorems (both basic and advanced) from any such set, even if the elements of the set could be articulated.

But, for our purposes, a more important question to consider is this:

Should we consider satisfaction of the above criteria as sufficient to allow us to conclude that the person understands the science of genetics (as I argued it would in the case of set theory), even if that person knows nothing about the intrinsic natures of the central entities theorized about in genetics? In other words, should the student be expected to know anything about the intrinsic natures of the central entities of the theory?

Certainly, chromosomes and genes are central entities of genetics, insofar as much of contemporary genetic theory is concerned with the description, characterization, mapping, functioning, and regulation of these things, so let us focus on their properties. It is clear, I believe, that the student would be expected to know a great deal about the properties of chromosomes (intrinsic and relational): she would be expected to know that chromosomes are to be found in the nuclei of cells, that chromosomes consist of DNA and proteins, that chromosomes are shaped like ribbons that can be coiled, folded, and cut, and that the behavior of chromosomes in the process called 'meiosis' suggests a mechanism for the Mendelian principles of segregation and independent assortment.

Consider now genes. Genes are said to be "on chromosomes". For example, it is said that the gene that Mendel hypothesized for smooth or wrinkled pods is on chromosome 4 of Mendel's pea plants. It is also said that sex-linked genes are always on the X-chromosome. Even popular publications use such terminology—in the 13 December 1999 issue of *Newsweek*, it was announced that scientists at Britain's Sanger Centre "had compiled a virtually complete list of the chemical letters that encode the genes on human chromosome 22".

Now what is this relation of "being on" that obtains between genes and chromosomes? Is this a relation, like the membership relation of set theory, that no one can classify as internal, external, or extrinsic? Well, is it the case that no one knows anything about the intrinsic natures of genes? Not at all. Much is known about the intrinsic properties of genes. For example, it is known that almost all genes are composed of DNA. Since it is also known that DNA is a double helix with its sugar-phosphate backbones on the outside and its bases on the inside, much would be known about the intrinsic physical and chemical properties of many specific genes. In the case of many specific genes, geneticists would know, for example, their length and their specific base sequence. (This is in sharp contrast to the case of sets, where essentially nothing is known about sets apart from the membership relationships that sets enter into.)

Now what about the relation of "being on" that some specific gene may bear to some specific chromosome? What does the geneticist know about that relation? She would know that to say that some gene is on some chromosome is to say simply that the gene is a physical part of that chromosome. It follows that the relation of "being on" is an internal relation. In short, unlike the membership relation of set theory, the "being on" relation can indeed be placed within the classifications utilized by van Inwagen and Lewis. Furthermore, the relation can be classified on the basis of what any competent geneticist could be expected to know. Those philosophers who believe that a student's knowledge of genetics can be tested in the same basic way we test a student's knowledge of set theory, requiring of the student no understanding of how the entities of the theory are related to actual things in the world, should ask: if that were the case, how could there be such a procedure as gene therapy, whereby specific genes are inserted into the actual cells of real people?[8]

---

[8] An article in the 28 April 2001 issue of the *San Francisco Chronicle* reported that researchers had been able to cure three dogs of blindness by injecting viruses that had been genetically engineered to contain the rpe65 gene directly into their eyes.

I conclude that the criteria of understanding described above for the case of set theory are not appropriate for the case of genetics.[9]

4. THE PROBLEM OF MULTIPLE REDUCTIONS

I would like to point out one nice consequence of taking the position indicated by my proposed resolution of the van Inwagen puzzle. Consider the so-called problem of multiple reductions of number theory that Paul Benacerraf made so famous. For those readers who are unfamiliar with his paper (Benacerraf, 1965), I will give the briefest of sketches of the basic points he makes that are relevant here. Benacerraf describes two imaginary sons of famous mathematicians: one a son of Zermelo and the other a son of von Neumann. The former takes the natural numbers to be the Zermelo finite ordinals; the latter takes the natural numbers to be the von Neumann finite ordinals. Each has been taught by his own illustrious father to believe that, unlike the vast majority of human beings, he truly knows what the natural numbers are. Benacerraf describes them as arguing over who is right. Of course, neither can give any decisive reason for picking his father's favored sets as the natural numbers.

The moral that Benacerraf wants us to draw from this story is that there is no correct answer to the question of which son is right. Both of the children's sets will do equally well as the natural numbers, but so also will countless other omega-sequences of sets.

Now, *from the point of view being set forth here, Benacerraf's moral is to be expected.* If we regard the Peano axioms of number theory as merely specifying a type of structure, then it is by no means odd that we can model this kind of structure in countless ways in set theory. No surprises here.

Similar considerations can lead to the view that the standard axioms of the real number system specify a type of structure. And again, we get multiple reductions of the real numbers to sets of various sorts. For example, the real numbers can be taken to be Dedekind cuts of rational numbers or they can be taken to be Cauchy sequences of rationals. In this case, as was the case of the natural number system, we have a type of structure being modeled in different ways in set theory. No surprise.

---

[9] Although my knowledge of Spinoza's writings is limited, I am confident that one can provide a convincing case that the criteria of understanding described for the case of set theory are also not appropriate for the case of Spinoza's metaphysics, especially if Charles Jarrett's partial formalizations of the theory (in Jarrett, 1978) are at all accurate.

5. AN INTUITIVE UNDERSTANDING OF SET

It might be argued that the structural account I have sketched runs counter to the fact that mathematicians and students of set theory have an intuitive understanding of 'set' and 'membership', which may be vague about certain aspects of what sets are and how they are related to one another by the membership relationship, but which nonetheless underlies the axioms of set theory that we now use. It is these background ideas about sets that guided the pioneers of set theory to formulate the particular axioms for set theory that we now have, and these background ideas still continue to guide set theorists when they attempt to prove new theorems or solve problems in their field. It seems, then, that set theory is about sets in a way that the structural account cannot accommodate.[10]

My response will take the form of accepting the facts upon which the objection is raised, while denying that these facts are incompatible with the structural account. Return to Hilbert's geometry. There is no doubt that Hilbert had an intuitive understanding of physical space and that his background ideas about physical space guided the formulation of the axioms of his geometry. Furthermore, it is safe to say that, even today, mathematicians proving theorems in Hilbert's geometry are guided, at least to some extent, by their understanding of physical space. So it can be said that, in some weak sense, Hilbert's geometry is "about physical space". But that does not imply that this geometry is about physical space in the way in which genetics is about heredity, genes, and chromosomes. Nor does it show that this geometry cannot be a pure mathematical theory that is structural (or model-theoretic).

Return to set theory. Yes, we can allow that set theory is "about sets" in the weak sense described above. But that does not commit us to maintaining that set theory is "about sets" in the much stronger way Gödel thought it was, so that not being able to answer the sort of questions van Inwagen and Lewis asked indicates a lack of understanding of the membership relation and of sets. We can still take set theory to be the kind of structural theory I have described it to be and, in this way, resolve the van Inwagen puzzle.

6. STRUCTURALISM?

Some readers may conclude that I have been giving an argument for the position in the philosophy of mathematics known as "structuralism".

---

[10] I owe this objection to one of the readers for Oxford University Press.

Actually, I think some aspects of structuralism are indeed supported by my account of the van Inwagen puzzle. But there is much in structuralism, especially as it has been developed by its foremost contemporary exponents Michael Resnik and Stewart Shapiro, that is metaphysical in nature and not supported by anything I have said in this chapter. I shall examine the views of these structuralists in the next chapter.

4

Structuralism

What has emerged from the preceding chapter is an inchoate view of pure mathematics that has much in common with the philosophical accounts of the nature of mathematics being developed and defended in great detail by Resnik and Shapiro—views that have been called "mathematical structuralism" or more simply "structuralism". In this chapter, I shall examine various aspects of the structuralist philosophies of mathematics of both Resnik and Shapiro, with the aim of clarifying, by way of contrast with their accounts, my own view of mathematics. I hope, in this way, to avoid having to defend my views against misguided objections that are only suitably directed against structuralism.

1. SHAPIRO'S CHARACTERIZATION OF STRUCTURES

Shapiro defines a *system* to be "a collection of objects with certain relations"—essentially what I mean by 'structure'—and a *structure* to be "the abstract form of a system, highlighting the interrelationships among the objects, and ignoring any features of them that do not affect how they relate to other objects in the system" (Shapiro, 1997: 73). Since the definite article is used in this characterization ("the abstract form"), it is implied that there is one and only one abstract form of a system of the sort described. He gives, as examples of systems, an extended family (a collection of people with blood and marital relations), a chess configuration (a collection of chess pieces under spatial and "possible move" relations), a symphony (a collection of tones under certain temporal and harmonic relationships), and a baseball defense (a collection of people with on-field spatial and "defensive role" relations).

How might one characterize a system that has the "abstract form" of a baseball defense? Presumably, it must have a domain that consists of baseball players, and the players must be positioned so as to satisfy the relevant "defensive role" relations that obtain in the particular baseball defense. Since a particular baseball defense will involve positioning the players in more or less particular places on the field, the question arises: when will that team play "that defense" again? Suppose, for example, that in the fourth inning, the shortstop plays six inches closer to the second base bag than he had in the first inning. Would it still be "that defense"? Would it still be "that defense" if the center fielder is positioned ten feet further to his left than he was in the first inning? What bothers me about Shapiro's characterization of systems and structures is the presupposition that, given some system—even one specified in the vague and imprecise way described above—there will always be one and only one "abstract form" that the system has. Might not such a vaguely specified system have many "abstract forms"? Could not such a system be described as having different forms, depending upon one's perspective and interests? Why should a particular baseball defense have some unique "abstract form"?

The principal primitive of Shapiro's structural philosophy is *exemplification*: a relation that holds between systems and structures. Since the term 'exemplification' is a primitive of Shapiro's theory, it is never defined. However, an intuitive idea of this relation can supposedly be obtained from the fact that a system exemplifies a structure *S* if the system has the abstract form of *S*. Thus, a particular collection of players from a specific professional team, appropriately arranged on the field of play during a World Series game, would presumably constitute an exemplification of some baseball defense. Another example Shapiro gives is that of the even natural numbers ordered in the standard way: such a system, according to Shapiro, constitutes an exemplification of the natural number structure studied in arithmetic.

2. MATHEMATICS VIEWED AS THE SCIENCE OF STRUCTURES

At one point in his book, Shapiro describes pure mathematics as "the study of structures, independently of whether they are exemplified in the physical realm" (Shapiro, 1997: 75), and he then quotes with approval the following description from Resnik:

In mathematics, I claim, we do not have objects with an "internal" composition arranged in structures, we have only structures. The objects of mathematics, that is, the entities which our mathematical constants and quantifiers denote, are structureless points or positions in structures. As positions in structures, they have no identity or features outside a structure. (Resnik, 1981: 530)

Parsons writes that such a view is a familiar one with "a long history, going back to the late nineteenth century". It is the view that "reference to mathematical objects is always in the context of some structure, and that the objects involved have no more to them than can be expressed in terms of the basic relations of the structure" (Parsons, 1995*b*: 74).

The above descriptions explain Shapiro's slogan that "mathematics is the science of structure" (Shapiro, 1997: 75). We can also see why Shapiro claims to be developing a "positive philosophical account" that attempts to say "what mathematics is about" (Shapiro, 1997: 8). Shapiro's structuralism "holds that the subject matter of, say, arithmetic is a single abstract structure, the natural-number structure" (Shapiro, 1997: 9). The natural numbers themselves, according to this view, are the places in the structure, so the expression 'the first even prime number' refers to one of these places.

A problem for the structuralist

Given that structuralism specifies the subject matter of all mathematical theories (namely, structures), it can be seen why structuralists maintain that all mathematical terms refer to positions in structures. This immediately generates a problem for the structuralist, namely, that of saying what a structure is. If one tries to explain what a structure is in the way Shapiro does (recall his characterization of structures in terms of systems, which in turn was characterized in terms of "collections" and "relations"), this seems to require that the theory being used in the characterization have an ontology containing such mathematical entities as collections and relations. But can structuralists take collections and relations to be fundamental mathematical entities? It is hard to see how they can, since they want to maintain that mathematics is always about structures (and indirectly about systems) and not other entities (so set theory is a theory about structures and not about *fundamental mathematical entities*—"sets" that are held to be non-physical entities to which we are not causally related). On the other hand, as Parsons has argued, if they take structures to be the fundamental mathematical entity and try to define what sets are in terms of structures, then they seem to be caught in a circularity (Parsons, 1996: 275). It is surprising that, in characterizing the notion of structure in terms of collections and relations as he does, Shapiro does not address this problem.[1] As will be pointed out later, Shapiro develops a theory of structures that seems to be troubled by just this problem.

---

[1] I say "surprising" because Shapiro is obviously thoroughly familiar with the article of Parsons's that I cite above, as he quotes with approval in his book a number of passages from it.

3. MATHEMATICAL AND ORDINARY STRUCTURES

Before leaving the topic of Shapiro's characterization of structures, it may be helpful to the reader to consider the question of what distinguishes mathematical structures from what Shapiro calls "ordinary structures". One distinguishing characteristic of mathematical structures, Shapiro tells us, is that the relations of mathematical structures are all "formal" (although he is not able to define precisely just what 'formal' means). Relations such as *being taller than* and *being heavier than* are not formal. Any relation involving such physical magnitudes is not formal. Shapiro suggests that if all the relations of a structure can be defined using only logical terminology and the objects and relations of the system, then these relations are all formal (Shapiro, 1997: 99).

There is another difference between mathematical and ordinary structures. Here's how Shapiro compares his baseball defense structure with mathematical structures:

Imagine a system that consists of a ballpark with nine piles of rocks, or nine infants, placed where the fielders usually stand. Imagine also a system of chalk marks on a diagram of a field, on which a baseball manager makes assignments and discusses strategy. Intuitively, neither of these systems exemplifies the defense structure. *A system is not a baseball defense unless its positions are filled by people prepared to play ball.* Piles of rocks, infants, and chalk marks are excluded. (Shapiro, 1997: 98, italics mine)

The positions in a mathematical structure are different: any object, according to Shapiro, can occupy any of the positions in a mathematical structure. For example, anything can occupy the place of the number 6 in the natural number structure: there is no restrictions to a particular kind of thing as there was in the case of the baseball defense. For this reason, Shapiro calls mathematical structures "freestanding" (Shapiro, 1997: 100).

4. SOME WAYS IN WHICH MY ACCOUNT WILL DIFFER FROM THE STRUCTURALISTS'

It should be emphasized that, although I am sympathetic to certain portions of what structuralists have espoused, my own view of mathematics will not attempt to provide an account of the content of mathematical assertions in the way their accounts do: I shall not attempt to describe what "reference to mathematical objects" in typical mathematical theories, such as number theory or set theory, consists in. Indeed, I will not claim to be giving, as structuralists do, some sort of semantical analysis of the actual meaning of the sentences of mathematical theories (literally construed). My position will be that, regardless of what may be actually asserted by contemporary mathematical theories, the fact that some sentence is an assertion or theorem of one of these theories provides us with information that is structural in content (it tells us what would have to be the case in structures of a certain type). Thus, let φ be some sentence of mathematical theory T. Then, the sentence

Any model of T will have to be a model of φ

will be said to express (or to give) the "structural content" of φ. (I shall give a more detailed and precise description of "structural content" in Chapter 9.) A proof of φ will justify asserting the structural content of φ. I will maintain that my account of how the assertions or theorems of contemporary mathematical theories provide us with such structural information should in no way be taken to be an account of what these theorems (literally construed) in fact assert. Hence, my claim that contemporary mathematical theorems provide such structural information will be significantly weaker than the claim that contemporary mathematical theorems in fact assert such and such. This can be seen in the fact that my claims will not require an appeal to detailed empirical studies of the linguistic practices of mathematicians in order to be confirmed or justified, as do the structuralists' claims. Thus, my basic position will be compatible with an analysis of the actual meaning of mathematical sentences that is Platonic, nominalistic, or even structuralistic.

It can be seen already that my structural account of mathematics will differ significantly from that of structuralism. As additional reasons emerge for distinguishing my views from those of the structuralists, it will become evident that there is much in the structuralists' views that I find unacceptable. Thus, serious confusions can arise from labeling my view of mathematics as a form of "structuralism". I prefer to refer to my view as "a structural account of mathematics", thereby distinguishing my views from those of such structuralists as Shapiro and Resnik.

5. ONTOLOGICAL ASPECTS OF SHAPIRO'S STRUCTURALISM

What, then, are the natural numbers, according to Shapiro? In some places, Shapiro tells us that each number is a "part" of the natural number structure (Shapiro, 1997: 76). However, in other places, Shapiro claims, as was suggested by the quote from Resnik, that a natural number is a "place" in the natural number structure (Shapiro, 1997: 77). Evidently, then, structures have parts, some or all of which are places or positions in the structure. Here, one might wish to ask if the logic of part–wholes (mereology) applies to a structure and its parts or places. Shapiro does not answer this question or explore the implications of taking the positions in a structure to be "parts" of that structure.

Consider the following quotation from Shapiro's book:

> [A] natural number is a place in the natural-number structure, a particular infinite pattern. ... The number 2 is the second place in that pattern. Individual numbers are analogous to particular offices within an organization. We distinguish the office of vice president, for example, from the person who happens to hold that office in a particular year ... Similarly, we can distinguish an object that plays the role of 2 in an exemplification of the natural-number structure from the number itself. The number is the office, the place in the structure. (Shapiro, 1997: 77)

In this passage, we find Shapiro's characterization of the "points" in a mathematical structure as the *places to be filled up* by objects in an exemplification of that structure. But Shapiro also wishes to treat the points in a structure as bona fide objects—objects that can constitute the domain of a system that exemplifies a structure (as the example above of the natural numbers shows). Thus, we get what he calls the "places-are-objects" perspective, from which the points in the natural number structure are viewed as genuine objects: "In mathematics," he tells us, "the places of mathematical structures are as bona fide as any objects are." This is why he feels he can say that "each structure exemplifies itself".[2] Of course, in saying that, Shapiro is implying that each structure is itself a system—that is, each structure is a system that has an abstract form, which is none other than itself.

The doctrine that each structure exemplifies itself brings to mind various conundrums of self-predication, such as the "third man regress",[3] which have been attributed to traditional Platonism. A structure is the abstract form of a system, and insofar as it exemplifies itself, it must be a system which has as its form the very form that it itself is. The implications of Shapiro's doctrine are

---

[2] Shapiro, 1997: 89. Later, he claims that "the places in the natural-number structure can be occupied by places in other structures (like finite von Neumann ordinals). Even more, the places in the natural-number structure can be occupied by the same or other natural numbers" (100).

[3] For a detailed discussion of the third man argument, see Malcolm, 1991. Michael Hand has developed a third man argument against structuralism in Hand, 1993. For Shapiro's reply to Hand, see Shapiro, 1997: 90.

similar to those of the Platonic proposition that the form of triangularity is itself triangular.[4]

There is something very strange about predicating triangularity of the form of triangularity. Does this form have three sides? Evidently it does, since a thing lacking three sides can hardly be triangular. But if it does have three sides, how long are its sides? And if the sides of the form of triangularity have no definite length, how can they be sides? Can a side with no definite length still be a side? It would seem that we have a situation that is bubbling over with paradox.

Traditional Platonists have wanted the form of triangularity to be something that particular triangles possess ("that plane figure you have drawn has the form of a triangle"). But they have also wanted the form of triangularity to be an object in its own right, as bona fide and real as the particular things that have that form.[5] Furthermore, they have held that these objects exist independently of the existence of any particulars that may instantiate (or have) that form. Similarly, on the one hand, Shapiro wants a structure to be the abstract form of some system, but on the other hand, he wants a structure to be an object with parts that are as real as any concrete object in the physical universe. One can see how Shapiro has come to adopt a kind of mathematical realism, according to which:

(1) Structures exist, whether or not they are exemplified in the physical world,

and

(2) The terms of a mathematical theory (such as arithmetic or set theory) refer to or denote genuine objects—the parts of the structures that form the subject matter of the theory.

He calls his view "*ante rem* structuralism", following in the tradition in which Plato's view of universals (his theory of forms) was called "*ante rem* realism". Traditional *ante rem* realism is taken by Shapiro to be the view that universals exist prior to and independently of any objects or systems that may

---

[4] I do not wish to attribute this doctrine to Plato himself, although my colleague Alan Code has informed me that the view is certainly "Platonist" and arguably Plato's: the self-predication assumption, he tells me, is applied to a wide range of forms in the middle dialogues. For example, there is a passage in the *Phaedo* (74) in which Socrates commits himself to the view that equality is equal and never unequal.

[5] The idea that a form is an object that is an instance of the form is closely related to the idea that a form is a paradigm case of the things that have that form. The paradigm case idea is discussed at great length in Malcolm, 1991: chs. 8–10.

instantiate or exemplify them (Shapiro, 1997: 84). Shapiro's own brand of *ante rem* structuralism takes "structures, and their places, to exist independently of whether there are any systems of [physical] objects that exemplify them" (Shapiro, 1997: 9). This type of realism is to be contrasted with "*in re* realism*", a view attributed to Aristotle, according to which universals have no existence independent of their instances.[6]

The third man objection

Michael Hand has criticized Shapiro's *ante rem* realism on the grounds that, as he sees it, Shapiro's structures must be nothing less than special kinds of universals or "forms" that are discussed in Plato's philosophy, and hence are open to various sorts of objections similar to the traditional objections, such as the third man regress, that have been leveled against the Platonic theory (Hand, 1993: 190–1). Hand's reasons for attributing to Shapiro a commitment to features of the Platonic doctrine of forms is to be found in the way Shapiro introduces his structures and in his description of his *ante rem* realism:

After all, abstract patterns, structures, are not entities newly posited by the structuralist to serve as the realm of mathematical reference. Instead, the structuralist is making use of things that we already know something about, and that we already put to use metaphysically in various ways. Since this is so, ... she is responsible to the metaphysical uses to which we already put them. (Hand, 1993: 188)

Shapiro has responded to Hand's objection by denying that his structures are Platonic forms; his structures, he tells us, can be taken to be a newly specified type of entity that is only similar in various respects to the Platonist's forms— they should not be taken to be just special kinds of Platonic forms. Shapiro tells us that his motivation for postulating these structures is different from the ones the traditional Platonists had for postulating their forms: he emphasizes, in particular, that his structures were not postulated to explain why "many are one", as were the Platonist's forms (Shapiro, 1997: 90).

Shapiro's reply to Hand, however, is by no means satisfactory. Assume that, by 'form' (or 'abstract form') he means to be referring to a new type of entity, distinct from Platonic forms and hitherto unknown to philosophers, and reexamine his explanation of what he means by 'structure'. Having explained what systems are, he tells us that a structure is the "abstract form" of a system. Now, how can he expect his readers to know what "the abstract form of a system" is, since he has not explained what he means by "abstract form" or

---

[6] See Shapiro, 1997: 84–5, for further details.

what this new type of entity is? How can he consider such an explanation to be logically satisfactory? Imagine that a philosopher, attempting to explain what he means by 'beaujum', tells us that "a beaujum is the glump of a game of chess"—having at no time explained what a glump is. Something is surely amiss here.

Also, what motivation has Shapiro supplied for postulating his *new kind of entity*—his "forms"? It might be thought that the existence of Shapiro's abstract forms are required by any structuralist philosophy of mathematics, so that it could be argued that the philosophical reasons for adopting a structuralist philosophy of mathematics gives us grounds for believing in Shapiro's forms. But it is evidently not the case that all structuralist philosophies presuppose the existence of such forms, since Resnik has developed a brand of structuralism (to be discussed in more detail shortly) that supposedly requires the existence of neither Shapiro's abstract forms nor entities like Shapiro's systems.[7]

6. WHY BELIEVE IN THE EXISTENCE OF STRUCTURES?

So what good reason does Shapiro have for postulating and believing in the existence of the abstract forms of systems? Even a superficial perusal of the history of philosophy will provide us with numerous examples of philosophers postulating or assuming the existence of all sorts of undetectable metaphysical entities for the flimsiest of reasons. Surely, today, we expect some sort of rational justification for such a postulation. In the particular case of Shapiro's postulation, what explanatory function do the postulated entities of Shapiro's theory of structures serve? Why should we suppose that there are such things? I do not deny that (in some sense) there are different systems that are strikingly similar in form, but why should I believe that *there is* some entity, a form, that all these systems have in common? I am here reminded of Quine's dispute with McX in "On What There Is". McX argues that "some

---

[7] Resnik writes: "One way of developing an ontological structuralism is to take it as positing an ontology of structures, themselves seen as universals that can exist prior to their instances ... I have distanced myself from such a view" (1997: 268–9). Cf. also: "Realism about mathematical objects does not commit one to realism about structures ... simply because current mathematics does not affirm the existence of structures" (1997: 261). Also, note: "Some philosophers ... have wanted to take structural properties construed as metaphysical universals, as primitive entities and interpret mathematics within a theory of universals. Their mathematical realism is a consequence of their being realists about properties first. I am a realist about mathematical objects without being a realist about properties at all" (1997: 261). I shall discuss Resnik's "no-structure" view in more detail later in this chapter.

houses, roses, and sunsets all have something in common, and this which they have in common is to be called 'an attribute'". Quine responds:

> One may admit that there are red houses, roses, and sunsets, but deny, except as a popular and misleading manner of speaking, that they have anything in common. The words 'houses', 'roses', and 'sunsets' are true of sundry individual entities ... but there is not, in addition, any entity whatever, individual or otherwise, which is named by the word 'redness' ... and it may be held that McX is no better off, in point of real explanatory power, for all the occult entities which he posits under such names as 'redness'. (Quine, 1961a: 10)

I cannot find anywhere in Shapiro's book anything like an *explicit* argument for believing in the kind of abstract forms or structures that he postulates. However, he does make an implicit case for accepting his *ante rem* view of mathematics. His basic strategy is to undermine the main nominalistic rivals to his realistic account of mathematics and then to argue that his is the most perspicuous account of mathematics that is available. Thus, he raises a number of specific arguments against what he considers the chief anti-realist accounts of mathematics. He also attempts to show that these nominalistic theories do not have any epistemological advantages over his realistic account. He then argues that "*ante rem* Structuralism is the most perspicuous account of contemporary mathematics" (Shapiro, 1997: 11).

I shall take up the objections that Shapiro raises to his nominalistic rivals in Chapter 7. In this chapter, I shall sow some seeds of doubt regarding his claim that his *ante rem* structuralism provides us with the most perspicuous account of contemporary mathematics, questioning the overall coherence of his own ontological views. But first, I shall take up a view of mathematics that is similar to, but quite distinct from, Shapiro's version of structuralism.

7. RESNIK'S EARLY THESES ABOUT STRUCTURES AND POSITIONS

Shapiro's *ante rem* realism has many points of similarity with a view of structures that Resnik once put forward. In a 1981 article, Resnik accepted the realist view of mathematics of his teacher Quine (to be discussed in the following chapter), according to which mathematics is about genuine objects that do not exist in physical space and time. As was pointed out in Chapter 1, such a position gives rise to the problem of how we humans, who exist in space and time, could gain knowledge of objects that do not exist in space and time. Mathematical objects appear to be completely isolated from any sort of causal interaction with us. Furthermore, since Resnik believed that no mathematical theory could do more than determine its objects up to iso-morphism, he had to face the paradox that the mathematician seems to be reasoning about specific things, even though it is impossible for the math-ematician to say definitely what these specific things are. These are philo-sophical problems that arise, according to Resnik, because of a "fundamental misconception" of the nature of mathematics:

[I]f we conceive of the numbers, say, as objects each one of which can be given to us in isolation from the others as we think of say, chairs or automobiles, then it is difficult to avoid conceiving of knowledge of a number as dependent upon some sort of interaction between us and that number. The same line of thought leads us to think that the identity of a number *vis-à-vis* any other objects should be completely determined. (Resnik, 1981: 529)

In fact, says Resnik, we are never given the objects of a mathematical theory in isolation but only in structures. For Resnik, mathematical objects are "points" or positions in structures which have "no identity or features outside of a structure" (Resnik, 1981: 530). For this reason, he suggests, mathematical objects are sometimes said to be "incomplete".

The above structuralist view of mathematics suggested to Resnik a way of dealing with Benacerraf's problem of multiple reductions which I discussed in Chapter 3. Benacerraf concluded, from the fact that there is no correct answer to the question of whether the natural numbers are the von Neumann finite ordinals, the Zermelo finite ordinals, or any of countless other systems of sets that form progressions, that the natural numbers are not sets of any sort. Indeed, he went on to argue that "numbers are not objects at all" (Benacerraf, 1965: 70), that "number words do not have single referents" (71), and that "there are no such things as numbers" (73). Benacerraf then took the position that such questions, as to whether the natural number 3 is identical to Caesar or to some particular set such as $\{\{\{\emptyset\}\}\}$, are simply senseless or "unseman-tical" (Benacerraf, 1965: 64–7).

Resnik's structuralist solution to the multiple reduction problem is partially in line with Benacerraf's views. Unlike Benacerraf, Resnik wished to maintain that numbers are bona fide objects: natural numbers are points in the natural number structure. However, he agreed with Benacerraf that it makes no sense to ask if some point in the natural number structure is or is not identical to some point in some structure studied in set theory. For Resnik, to ask whether the natural numbers are von Neumann finite ordinals or Zermelo finite ordinals is to ask a meaningless question with no sensible answer.

Thus, Resnik arrived at the following sort of ontology for mathematics.[8] Mathematics deals with two sorts of things: structures and positions in structures. About these two sorts of things, Resnik advanced the following theses:

[1] If α and β are structures, then it makes no sense to say either that α is identical to β or that α is not identical to β. As he sometimes put it, there is simply no fact of the matter as to whether identity or non-identity holds between structures—structures are simply excluded from the field of the identity relation.

[2] If x is a position in one structure and y is a position in a different structure, then it makes no sense to say either that x is identical to y or that x is not identical to y.

Thesis [2] is what I once called "the doctrine of the nonsensicality of trans-structural identity" (Chihara, 1990: 133). Of course, there is something paradoxical in the very statements of these theses, since the second thesis makes reference to the non-identity of structures—something that is supposedly senseless according to thesis [1].[9] Not surprisingly, Resnik has wavered in his advocacy of these theses. For example, in a 1988 article, Resnik asserts: "I count patterns as identical when they are generated by the same positions and the same relations. I take relations to be identical when they are coextensive" (Resnik, 1988a: 411 n. 16). But even in that article, Resnik runs into serious difficulties in trying to sketch a coherent and workable view of structures.[10] In his more recent works, Resnik has attempted to extricate himself from the difficulties of his earlier positions by taking the more radical position that there are no structures at all. (Resnik's more recent views on the nature of structures will be taken up later in this chapter.)

8. SHAPIRO AND THESIS [2]

Returning to Shapiro's account of structures, he agrees with both Benacerraf and Resnik on one point (to a degree): he agrees that identifying positions in different structures is generally meaningless, writing,

[I]t makes no sense to pursue the identity between a place in the natural-number structure and some other object, expecting there to be a fact of the matter. Identity

---

[8] In Resnik, 1981, which it should be noted was subtitled "Ontology and Reference".

[9] I present a critical discussion of this view of structures in Chihara, 1990: ch. 7, sect. 5.

[10] See, for example, the discussion on 413–14.

between natural numbers is determinate; identity between numbers and other sorts of objects is not, and *neither is identity between natural numbers and the positions of other structures*. (Shapiro, 1997: 79, italics mine)

It is clear from what Shapiro says in this context that not only does he think that such questions as 'Is 2 identical to {{∅}}?' (positions from different structures) make no sense, but he also believes that such questions as 'Is 2 identical to Julius Caesar?' have no determinate answers to be discovered (Shapiro, 1997: 79). In this way, Shapiro's form of structuralism is supposed to resolve the "Caesar problem" (which arises when one asks if a natural number is identical to Julius Caesar) that supposedly bedeviled Frege's account of the natural numbers (see Shapiro, 1997: 78–9). Also, on a similar point, he writes:

[O]ne can ask about *numerical* relations between numbers, relations definable in the language of arithmetic, and one can expect determinate answers to these questions. . . . But if one inquires whether 1 is an element of 4, there is no answer waiting to be discovered. It is similar to asking whether 1 is braver than 4, or funnier. (Shapiro, 1997: 79)

In other words, both are nonsense.
He continues:

It is nonsense to ask whether *the shortstop* is identical to Ozzie Smith—whether the person is identical to the position. Ozzie Smith is, of course, a shortstop and, arguably, he is (or was) the quintessential shortstop, but is he the position? (Shapiro, 1997: 79–80)

It can be seen that Shapiro accepts some form of the nonsensicality of trans-structural identity. Notice that this nonsensicality principle applies not only to the positions in mathematical structures but also to the places in ordinary structures (such as a baseball defense). Obviously, Shapiro is not merely concerned with ruling out the meaningfulness of certain sentences of mathematics (such as '2 = {{∅}}'). He also wishes to show the illegitimacy or absurdity of asking certain sorts of questions or making certain sorts of inquiries.

Shapiro later adds a caveat to his acceptance of thesis [2]:

I do not wish to go as far as Benacerraf in holding that identifying positions in different structures . . . is always meaningless. On the contrary, mathematicians sometimes find it convenient, and even compelling, to identify the positions in different structures. . . . cross-identifications like these are matters of *decision*, based on convenience, not matters of discovery. (Shapiro, 1997: 81)

If mathematicians are able to decide, based on convenience, that the "whole positive numbers" of the real number structure are the numbers of the natural number structure, it would seem that these identities are matters of stipulation. I shall call Shapiro's thesis that mathematicians are able to make such stipulations 'the stipulation thesis'. Then we can say that Shapiro accepts a form of thesis [2] that is qualified by the stipulation thesis.

9. SHAPIRO'S REJECTION OF THESIS [1]

One important respect in which Shapiro's view of structures deviates from that of Resnik's is in its rejection of Resnik's thesis [1] (recall that this was the doctrine that there is no fact of the matter as to whether identity or non-identity holds between structures). Shapiro notes that his doctrine that structures are genuine objects, when conjoined with [1], conflicts with Quine's principle "no entity without identity", which is amplified as the doctrine that if "we do not have an identity relation, then we do not have bona fide objects" (Shapiro, 1997: 79):

Quine's thesis is that within a given theory, language, or framework, there should be definite criteria for identity among its objects. There is no reason for structuralism to be the single exception to this. If we are to have a theory of structures, we need an identity relation on them. ... [I]f one is to speak coherently about structures, then such a theory is needed. (Shapiro, 1997: 92)

Shapiro then sets out to provide at least a sketch of such a theory of structures. What he gives us is a list of its main axioms. Here are the first four of these axioms:

Infinity: There is at least one structure with an infinite number of places.

Subtraction: If S is a structure and R is a relation of S, then there is a structure S' which is isomorphic to the system that consists of the relations and places of S except for R.[11]

Subclass: If S is a structure and c is a subclass of the class of places in S, then there is a structure that is isomorphic to the system that consists merely of c without any relations.

Addition: If S is a structure and R is any relation on the places of S, then there is a structure isomorphic to the system that consists of the places and relations of S together with R.

---

[11] I have stated Shapiro's axiom without bringing in functions as he does, primarily for the sake of simplicity, since functions can be regarded as a special sort of relation.

These are followed by two more that bring to mind the Zermelo-Fraenkel set theory upon which Shapiro consciously models his theory of structures:

> *Power Set*: For any structure S, there is a structure T such that each subset of the places of S is related to a place of T in such a way that there must be at least as many places of T as there are subsets of the places of S.
>
> *Replacement*: Suppose that S is a structure and f is a function such that, for every place x of S, fx is a place of a structure S_x. This axiom tells us that, for any structure S, there is a structure at least as large as the structure one would get by "replacing" each place in S with the places in a structure S_x.

The system sketched here is said by Shapiro to be a "variant of second-order ZFC" (Shapiro, 1997: 96). Presumably, the system is envisaged as some sort of formal system formulated in a many-sorted second-order logic, with distinct variables for structures, places, relations, and functions (Shapiro, 1997: 93).

Identity between structures is defined as follows:

Structure S is identical to structure T iff S is isomorphic to T.

How identity *within* structures is handled formally is not at all clear to me, especially in view of Shapiro's acceptance of a qualified form of Resnik's thesis [2]. The reason I have problems with Shapiro's account above is this: if the above axioms are formalized in a second-order logic, as is suggested in the passage quoted, then it would seem that we need to quantify over places in structures. Since Shapiro thinks that these places are bona fide objects, and since he accepts Quine's thesis about the need for a criterion of identity for all bone fide objects, his theory is committed to there being a general criterion of identity for places in structures. But he never supplies the reader with any such criterion. And it is doubtful that his commitment to such a criterion of identity is even consistent with his acceptance of his version of thesis [2]. Note: there is no axiom of the object theory that provides us with such a criterion of identity, and there is no explanation in the metatheory that fills this lacuna. Furthermore, it seems to be no trivial task to come up with a really workable criterion of identity of places.

10. SOME PROBLEMS WITH SHAPIRO'S
ANTE REM STRUCTURALISM

A problem with Shapiro's stipulation theory

Let me first say that I have difficulty making sense of Shapiro's stipulation thesis. According to his *ante rem* realism, the parts or positions of a structure are genuine entities that exist independently of our thoughts and theories and also of any objects or systems that may instantiate or exemplify the structure. Now suppose that, at one point prior to any identity stipulations, x is a position of structure A and y is a position of structure B and $A \neq B$ (according to the isomorphism criterion). Then, prior to any stipulations about these entities, thesis [2] is in force, so there is no fact of the matter as to whether $x = y$. Then it would be incorrect to say that identity holds of the ordered pair $\langle x, y \rangle$. At some later time, the mathematicians stipulate that $x = y$. Is it then correct to say that the identity relation now holds of the ordered pair $\langle x, y \rangle$? If so, then it would have to be the case that, at one time, the identity relation did not hold of the ordered pair $\langle x, y \rangle$, and that, at another time, the identity relation did hold of this ordered pair. Does that make any sense at all? How can anyone bring it about by mere stipulation that what does not hold of an ordered pair at one time, does hold at another time? I can see how we can stipulate how we are to talk in such and such situation or how we are to use such and such an expression. But I don't see how stipulation can bring it about that there is such a fact of identity when, before, there was no such fact.

Here is one way of making sense of Shapiro's stipulation theory.[12] Let us say that, where bivalence holds for a declarative sentence S (i.e. S is either true or false), then there is a fact of the matter as to whether S is the case. Now, in those cases where Shapiro wants to say that it makes no sense to say that S or not-S (i.e. where he asserts that bivalence does not hold), we shall say that there is no fact of the matter as to whether S is the case. We can take Shapiro's stipulation thesis not to be the assertion that the mathematician's stipulation brings it about that there is a fact of the matter, where before the stipulation there was no fact of the matter. Instead, we can interpret Shapiro's claims to be that the mathematician's stipulation merely expresses a decision about how they are going to talk or use language. Thus, the stipulation that the positive whole numbers of the real number structure are to be identified with the positions in the natural number structure is, under this interpretation, the stipulation that the mathematician is going to talk and write as if the positions in the respective structures specified are one and the same. It is a stipulation about how they are going to talk and use language, and as such there are no metaphysical

---

[12] This was suggested to me by John MacFarlane.

mysteries about it. Seen in this light, Shapiro's version of thesis [2] is little different from Resnik's.[13]

Difficulties with Shapiro's views about ordinary structures

Consider what Shapiro says about "ordinary structures", such as a baseball defense: "[O]ne might say that a baseball defense consists of four infielders, arranged thus and so, three outfielders, and so on" (Shapiro, 1997: 74). Shapiro makes it clear that for any system to be an exemplification of a baseball defense, the domain of the system must consist of real human beings—a collection of stones or flags will not do, even if the elements of the collection are arranged on a baseball field in exactly the appropriate places. Thus the above description of the structure makes it clear that positions of the "baseball defense" must be occupied by players ("infielders", "outfielders", etc.).[14] Furthermore, the relation of the structure must be one involving relative spatial relationships (the "defensive role" relation). In particular, since the relation of the structure involves a physical magnitude such as distance, the places of this structure cannot be filled by abstract objects, "because such objects do not have distance relations with each other" (Shapiro, 1997: 100). But how is such a requirement compatible with Shapiro's doctrine that every structure is self-exemplifying? The places or positions in the structure (the baseball defense) are surely not real human beings. So how can they be baseball players? Is Shapiro's account of structures intelligible? I do not think it is.

Problems involving the identity of positions

Consider the following simple system S that has a domain consisting of the two persons Al and Jo, and whose only relation in the system (other than identity) is *being citizens of the same country and having the same height*. It turns out that Al and Jo are both citizens of the United States and they have the

---

[13] I have learned from Jonathan Kastin that, at the 1999 Eastern Division American Philosophical Association meeting, Shapiro responded to criticisms of the stipulation thesis in a paper by Kastin by accepting the criticism and changing his view about cross-structural identities. According to the revised view, there simply are no cross-structural identities at all—not even stipulated ones. However, in a more recent work, Shapiro still maintains that "it is surely wise to identify positions in the natural number structure with their counterparts in the integer, rational, real, and complex number structures" (Shapiro, 2000: 267).

[14] "A system is not a baseball defense unless its positions are filled by people prepared to play ball. Piles of rocks, infants, and chalk marks are excluded" (Shapiro, 1997: 98).

same height. Obviously, neither has the same height as anyone else in the domain apart from the other.

Now Shapiro holds that we can *apprehend particular structures* "through a process of pattern recognition, or abstraction":

One observes a system, or several systems with the same structure, and focusing attention on the relations among the objects—ignoring those features of the objects that are not relevant to these relations. For example, one can understand a baseball defense by going to a game (or several games) and noticing the spatial relations among the players who wear gloves, ignoring things like height, hair color, and batting average, because these have nothing to do with the defense system. (Shapiro, 1997: 74)

We can do the same with the system S above. The abstract form we seek is not hard to make out. It is the structure S^* with a domain consisting of two positions α and β, and a binary relation R which is such that xRy iff x and y are citizens of the same country and x has the same height as y. Furthermore, we have $\alpha R\beta$, $\alpha R\alpha$, $\beta R\beta$, and $\beta R\alpha$. Thus, any system that exemplifies S^* must have a domain consisting of two persons and these two persons must be citizens of the same country and have the same height.

Shapiro is also committed to the view that the two positions in the domain of S^* are distinct entities that actually exist. But have I specified two distinct entities α and β? My intuition is that I have not. If α is distinct from β, what makes it a distinct entity? What property does α have that β does not have? Well, we can say that α has the property of being distinct from β, and that is a property that β does not have. Of course, α has the property of being distinct from β only if α is distinct from β. Must not α have some qualitative property in virtue of which it is distinct from β? Must not this "being distinct from β" be grounded in something other than just being distinct from β? Graeme Forbes, for one, has argued strenuously for a *principle of grounding*, according to which "it is part of the concept of identity ... that there are no ungrounded facts about such identities" (Forbes, 1985: 128). According to his principle, there must be some fact (distinct from the mere fact of the identity or difference) in virtue of which that identity or difference obtains. If one were to accept some such principle, then Shapiro's position would be unacceptable.

In what follows, I would like to focus on three features of Shapiro's realism: first, his acceptance of the Quinean principle "no entity without identity"; second, his thesis that the positions in structures are genuine entities; and third, his structuralist doctrine that the positions in structures have no identity or features outside a structure (that is, the view that positions in structures have no more to them than can be expressed in terms of the basic relations of the structure).

It is clear from the previous paragraph that there is nothing in the structure S^* that would enable us to distinguish α from β. But it is Shapiro's view of structures that α and β have no identity or distinguishing features outside S^*. So how could they be genuine entities that are distinct from each other? Thus, according to (2), the objects α and β of structure S^* are bona fide objects. By (1), there must be a criterion of identity of these two objects, and in view of (3), there must be a criterion by which α could be distinguished from β just using the relation R. It is obvious, however, that R alone could not possibly allow us to distinguish α from β, so we see that Shapiro's structuralist philosophy gives rise to a serious paradox.[15]

Another problem with Shapiro's realism

A structure is, according to Shapiro, "the abstract form of a system, highlighting the interrelationships among the objects, and ignoring any features of them that do not affect how they relate to other objects in the system". Now consider the following system:

It has a domain consisting of two distinct apples x and y, and it has, in addition to identity, the binary relation $R^\wedge$ of *being the fruit of the same apple tree and having the same mass as*, and $xR^\wedge y$.

By abstraction, we obtain the structure S% which we can describe as follows:

S% has a domain consisting of two distinct places α and β. It has, in addition to identity, the binary relation $R^\wedge$ described above. Furthermore, we have $\alpha R^\wedge \beta$, $\alpha R^\wedge \alpha$, $\beta R^\wedge \beta$, and $\beta R^\wedge \alpha$.

It seems to me that S^* (specified earlier) and S% are two distinct structures—they are distinct since they have different relations. On the other hand, by the isomorphism criterion of identity of Shapiro's theory of structures, S^* is identical to S%.[16] But if S^* and S% are identical structures, then any exemplification of one is an exemplification of the other. But any exemplification

---

[15] The principle idea for this objection was suggested to me by my colleague John MacFarlane, who had obtained the idea from an unpublished paper written by Jukka Keränen. I have no idea, however, whether Keränen would agree with the objection I have raised here.

[16] Shapiro also says: "a categorical theory characterizes a single structure if it characterizes anything at all" (Shapiro, 1997: 133).

of S* must have people in the domain; and any exemplification of S% must have apples in the domain; so, any person must be an apple![17]

11. A Problem with Shapiro's Acceptance of Thesis [2]

Shapiro's realism, coupled with his acceptance of a form of Resnik's thesis [2] above, gives rise to another problem. The number 3 is a place in the natural number structure; the singleton $\{\{\{\emptyset\}\}\}$ is a place in a different structure, a set-theoretical structure. In general, according to Shapiro, it makes no sense to identify positions in different structures. According to Shapiro's *ante rem* realism, the number 3 is a bona fide object as is the singleton set $\{\{\{\emptyset\}\}\}$; they are both as real and genuine as tables and chairs. Yet, he is committed to the claim that the identity assertion is meaningless, so he is committed to the position that it is not true that the natural number 3 is identical to $\{\{\{\emptyset\}\}\}$, for if it were true then it would make sense to assert the identity of the number with the singleton, contradicting the above doctrine. On the other hand, his doctrine that the number 3 is a genuine object seems to commit him to the view that the number 3 is identical to the number 3, since it is hard to see how a genuine object could fail to be identical to itself. Thus, Shapiro seems to be committed to the following two propositions:

(1) It is not true that the number 3 is identical to $\{\{\{\emptyset\}\}\}$.
(2) It is true that the number 3 is identical to the number 3.

It seems to follow, by reductio, that 3 is not identical to $\{\{\{\emptyset\}\}\}$. Thus,

(3) Assume that 3 is identical to $\{\{\{\emptyset\}\}\}$.
(4) Then, (1) is false.
(5) Thus, 3 is not identical to $\{\{\{\emptyset\}\}\}$.

---

[17] Shapiro has suggested to me (in an email message) that he might be able to avoid this difficulty by abandoning his isomorphism criterion of identity of structures for non-mathematical structures. In that case, he would no longer have a general criterion of identity of structures, but only a criterion of identity for structures that are "freestanding" structures. Then, he would no longer have the kind of theory of structures that his axiomatized version was supposed to be. It was put forward in response to his belief that "if one is to speak coherently about structures and avoid the ontological and modal options, then such a theory [with a criterion of identity of structures] is needed at least at some stage of analysis" (Shapiro, 1997: 92).

Of course, the above line of reasoning can be blocked by rejecting step (3). It is being incorrectly assumed, in the above reasoning, that it makes sense to assert that 3 is identical to $\{\{\{\emptyset\}\}\}$. So let us proceed as follows.

From (1) and (2), we can infer that the open-sentence 'the number 3 is identical to x' is not satisfied by $\{\{\{\emptyset\}\}\}$ and that the open-sentence is satisfied by 3. Thus, one can conclude, by a principle known as 'the indiscernibility of identicals', that $\{\{\{\emptyset\}\}\}$ is not identical to 3 after all—a conclusion that conflicts with Shapiro's doctrine that it is meaningless to assert such things.[18]

Thus, we have the following problem. Shapiro maintains that it is meaningless to assert either the identity of 3 with $\{\{\{\emptyset\}\}\}$ or the non-identity of 3 with $\{\{\{\emptyset\}\}\}$. However, we saw above, using the indiscernibility of identicals, that 3 is not identical to $\{\{\{\emptyset\}\}\}$. But then how can it be meaningless to assert what is the case? We seem to be landed in another paradox.

One way out of this conundrum is to adopt the Benacerrafian position that 3 and $\{\{\{\emptyset\}\}\}$ are not objects at all, for this move would block the inference to the conclusion that the open-sentence 'the number 3 is identical to x' is not satisfied by $\{\{\{\emptyset\}\}\}$, but that it is satisfied by 3. That is, we can avoid this puzzle if we reject the realism about structures and places in structures that gave rise to the paradox. If we deny that there are such things as structures and places in structures, then the above reasoning cannot get off the ground. We would then have a kind of structuralism without structures. The kind of structuralism that would result has been called "eliminative structuralism" by Charles Parsons.[19] It has a number of advantages over its *ante rem* rivals: besides

---

[18] The principle of the indiscernibility of identicals states: if object α is identical to object β, then any open-sentence true of α is true of β. It is hard to see how this principle can be seriously doubted, for if α and β are one and the same thing, and if an open-sentence θ is true of α, how can θ not be true of β? It might be objected that the principle can be refuted by using open-sentences containing "referentially opaque contexts", such as the following: 'Sue wonders if x murdered Tom'. Suppose that Mary is the murderer of Tom. Then is that open-sentence true of Mary but not true of the murderer of Tom? No, I would say that the open-sentence is not true of the person Mary at all. Of course, there is an enormous literature on this topic and I do not wish to get involved in a general discussion of it. However, no matter how such examples of "referentially opaque" open-sentences should ultimately be treated, they should not undermine the objection I am raising to Shapiro's doctrines. For the open-sentences I use above ('the number 3 is identical to x') clearly does not involve "referentially opaque contexts". And I am inclined to agree with David Kaplan's assessment that "whatever you may think about the identity of indiscernibles, no sensible person would deny the indiscernibility of identicals" (Kaplan, 1979: 99).

[19] In Parsons, 1996. There, Parsons notes that there is a plausible reading of a passage in Richard Dedekind's *Was sind und was sollen die Zahlen?* according to which Dedekind is holding that "statements about natural numbers are implicitly general, about *any* simply infinite system" (Parsons, 1996: 276). However, he gives reasons on the following page for concluding that such a reading is probably not what Dedekind intended.

avoiding the above paradoxical doctrines, Shapiro himself notes that eliminative structuralism is not saddled with the third man type problem which was discussed earlier (Shapiro, 1997: 101). Why, then, does Shapiro favor his more ontologically robust *ante rem* version?

12. The Main Problem with Eliminative Structuralism

Shapiro believes (following Parsons) that eliminative structuralism is vulnerable to a formidable objection, which Parsons calls "the problem of avoiding vacuity in the conditionals used to interpret mathematical statements".[20] Consider a statement Φ of number theory. This must be understood, according to his version of eliminative structuralism, to be some such statement as:

[Φ]*: For every system S, if S exemplifies the natural number structure, then $\Phi[S]$,

where $\Phi[S]$ is the sentence one obtains from Φ by restricting the variables in Φ to the objects in S and interpreting the non-logical constants in Φ in terms of the objects in S (and the phrase 'exemplifies the natural number structure' is expressed in a way that does not presuppose the existence of structures). Now if the eliminative structuralist's background ontology is finite, she will be forced to hold that there is no system that exemplifies the natural number structure; in which case, both [Φ]* and [$-\Phi$]* would have to be considered to be true. This is unacceptable, since we "do not end up with a rendering of arithmetic if the background ontology is finite" (Shapiro, 1997: 86). This problem is what Shapiro calls the "main stumbling block of the eliminative program" (Shapiro, 1997: 86).

What prevents an eliminative structuralist from postulating an ontology of infinitely many objects? Hartry Field argued (in Field, 1980: 31–4) that spacetime points are physical objects, and concludes that there are good empirical grounds for postulating the existence of infinitely many physical objects. Such a response to this problem, however, runs into the problem that even an infinity of such "physical objects" would not be enough for many mathematical theories such as set theory (Shapiro, 1997: 86). Parsons concludes that set theory would require a domain "of pure abstract objects, of

---

[20] Parsons, 1996: 288. For his discussions of this problem, see especially Parsons, 1996: sects. 3 and 7. See also (Parsons, 1995b: 75). I shall discuss Parsons's views on this topic in more detail in Chapter 9.

which pure mathematical objects are a paradigmatic instance" (Parsons, 1995*b*: 75).

Much more needs to be said about this particular objection to eliminative structuralism. However, I feel that, before I can effectively and adequately discuss this reasoning, I shall need to provide some additional development of theory, which would best be given in a later chapter. Let us then put off, for further theoretical developments, a more thorough discussion of this topic and go on to an examination of another form of structuralism.

13. Resnik on Structures

I turn now to Resnik's more recent views on the nature of structures and positions in structures, as developed in his book *Mathematics as a Science of Patterns* (Resnik, 1997). The following quotation shows that his basic position on structures has not significantly changed:

The underlying philosophical idea here is that in mathematics the primary subject-matter is not the individual mathematical objects but rather the structures in which they are arranged. The objects of mathematics, that is, the entities which our mathematical constants and quantifiers denote, are themselves atoms, structureless points, or positions in structures. And as such they have no identity or distinguishing features outside a structure. (Resnik, 1997: 201)

Recall that Resnik had espoused earlier the thesis that if *x* is a position in one structure and *y* is a position in a different structure, then it makes no sense to say either that *x* is identical to *y* or that *x* is not identical to *y* (thesis [2]). That Resnik continues to maintain a form of this thesis can be inferred from the following:

[W]hat would strike one as a problematic oversight if one thought of mathematical objects along the lines of ordinary objects seems quite natural when it comes to positions in patterns. For restricting identity to positions in the same pattern goes hand in hand with their failure to have any identifying features independently of a pattern. (Resnik, 1997: 211)

(A reminder: in the above passage and throughout his book, Resnik uses the term 'pattern' more or less interchangeably with 'structure'.) Again, he writes: "[T]aking the language of number theory at face value, we can conclude that its variables range over numbers, but there is no fact of the matter as to whether they range over sets" (1997: 221). He later explains that, "in saying that there is no fact of the matter as to whether certain mathematical objects (or positions) are identical to others, I was tacitly denying truth-values to ['numbers are sets'] and related sentences" (1997: 245).

What are structures for Resnik?

Recall from Chapter 3 that Resnik explained what he means by 'structure' or 'pattern' with the following words: "I take a pattern to consist of one or more objects, which I call positions that stand in various relationships" (1997: 202–3). From this characterization, structures for Resnik seem to be, initially, what Shapiro called 'systems'. What distinguishes Resnik's structures or patterns from Shapiro's systems is a feature that Resnik attributes to the positions in a structure: "they have no identity or distinguishing features outside of a structure" (1997: 201). Thus, a group of football players arranged in an old-fashioned single wing formation would be regarded by Shapiro as forming a system, whereas such a concrete arrangement would not qualify as a structure for Resnik, since the particular players in the formation would have many distinguishing qualities which are "outside of a structure". He refers to such a formation as an "arrangement of concrete objects".

What then corresponds, in Resnik's philosophical theory, to the relation of *exemplification* that Shapiro postulates to hold between his systems and structures? Consider an "arrangement of concrete objects" such as a stack of ten pennies. Such an arrangement (related by the relation of being immediately on top of) is isomorphic to an abstract mathematical structure, for example, the set of natural numbers less than 10 (related by the immediate successor relation). In that case, the arrangement is said to be an *instantiation* of the mathematical structure. Thus, Resnik's instantiation corresponds to Shapiro's exemplification. For Resnik, instantiation is just a special case of structural isomorphism—one that holds between a structure and an arrangement, so that the positions in the arrangement have identifying features (such as being made of copper, being disc-like, etc.) over and above the structural features of the arrangement (Resnik, 1997: 204).

A significant difference between Shapiro's view of structures and Resnik's can be discerned from their differences over whether isomorphism can be regarded as giving an appropriate criterion of identity of structures. As noted above, Shapiro thinks it can. Resnik thinks it cannot. Resnik supports his position as follows: The natural number structure is isomorphic to a substructure whose domain consists of the even natural numbers. If we took this substructure to be identical to the natural number structure, then according to Resnik, "we would be forced to take the even natural numbers to be identical to the natural numbers" (1997: 209), which would be absurd.

Shapiro, however, is not forced to this absurdity by his position. Remember that Shapiro distinguishes structures from systems. Working within Shapiro's theory, we would have to distinguish the *system* consisting of the even natural numbers related by the successor of the successor relation from the abstract form of this system (the *structure* of this system). This abstract form, he can argue, is indeed isomorphic to the natural number structure. Thus, he notes that "the even-number places of the natural number structure constitute a system, and on this system, a 'successor' function could be defined that would make the system exemplify the natural number structure" (Shapiro, 1997: 94). However, we cannot conclude from this identity that the even numbers are identical to the natural numbers. Drawing such a conclusion would undoubtedly result from confounding the system of even natural numbers (described above) with its abstract form.[21]

14. RESNIK'S NO-STRUCTURE THEORY

Recall that, earlier, Resnik had defended two theses: in addition to thesis [2], a form of which he continues to accept, he had also adopted the position (thesis [1]) that there is simply no fact of the matter as to whether identity or non-identity holds between structures. This is the thesis that Shapiro rejects in his theory of structures by, in effect, going with the Quinean idea of "no entity without identity" and defining the identity of structures in terms of isomorphism. In his more recent book, Resnik thinks that there are strong reasons for not taking the route Shapiro advocates. Resnik holds that "mathematics is largely a conglomeration of theories each dealing with its own structure or pattern and each forgoing identities leading outside its pattern" (Resnik, 1997: 211). The typical mathematical theory excludes the structure it deals with from its universe of discourse. Thus, treating structures as entities, as Shapiro does in his theory of structures, "would undo these parallels with mathematics, since it would allow identities between patterns, and this in turn would permit identities between their positions" (Resnik, 1997: 211).

I find this argument rather unconvincing, based as it is on features of typical mathematical theories. Shapiro's theory of structures is not supposed to be just a typical mathematical theory—it was put forward as an all encompassing ontological theory—so why should we expect it to have only those features one expects to find in typical mathematical theories?

---

[21] See, in this regard, Shapiro, 1997: 93 n.

Why should Shapiro care if his theory undoes the parallels with mathematical theories? In particular, why should we conclude that all theories (even theories of structures) are precluded from treating structures as entities, just because typical mathematical theories exclude the structures they deal with from their universes of discourse?

Resnik has a second reasons for not allowing us to speak meaningfully about structures being identical or not identical to one another. He expresses this reason as follows:

[S]peaking of patterns as identical or distinct is to treat them as individuals, *since identity is a relation between individuals*. But ... there are reasons to avoid being forced into treating patterns as individuals or even as entities of any kind. (Resnik, 1997: 210, italics mine)

Of course, Resnik cannot allow there to be facts concerning identities between positions from all patterns, since this would require there to be identities between positions from different patterns—something not allowed by thesis [2].

But what are Resnik's grounds for maintaining that identity is a relation that holds only between individuals, as the italicized portion of the above quote suggests? Shapiro is not alone in thinking that identity can hold between things that are not "individuals". For example, Russell and Whitehead certainly thought that identity held between propositional functions (which they distinguished from "individuals").[22] The doctrine that the field of identity is restricted to individuals is certainly not self-evident. So one would expect Resnik to provide some plausible reason for adopting such a position on identity. But he doesn't.

As can be seen from the above quotation, Resnik feels that he cannot affirm that structures are entities. Resnik continues to subscribe to a form of thesis [2] of his previous position on structures, but he no longer accepts thesis [1] in quite the form given earlier. He no longer holds that structures are excluded from the field of the identity relation, since he no longer believes that structures are entities at all, asserting: "Realism about mathematical objects does not commit one to realism about structures, even when one maintains that objects are positions in patterns or structures and that mathematical objects are positions in patterns" (Resnik, 1997: 261). Resnik then goes on to assert that "there is no fact of the matter as to whether patterns are mathematical objects or entities of any kind" (1997: 248). Thus, one might attribute

---

[22] See Russell and Whitehead, 1927: *13.

to Resnik a form of eliminative structuralism, that is, a structuralism without structures. But if Resnik is defending a form of eliminative structuralism, it is a very strange form of the view in so far as it maintains that the points in a structure exist even though the structure itself does not exist.

I find it hard to reconcile Resnik's "no-structure" doctrine with much of what he says about structures, especially in earlier parts of his book. It is almost as if we have two philosophers at work: a philosopher who believes in structures and bases his realistic view of mathematics upon the existence of structures, and another philosopher who tries to get along without any structures (only positions in structures).

The following quotations from Resnik's book are just a small sample of the many statements he makes that strongly suggest that he is presupposing the existence of structures:

[I]n mathematics the primary subject-matter is not the individual mathematical objects but rather the structures in which they are arranged. (201)

Pattern-congruence is an equivalence relation whose field I take to include both abstract mathematical structures and arrangements of more concrete objects. (204)

The natural number sequence, *qua* pattern, has multiple occurrences within iterative set-theoretic hierarchies, *qua* pattern. (216)

[I]n describing the pattern the detective simultaneously develops a theory of it and of the data fitting it. (225)

On the structuralist view, the theorems of a branch of mathematics are supposed to be true of the structure (or structures) they describe. (240)

Now how does Resnik try to reconcile the statements he makes about structures, many of which evidently refer to or quantify over (and hence presuppose the existence of) structures, with his paradoxical thesis that there are no such entities as structures? He suggests that he could do the sort of thing nominalists do in analyzing away familiar informal talk that seems to presuppose the existence of properties and numbers, writing: "Philosophers explicate examples of this sort by preserving the apparent logical form of some of the cases while paraphrasing away the conflicting ones" (1997: 249).

The general idea of such paraphrasing is clear enough. We have the example of Russell's "no-class" theory, according to which statements which appear to be about classes (or sets) are actually abbreviations of statements about propositional functions. In *Principia Mathematica*, Russell supplies a relatively precise set of rules for paraphrasing the statements about classes that can be made in the formal language of the system into statements about propositional functions.[23] Unfortunately, Resnik does not undertake to make the sort of detailed paraphrases of his many statements about structures that one would expect from his suggestion. He seems to think that producing such a detailed paraphrasing would require that he develop a *mathematical* theory of structures. But he says that he has neither the talent nor the taste for carrying out such a project. And he sees no current mathematical need for doing so (Resnik, 1997: 257).

Why he thinks he needs to develop a *mathematical* theory of structures in order to supply the needed paraphrasing is by no means clear. Why he thinks a *philosophical* theory of structures will not do what is wanted is never explained. And the fact that there is no current mathematical need for such a theory seems to be beside the point. I should think that he would acknowledge a pressing philosophical need for such a theory, given his overall position.

Despite Resnik's suggestion that such a paraphrasing can be carried out, I have my doubts. Consider one of his most fundamental of principles: the thesis that positions in a structure have no distinguishing features that are "outside the structure". How are we to make sense of this principle? He seems to be maintaining that positions "in a structure" (whatever that may mean) have no distinguishing features that are "outside" the *structure*—something that does not even exist. This is a very paradoxical principle indeed.

How is this principle to be paraphrased in a way that does not refer to or presuppose in any way the existence of structures? Perhaps this talk of structures can be understood in terms of classes of positions and relations among the positions? There are obvious problems with this idea. For as I pointed out earlier, Resnik does not believe in the existence of properties and hence of relations. (Recall his words: "I am a realist about mathematical objects first without being a realist about properties at all" (Resnik, 1997: 261)). Thus, the suggested paraphrase of Resnik's fundamental principle would have to involve some method of avoiding all reference to relations and properties. How this could be done is not at all obvious.

Of course, Resnik could define relations to be classes of ordered pairs. But classes are not things that Resnik could, in any obvious way, make use of for the present purposes, since classes themselves are held to be positions in a structure, and the classes required for the suggested paraphrase are classes of positions in *arbitrary* structures. Hence these classes cannot have only

---

[23] See Chihara, 1973: ch. 1, sect. 4, for details.

distinguishing features in the structure of classes—they would have to be related to points of arbitrary structures.

Could a clever use of modal concepts enable Resnik to overcome some of these difficulties? This pathway seems to be completely blocked by his insistence that there are no modal facts (not even logical facts). I think Resnik would face some truly daunting problems if he attempted to carry out the suggested program of paraphrasing away all the apparent references to structures.

15. PROBLEMS WITH RESNIK'S NO-STRUCTURE THEORY

A problem involving identity

Here's what Resnik says about the natural number structure: "I take this to be a pattern [or structure] with a single binary relation (successor) and the natural numbers to be its positions" (Resnik, 1997: 203). Let us turn our attention to a simpler structure S_1: it has a single binary relation (which I'll call 'R_1') and the two positions α and β, which satisfy the following law:

$$(x)(y)(R_1xy \leftrightarrow x \neq y)$$

According to Resnik's structural version of realism, α and β are genuine objects that have no features or qualities other than what I have stated above. By stipulation, $\alpha \neq \beta$.

Now let S_2 be specified as follows: it has a single binary relation (which I'll call 'R_2') and the three positions χ, δ, and ε, which satisfy the following law:

$$(x)(y)(R_2xy \leftrightarrow x \neq y)$$

We can infer that $\chi \neq \delta$, but we cannot infer $\chi \neq \alpha$. '$\chi \neq \alpha$' supposedly has no truth value, since χ and α are positions in different structures.

Let us assess these doctrines about positions in the light of Resnik's no-structure thesis. According to Resnik's realism, these positions are held to be genuine entities that exist independently of our beliefs and wishes. This gives rise to the problem of trying to make sense of the view that $\chi \neq \delta$ even though neither χ nor δ have any features in virtue of which one could infer that they are different objects. Do we not have here the sort of bare (ungrounded) non-identity that was discussed in connection with Shapiro's doctrines? (Does that make sense?)

Resnik's doctrine that there is no fact of the matter as to whether $\chi = \alpha$ gives rise to an even more puzzling mystery. When thesis [2] was first advanced, it was assumed that these strange features of positions were somehow due to what structures were like—that is, to the nature of structures. One imagined that different structures were like different worlds, so that the positions of different structures were like objects in different worlds, and the non-sensicality of the transtructural identity of positions could be likened to the "incommensurability" of objects in different worlds. However, when Resnik put forward his no-structure theory, implying that there are no structures, the worlds analogy completely dissolved. We now have just the positions and no structures. Now, the identity and diversity features described above must stem from the natures alone of the entities concerned. There must be something about the natures of the positions χ and α themselves in virtue of which there is no fact of the matter as to whether $\chi = \alpha$. But by Resnik's own account, these entities have no features or qualities that are not given in the above structural descriptions; in other words, their natures are completely given by the structural descriptions. But the features that the structural descriptions attribute to χ and α do not allow us to infer that there is no fact of the matter as to whether or not they are identical. Yet, Resnik insists that there is no such fact of the matter. How can that be? All of this strikes me as very implausible, if not downright unintelligible.

Typosynthesis again

Reconsider the structure S_1 described above. According to Resnik's views, the above description of the structure amounts to a specification of a relation R_1 that obtains between the points α and β. But have I specified a genuine relation? Is it an internal relation? That is, is it in virtue of the intrinsic properties of α and β that the former is in the relation R_1 to the latter? But we know nothing about the intrinsic properties of these points. Taking Resnik's position on such mysterious relations is tantamount to maintaining that a genuine relation has been specified by the structural description of the typosynthesis theory discussed in Chapter 3. This would set one right back in the van Inwagen mess.

16. Resnik's Rejection of Classical Logic

In the earlier discussion of Shapiro's radical views, I noted that the acceptance of thesis [2] brings on a daunting philosophical difficulty involving the logic of identity. Resnik is certainly aware of the fact that his position faces serious problems. He writes:

I am committed to more than denying truth-values to whole sentences such as 'each number is a set'. For I have also held that for any number (or more generally, any thing) there may be no fact as to whether it is a set, i.e. whether 'set' is true of it. Unfortunately, it is not clear that I can satisfactorily follow through on this by merely restricting my truth-predicate or truth-theory. (Resnik, 1997: 245)

His response to this sort of problem is to "restrict logic so that excluded-middle does not apply generally." He concludes that "implementing my claim that there is no fact of the matter as to whether numbers are sets involves rejecting the universal applicability of classical logic" (Resnik, 1997: 245).

Resnik opines that such a position is not so radical as it might at first appear. He suggests that intuitionistic logic could be used in place of classical logic for this area of discourse (1997: 245 n). But what is "this area of discourse"? Has he circumscribed any specific area of discourse for which intuitionistic logic could be used in place of classical logic? Not at all. He tells us that he is not espousing using intuitionistic logic *within* individual mathematical theories, but that does not tell us where we are to use it.

Resnik seems to think that it is the rules of classical logic (with its law of excluded middle) that are responsible for the troubling position to which he had been driven. Thus, he treats the sort of paradoxical conclusions we drew from his thesis [2] as resulting from the employment of rules of inference (classical logic) that sometimes lead us from true premises to an unwanted position. All we need to do, he seems to be thinking, is to reject the rules that led us from [2] to the unwanted position and we shall avoid the paradox. It's as if we had a game in which, at certain times, a troubling situation can arise, and all we need to do is to change the rules to block such situations from arising. Just switch from classical to intuitionistic logic, Resnik suggests, and we can avoid the problem.

Now this way of looking at the paradoxical situation arising out of thesis [2] is, in my judgment, quite misleading. The rules of logical inference we used to draw the paradoxical conclusions are not like the rules of a game, which we players of the game can more or less arbitrarily change to make the game more satisfactory. The rules of logic we used are tied to, and answerable to, semantical matters—in particular, to what the premises in question mean—which are not up to us to change at will. Consider, in the light of these last comments, the case of intuitionistic logic.

Intuitionistic logic

Let us briefly review some of the principal features of the version of intuitionistic logic that was developed and explained by Arend Heyting (Heyting, 1956). The most salient feature of this logic is how its logical constants are defined: not in terms of truth, as is standardly done in classical logic, but rather in terms of assertability. Roughly, one can say that a mathematical proposition is assertable iff one has a proof of it. Thus, consider the connectives: '−' and 'v' of intuitionistic logic:

> The Intuitionistic negation should not be read as just 'not true'. One can assert −φ only when one has proved that, from the assumption that φ is assertable (i.e. there is a proof of φ), one can prove a contradiction. One might put it this way: one can assert −φ if one can prove that, from the assertability of φ, one can assert a contradiction. This amounts to the following: we have a construction such that, when added to a proof of φ, it would yield a proof of a contradiction.

> The Intuitionistic disjunction can be understood as follows: one can assert 'φ v θ', if one can assert φ or one can assert θ.

Intuitionistic logical laws

In classical logic, one can assert a logical law if one can prove that *every instance of the formula will be true*. Thus, we can assert that (P v −P) is a logical law since, no matter what truth value P may have, the disjunction must have the truth value true. In intuitionistic logic, one can assert a logical law only when one can prove that every instance will be assertable. Thus, one can show that some formula of propositional logic is not assertable by constructing an instance of the formula that is not assertable. For example, the proposition φ that asserts that there are seven consecutive sevens in the decimal expansion of π cannot now be asserted; furthermore, we cannot assert −φ (since we cannot construct a proof showing that the assertion of φ leads to a contradiction). This shows that we are not in a position to assert that the formula 'P v −P' expresses a law of intuitionistic logic.

Regarding Heyting's views about logic, the following should be noted:

(1) For Heyting, the formula 'P v −P' does not express a logical law of Intuitionistic logic. He does not claim that the formula should not be taken to express a law of classical mathematics, nor does he claim that the classical law of excluded middle is "false" or "invalid".[24] He is content to claim only that excluded middle does not hold in his own system of logic.

(2) Heyting specifically warns his readers that the logic he has developed is meant only to apply in his intuitionistic *mathematics*: he did not design his

---

[24] In this respect, Heyting's position differs from that of L. E. J. Brouwer (in Brouwer, 1967). There can be little doubt that Heyting was clearer-headed than Brouwer about the dispute between intuitionists and classical mathematicians.

logic to codify logical reasoning carried out in ordinary language about scientific or everyday matters.

(3) Heyting's logical laws are justified by appeal to the semantics of his mathematical language. In particular, the laws of Heyting's intuitionistic propositional logic are justified by appeal to his definitions of the logical connectives.

Thus, when Resnik suggests that intuitionistic logic could be used for "this area of discourse" (which, I assume, includes discourse about numbers being or not being sets), is he suggesting that the logical constants that are used in "this area of discourse" should be understood to be defined in the intuitionistic manner in terms of assertability (instead of truth conditions)? What grounds could he possibly supply in support of such a suggestion?

Michael Dummett once put forward grounds which might be taken to support Resnik's suggestion. In his famous article on truth, he wrote:

We learn the sense of the logical operators by being trained to use statements containing them, i.e., to assert such statements under certain conditions. Thus, we learn to assert ... 'P or Q' when we can assert P or can assert Q. (Dummett, 1964: 109)

He then went on to say:

We no longer explain the sense of a statement by stipulating its truth-value in terms of the truth-values of its constituents, but by stipulating when it may be asserted in terms of the conditions under which its constituents may be asserted. The justification for this change is that this is how we in fact learn to use these statements. (1964: 110)

However, these claims about how we in fact learn the logical operators and how we in fact learn to use such statements seem to me to be quite implausible. Do English-speaking parents correct their children when they say such things as "I know the cat is in the house somewhere, so it is either upstairs or it is downstairs in the kitchen"? Do they object to such assertions by saying "You oughtn't to say that it is either upstairs or it is downstairs in the kitchen, unless either you know that it is upstairs or you know that it is downstairs in the kitchen". It should be noted that Dummett did not support his claims about how we in fact learn the relevant elements of the language by citing any empirical studies, and so far as I know few (if any) philosophers of language or mathematics have adopted Dummett's view of how we learn them. It is also striking that Dummett has never, to my knowledge, repeated these early claims about how we learn to use statements containing the logical connectives.

17. THE MAIN PROBLEM WITH RESNIK'S VERSION OF THESIS [2]

Resnik maintains, roughly, that if x is a position in a structure, and y is a position in a different structure, then there is no fact of the matter as to whether x is or is not identical to y. I say "roughly" here, because strictly speaking Resnik does not allow there to be any fact of the matter as to whether a structure A is or is not identical to a structure B. The "rough" statement requires A to be distinct from B, which is not allowed by Resnik. So we find Resnik affirming not the general "rough" claim above, but rather such particular propositions as "there is no fact of the matter as to whether numbers are sets".

What does it mean to say that "there is no fact of the matter" as to whether say, 3 is identical to $\{\{\{\emptyset\}\}\}$? Resnik makes several attempts to clarify his use of the expression. Here's one:

[W]hen I deny that there is a fact of some matter, for example, as to whether numbers are sets, ... I am tacitly referring to certain sentences in our language, and at the very least I am denying that they have truth-values. (Resnik, 1997: 244)

We can conclude that, in claiming that there is no fact of the matter as to whether 3 is identical to $\{\{\{\emptyset\}\}\}$, Resnik is denying, among other things, that the sentence '3 is identical to $\{\{\{\emptyset\}\}\}$' has a truth value.

We saw in Section 11 that, from the thesis that both '3 is identical to $\{\{\{\emptyset\}\}\}$' and also '3 is not identical to $\{\{\{\emptyset\}\}\}$' have no truth values, and using the indiscernibility of identicals, one can infer that 3 is not identical to $\{\{\{\emptyset\}\}\}$. Of course, if 3 is not identical to $\{\{\{\emptyset\}\}\}$, then the sentence '3 is not identical to $\{\{\{\emptyset\}\}\}$' is true—a conclusion that contradicts Resnik's thesis that the sentence '3 is not identical to $\{\{\{\emptyset\}\}\}$' has no truth value.

Now how is Resnik's limitation of classical logic supposed to enable him to avoid the paradoxical conclusion that $\{\{\{\emptyset\}\}\}$ is not identical to 3? Evidently, by allowing him to reject the applicability of the principle of the indiscernibility of identicals. It is easy to see how this simple move would block the inference that $\{\{\{\emptyset\}\}\}$ is not identical to 3, but how reasonable is the move? Does such a move make sense? After all, assessing the validity of the inference in question is not just a matter of deciding whether one should allow the use of some syntactic rule of inference which some classical logician made up. Supposedly, we have here a genuine entity 3 and a genuine entity $\{\{\{\emptyset\}\}\}$ which are such that the open-sentence 'the number 3 is identical to x' is true of 3 but not true of $\{\{\{\emptyset\}\}\}$. Now how in the world could 3 be the very same entity as {{{∅}}}? It seems to me clear that 3 couldn't possibly be the very same thing as {{{∅}}}. If 3 were one and the same entity as {{{∅}}}, and if the open-sentence 'the number 3 is identical to x' were true of 3, it would also have to be true of {{{∅}}}. Thus, it would seem that any rational person would be driven to the conclusion we arrived at earlier: that 3 is not identical to {{{∅}}}.

Why realism?

Must Resnik abandon the principle of the indiscernibility of identicals in order to save his theory from absurdity? Resnik does not seem to consider (seriously) the possibility of obviating the above problem by giving up one of the basic theses underlying his version of structuralism. He is willing to question the idea that structures are genuine entities, but he never seems to question seriously the belief that the positions in structures are genuine entities. Is that belief more strongly supported than the principle of the indiscernibility of identicals? We need to investigate the reasons Resnik gives for being a mathematical realist. These reasons will be the topic of the next chapter.

18. A CONCEPTUAL OBJECTION TO STRUCTURALISM

Before leaving the topic of structuralism, I shall take up briefly a type of objection to the structuralist's view of set theory that Parsons has examined in great detail (Parsons, 1995b). Consider, for example, the idea that a set is ontologically dependent upon its members—the idea that a set wouldn't exist were its members not in existence—whereas the members of a set are not (in the same way) ontologically dependent upon the set. Such an idea is generally used in explaining certain features of the "iterative conception of sets", and this implies (according to the objection) that the concept of set has more content than the structuralist view allows.

James Brown has recently advanced a variation on this argument as follows:

The notion of set, however, seems quite contrary to the spirit of structuralism. The members (objects) have a kind of priority over the sets (structures) that they con-stitute. If a set was just a structure, then changing its members would not affect it any more than changing the first baseman changes the structure of the infield. But a set's identity is wholly dependent upon its members—change its members and you change the set. Structuralism doesn't do justice to this basic fact about set theory. ... the support for a descriptive view of mathematics offered by structuralism cannot be upheld. (Brown, 1999: 61)

Note the two sentences: "If a set was just a structure, then changing its members would not affect it any more than changing the first baseman changes the structure of the infield. But a set's identity is wholly dependent upon its members—change its members and you change the set." Brown's strategy is to reduce structuralism to a kind of absurdity by attempting to show that the structuralist's view that *a set is just a structure* conflicts with the idea that a set's identity is dependent upon its members. The trouble with this strategy is that the view being attacked is not a view that the structuralist has adopted. Structuralists do not claim that a set is a structure (certainly not the structuralists Brown is discussing in this section—notably Shapiro and Resnik). Recall that, for Resnik,

[I]n mathematics the primary subject-matter is not the individual mathematical objects but rather the structures in which they are arranged. The objects of mathematics, that is, the entities which our mathematical constants and quantifiers denote, are themselves atoms, structureless points, or positions in structures. (Resnik, 1981: 201)

In short, sets for Resnik are not structures but positions in structures. Similarly, we saw that Shapiro, too, holds that mathematical terms refer to positions in structures. So this part of Brown's objection fails; he is attacking a straw man.

Brown goes on to say: "Structuralism doesn't do justice to this basic fact about set theory." Is it true that structuralism fails to do justice to the fact that a set's identity is wholly dependent upon its members? Take a particular set theory, say Zermelo-Fraenkel set theory. It gives a criterion of identity of sets: $\alpha = \beta$ iff $(x)(x \in \alpha \leftrightarrow x \leftrightarrow \beta)$. In effect, this criterion tells us that if you change a set's members, you change the set. Any structure that is a model of this set theory will reflect that fact. Why does structuralism fail to do justice to the basic fact that if you change a set's members you change the set?

Perhaps Brown got carried away in raising his objection to structuralism and intended, initially, only to raise the objection that Parsons considers (Parsons, 1995*b*), namely, that the concept of set has more content than the structuralist view allows. Since Parsons discusses this version of the objection in a thoroughly convincing fashion, I see no need to review his reasoning here. His conclusion is: "This discussion of the arguments that are actually in the literature should make plausible that there is not a set of persuasive direct 'intuitive' considerations in favor of the axioms of ZF that are incompatible with the structuralist conception of what talk of sets is."[25]

---

[25] Parsons, 1995*b*: 87–8. Shapiro agrees with the assessment that Parsons gives of the above argument. See Shapiro, 1997: 103–4.

It should be noted that, even if Parsons's version had raised serious difficulties for structuralism, it would not raise a serious difficulty for my structural account. For as I emphasized early in this chapter, my account is not concerned with the actual content of set-theoretical assertions—with what the sentences of set theory actually mean or say. Whether or not the actual content of the theorems of set theory can be cashed out in the way structuralists claim is not something upon which my own structural account depends.

5

Platonism

This chapter is concerned with the dispute in the philosophy of mathematics between mathematical realists (or Platonists) and mathematical nominalists, which has been simmering in the Anglo-American world of philosophy these past fifty or so years. A close approximation to what the terms 'mathematical realist' and 'mathematical nominalist' mean can be given as follows: a mathematical realist is one who maintains that *mathematical objects exist*, whereas a mathematical nominalist is one who opposes the realist's position.[1] Thus, there are two types of nominalists: those who deny that there are mathematical objects, and those who adopt the weaker thesis that we lack compelling grounds for believing in the existence of mathematical objects. (Many nominalists accept both theses, using the weaker to support the stronger.) As we saw in the previous chapter, both Shapiro and Resnik espouse a form of mathematical realism.

The dispute between mathematical realists and nominalists is a special case of the long-standing and more general dispute between Platonists and nominalists over the existence of universals, which goes back over two thousand years.[2] In this century, the controversy appeared in the philosophy of mathematics in a revitalized form, when it was given a powerful new logical flavor by the highly respected philosopher-logicians Kurt Gödel, Willard Quine, and Hilary Putnam, each of whom espoused some form of Platonism.

1. GÖDELIAN PLATONISM

Unlike the logical positivists, who held that traditional philosophical questions (such as "Do mathematical objects exist?" and "Do physical objects

---

[1] More accurate characterizations of these terms will be given later in this chapter.

[2] This might be contested on the ground that objects are not universals. However, the notion of "object" is not sufficiently precise to rule out identifying mathematical objects with universals. Thus, Penelope Maddy has argued that numbers are properties and that properties are universals. See Maddy, 1990: ch. 3, sects. 2 and 3.

exist?") were meaningless, Gödel was a thoroughgoing metaphysician.[3] For Gödel, the question of the existence of mathematical objects was "an exact replica of the question of the objective existence of the outer world [the world of physical objects]" (Gödel, 1964a: 220), and both these traditional metaphysical questions were, from his point of view, sensible and answerable. He believed that the assumption of mathematical objects "is quite as legitimate as the assumption of physical bodies", the reason being that the postulation of such mathematical objects as sets is, for him, "in the same sense necessary to obtain a satisfactory system of mathematics as physical objects are necessary for a satisfactory theory of our sense perception" (1964a: 220). Thus, Gödel held that we have rational grounds for believing in both physical objects and mathematical objects: in both cases, the postulation of such entities is justified by something like "inference to the best explanation".[4]

Why would Gödel, a world-renowned mathematician whose fame was inextricably tied to his extraordinary accomplishments in mathematical logic, publish articles which give philosophical arguments defending the rationality of belief in the existence of mathematical objects? One reason is to be found in his attitude toward the continuum hypothesis (Cantor's conjecture that there is no cardinal number greater than aleph-null but less than the cardinality of the continuum.) Gödel worked for many years attempting to resolve the question and was convinced that Cantor's conjecture had a truth value that was independent of whether or not the conjecture was formally decidable from the axioms of standard versions of set theory. This conviction was tied to his belief in the existence of sets. Thus, in his paper (Gödel, 1964b), we find a discussion of the idea which had been gaining some currency that, if the continuum hypothesis were proven to be independent of the standard axioms of set theory, then the question of its truth or falsity would lose its meaning, just as the question of the truth or falsity of the fifth postulate of Euclidean geometry was thought to have lost its meaning with the discovery of its independence

---

[3] There is evidence that Gödel had maintained a Platonic view of mathematics "since his student days in Vienna" (Feferman, 1998e: 168). Feferman also notes: "There is certainly no questioning Gödel's unremitting espousal in print of full-fledged platonism, beginning with his 1944 article on Russell's mathematical logic and continuing (especially in his 1947 article on Cantor's continuum problem) until his death" (ibid.). For additional details of Gödel's philosophical development, see Feferman, 1998e.   [4] See Harman, 1965 for a discussion of such inferences.

from the other postulates.[5] Gödel argued in this paper that such an independence result in set theory would render the question of the truth or falsity of the continuum hypothesis meaningless only if set theory were regarded as a hypothetico-deductive system in which the meanings of the primitives of set theory are left undetermined (1964b: 271). But, as was noted earlier, for Gödel, set theory is not that sort of system. According to Gödel:

(1) The objects of set theory "exist independently of our constructions";
(2) We have "an intuition of them individually" (an intuition that is something like a "perception" of individual sets); and
(3) The general mathematical concepts we employ in set theory are "sufficiently clear for us to be able to recognize their soundness and the truth of the axioms concerning" these objects (Godel, 1964b: 262).

He concluded that "the set-theoretical concepts and theorems describe some well-determined reality, in which Cantor's conjecture must either be true or false" (263–4), even if the conjecture is independent of the other axioms.

When Gödel maintained that the objects of set theory exist independently of our constructions, he was adopting a metaphysical view of sets that has many features common with the Platonic doctrine of forms. For Gödel, sets, like the forms of Plato, do not exist in the physical world, and they are completely independent of human beliefs and desires. Given this independence feature of sets, one may well wonder how we can gain any knowledge of their existence and properties. This is where mathematical intuition comes in. We are supposed to have something like a faculty by means of which we have something like a perception of sets.

Why did Gödel think that we have such a faculty? Why did he think that we have something like a perception of the objects of transfinite set theory— which gives meaning to the question of whether the continuum hypothesis is true or false?[6] That we have such a faculty of mathematical intuition, he

[5] Cf. "Probably we shall have in the future essentially different intuitive notions of sets just as we have different notions of space, and will base our discussions of sets on axioms which correspond to the kind of sets we wish to study. ... everything in the recent work on foundations of set theory points toward the situation which I just described" (Mostowski, 1967: 94).

[6] Paul Cohen proved the independence of the continuum hypothesis from the standard axioms of set theory. See Cohen, 1966 for a detailed presentation of the proof for non-specialists.

argued, can be seen from the fact that the axioms of set theory "force themselves upon us as being true". He continued:

I don't see any reason why we should have less confidence in this kind of perception, i.e. in mathematical intuition, than in sense perception, which induces us to build up physical theories and to expect that future sense perceptions will agree with them and, moreover, to believe that a question not decidable now has meaning and may be decided in the future. (Godel, 1964b: 271)

The idea here seems to be that mathematical intuition plays a role in mathematics analogous to the role that sense perception plays in the empirical sciences. In both cases, certain sorts of beliefs are thought to be forced upon us by these respective sorts of "perception". In both cases, we are pictured as constructing theories that have implications about future "perceptions" which are, in favorable instances, confirmed by "perceptions". All of this suggested to him the possibility of devising some new set-theoretical axioms that have implications that our mathematical intuitions may "confirm" in some way analogous to the way sense perceptions may confirm a new scientific law or general hypothesis:

There might exist axioms so abundant in their verifiable consequences, shedding so much light upon a whole field, and yielding such powerful methods for solving problems ... that, no matter whether or not they are intrinsically necessary, they would have to be accepted at least in the same sense as any well-established physical theory. (Gödel, 1964b: 265)

The simplest case of an axiom that would be "successful" in the above way would be one that had number-theoretic consequences that could be verified by straightforward computation up to arbitrarily high integers (Gödel, 1964b: 272). The above ideas have stimulated much mathematical research, and they have been mined by a number of set theorists to obtain interesting results. Certainly, these suggestions of Gödel's have been most influential.[7]

On the philosophical front, although Gödel's belief in the existence of sets has an initial plausibility and may appeal to set theorists who are Platonically inclined, I believe that it cannot sustain a close critical examination and is

---

[7] In one of his Alfred Tarski lectures given at Berkeley on 16 April 2001 (entitled "On the Philosophical Foundations of Set Theory"), Ronald Jensen described the views I attribute above to Gödel as being the most influential of Gödel's philosophical views, at least among set theorists. For a detailed discussion of these views of Gödel's, especially those concerning the search for new axioms of set theory, see Feferman, 1998d: 69–73, and Maddy, 1990: ch. 4. But compare Maddy's more recent views on this topic in Maddy, 1997: especially chs. 5 and 6.

open to many objections.[8] However, as I have already subjected Gödel's reasoning about these matters to many criticisms in several of my previous works, I shall not repeat my objections here.[9]

Maddy's version of realism

Gödel's brand of Platonism was given a kind of philosophical rebirth by Penelope Maddy, who interpreted Gödel's claim that we have a mathematical intuition ("this kind of perception") of sets to be the claim that we can literally see sets.[10] According to Maddy, a "mixed set"—a set whose transitive closure contains urelements (or non-sets)—has a location in physical space (namely, the exact place in space at which the urelements are located) and can be perceived. The set of keys in my pocket, for example, is held to be located in exactly the place that the keys themselves are to be found. Thus, by taking mixed sets to be locatable in physical space, Maddy believed that she could give a reasonable explanation both of how we can gain knowledge of them and their properties.[11] Making use of the theories of perception of the philosopher George Pitcher and the neurophysiologist Donald Hebb, she tried to explain both how we are able to perceive sets and also how we arrive at our knowledge of certain intuitively obvious "truths" about sets.[12] Maddy made a

---

[8] For a scholarly and sympathetic discussion of Gödel's Platonism, see Parsons, 1995a. Parsons gives the above passage quoted from Gödel's 1964 article an unusual interpretation, writing: "We can see that by 'the objects of set theory' Gödel means not just sets but the primitive *concepts* of set theory, 'set' itself, membership, what he calls 'property of set' (264 n. 18)" (65). I find Parson's interpretation of this passage questionable, given that, immediately preceding the quoted passage, we find Gödel saying that "the objects of transfinite set theory, conceived in the manner explained on [pp. 262–3 of Gödel, 1964b] and in footnote 11, clearly do not belong to the physical world and even their indirect connection with physical experience is very loose (owing primarily to the fact that set-theoretical concepts play only a minor role in the physical theories of today)". There then follows the passage that Parsons calls "the most quoted in all [Gödel's] philosophical writing": "But, despite their remoteness from sense experience, we do have something like a perception of the objects of set theory".

[9] See in this connection Chihara, 1973, Chihara, 1982, and Chihara, 1990. I should add that, unlike Quine's "indispensability argument" (to be taken up in the next section), Gödel's reasons for believing in mathematical objects have not generated much of a following among philosophers of mathematics. This is another reason why I have not gone into a detailed discussion of my objections to Gödel's reasoning.

[10] Maddy, 1980. I criticized her interpretation of Gödel in Chihara, 1982.

[11] Maddy put forward these views in Maddy, 1980, and later revised the view in her book Maddy, 1990.

[12] See Maddy, 1980. Later (Maddy, 1990: ch. 2), she gave a much more detailed account of how we are able to gain knowledge of the apparently obvious truths of set theory, calling upon the theories of Hebb.

valiant effort to fit Gödel's ontological views about sets into a coherent framework in which contemporary philosophical and scientific views about perception and concept formation were placed. Certainly, one can regard her work as an attempt to fit Gödel's Platonic view into what I called earlier the Big Picture.

Subsequently (in Maddy, 1990: ch. 2), she adopted the view that her doctrines regarding our ability to see sets should not be taken to constitute an interpretation of Gödel's views, but rather part of her alternative form of mathematical Platonism (which she called "set-theoretical realism")—a view that she has more recently come to reject. I shall not repeat the many criticisms of her Platonic views that I have given in previous publications.[13]

2. QUINE'S CHALLENGE TO THE NOMINALIST

It was Quine who advanced the well-known "indispensability argument" for the existence of mathematical objects, by making such declarations as:

Mathematics—not uninterpreted mathematics, but genuine set theory, logic, number theory, algebra of real and complex numbers, differential and integral calculus, and so on—is best looked upon as an integral part of science, on a par with the physics, economics, etc., in which mathematics is said to receive its applications. (Quine, 1966c: 231)

With such a view of mathematics, it is not surprising that Quine would maintain that the mathematical nominalist "*is going to have to accommodate his natural sciences unaided by mathematics*" (Quine, 1960: 269, italics mine). Quine had advanced a *criterion of ontological commitment*, according to which a theory T, expressed in a first-order logical language, is ontologically committed to entities of kind K iff such entities would have to be in the range of the bound variables of T in order that T be true.[14] Thus, when he asserted that "mathematics, except for some trivial portions such as very elementary arithmetic, is irredeemably committed to quantification over abstract objects",[15] he was in effect claiming that science itself was ontologically committed to the existence of mathematical objects. If the truth of all but trivial portions of mathematics requires that the quantifiers of the

---

[13] I give critical analysis of her set theoretical realism in Chihara, 1990: ch. 10, and in Chihara, 1998: ch. 9. For her more recent abandonment of realism in favor of "naturalism", see Maddy, 1995.

[14] See Chihara, 1973: ch. 3, sect. 3, for a detailed discussion of Quine's criterion.

[15] Quine, 1960: 269.

mathematical language range over mathematical objects or logically similar abstract objects, then it is hard to see how a nominalist can accept the pronouncements of the natural sciences without implicitly accepting the existence of mathematical objects. It is no wonder, then, that Quine thought that the nominalist must *"accommodate his natural sciences unaided by mathematics"* (1960: 269).

In *Ontology and the Vicious-Circle Principle* (Chihara, 1973), I took Quine to be proposing not so much a direct argument for the existence of mathematical objects, as a kind of challenge to the nominalist to produce a non-trivial system of mathematics which would be adequate for the needs of the natural scientist, but which would not require its quantifiers to range over mathematical objects.[16] I regarded him arguing roughly as follows:

> Let's see you nominalists accommodate science without committing yourself to mathematical objects. Let's see you produce a system of mathematics which would be adequate for the needs of the natural scientist, but which would not require its quantifiers to range over abstract objects.

If this challenge cannot be met, then it can plausibly be argued, by a sort of inference to the best explanation, that we have good grounds for believing in the existence of mathematical objects, that is, for being mathematical Platonists. John Burgess and Gideon Rosen have recently put the argument as claiming that "we should believe in abstract entities, but only because nominalistic alternatives to standard scientific theories cannot be developed" (Burgess and Rosen, 1997: 64).

Of course, Quine was convinced that no such nominalistic system of mathematics could be devised—the reason being that he believed that mathematical objects were *indispensable* to the practice of science—so he reluctantly adopted the Platonic view of mathematics for which he is now famous: "reluctant Platonism".[17]

Mark Steiner has suggested that what Quine was getting at with his indispensability argument can be expressed in the following way:

> [T]o describe the experience of diversity and change requires mathematical entities. Imagine defining *rate of change* without the resources of analysis. We cannot say what

---

[16] I was willing to grant Quine a version of his "criterion of ontological commitment". See my reconstruction of his argument in Chihara, 1973: ch. 3, sect. 5.

[17] It is possible that one reason Quine was so sure that no nominalist could meet his challenge is that he also adopted a strong thesis about the kind of language that science can legitimately be expressed in. This is the thesis that the language of science ought to be a logical, first-order, extensional language. For a discussion and criticism of this thesis, see Chihara, 1990: 8–14.

the world would be like without numbers, because describing any thinkable experience (except for utter emptiness) presupposes their existence. (Steiner, 1978: 19–20)

Steiner's formulation can also be regarded as a form of a challenge to the nominalist. Defenders of the argument can be understood to be saying: "Let's see you describe any thinkable experience without making reference to or quantifying over mathematical objects."

There were philosophers who believed that Quine's challenge had already been met by *Principia Mathematica*, since it contained a type-theoretical version of classical mathematics, in which classes were regarded as "logical fictions" and statements containing class-referring expressions were understood to be abbreviations for statements that referred only to individuals and propositional functions.[18] But Quine dismissed such beliefs:

The nominalist program seems already and painlessly achieved if Whitehead and Russell's elimination of classes by a theory of incomplete symbols is thought to bear. But it does not; it only eliminates classes in favor of attributes.[19]

For Quine, to "have reduced classes to attributes is of little philosophical consequence, for attributes are no less universal, abstract, intangible, than classes themselves" (Quine, 1966d: 22).

With respect to the Platonism–nominalism dispute, Quine is surely correct. Since the attributes (or "propositional functions"), to which classes had been reduced in *Principia Mathematica*, turn out to be little different in ontological features from the classes themselves, Quine was certainly justified in his skeptical appraisal of the significance of the "no-class" theory for the dispute.[20] However, his claim that Russell's reduction "is of little philosophical consequence" is unfair and misleading. This can be seen by comparing the abstract ontology of Russell's logical system with Frege's. Frege's abstract ontology contains both concepts and extensions of concepts, whereas Russell's requires only concepts ("propositional functions"), extensions of concepts being eliminated by the no-class theory. Eliminating all the classes from one's ontology should not be considered to be philosophically insignificant, even by Quine, given that he trumpeted, in several places, the importance of the reduction of all mathematical objects (such as numbers,

---

[18] See Chihara, 1973: ch. 1 for a presentation of the basic ideas of that classic work.

[19] Quine, 1960: 269 n. See Quine, 1966d: 19 for an explanation of why he concludes that the Russell elimination reduces classes to attributes.

[20] For a more detailed discussion of this last point, see Chihara, 1973: ch. 1.

functions, ordered pairs, etc.) to sets. If the reduction of functions to sets is ontologically significant, as Quine implies, then why is not the reduction of extensions to propositional functions?

3. NOMINALISTIC RESPONSES TO QUINE'S CHALLENGE

How is a nominalist to respond to Quine's challenge?

It is important to see that a nominalist is not required, by Quine's challenge, to show that the scientific theories *in actual use* are nominalistically acceptable—that when properly analyzed, these theories do not, in fact, make reference to or quantify over mathematical entities. Quine challenged the nominalist to produce a version of science that is both ontologically acceptable to the nominalist and also scientifically (in some rough sense) "on a par with" our present-day scientific theories. He did not expect the nominalist to show that our actual scientific theories are ontologically acceptable to the nominalist, since he was convinced that these theories are ontologically committed to mathematical objects. In the following, I shall indicate how some nominalists have responded to this challenge.

Constructibility quantifiers and predicativity

My own initial response to Quine's challenge (in Chihara, 1973) was to produce a *predicative* system of mathematics, which utilized *constructibility quantifiers* (to be discussed in detail shortly) in place of the existential quantifiers of standard logic, thus avoiding existential quantification over mathematical objects.[21] The idea was to undercut Quine's claim that the nominalist would have to give up all but trivial portions of arithmetic, by showing how a nominalistic version of predicative mathematics could be developed.[22] One problem with this approach, however, was the unresolved worry that significant portions of the mathematics of contemporary science,

[21] Roughly, a predicative system of mathematics is a system that obeys the vicious-circle principle. One can find a detailed discussion of this principle in Chihara, 1973: ch 1.

[22] Philip Kitcher and William Aspray get the point of my reconstruction in writing: "Chihara proposed to reevaluate the Quinean argument for Platonism by showing how the mathematics needed for science could be articulated without commitment to abstract entities" (Kitcher and Aspray, 1988: 13).

such as quantum physics, would be unreproducible within a predicative framework.[23]

My use of constructibility quantifiers to meet Quine's challenge illustrates one of the principal tools that nominalists have adopted to avoid "ontological commitments" to mathematical entities, namely the use of *modal concepts*. What are modal concepts? Here are a couple of examples: *necessity* and *possibility*. Why are they called "modal concepts"? What's *modal* about necessity and possibility? Statements are said to be true or false. Now sometimes we notice that some statements are not only true but are also necessarily true. Necessity seems to be a *way or mode in which the statement is true*. Similarly, possibility has been taken to be a mode in which a statement is true. Necessity and possibility are thus called "modalities", and the logics of these modalities are classified as modal logics.

Philosophers have reasoned using modal concepts for thousands of years. Aristotle, for example, frequently used modal reasoning in his philosophizing. In recent years, modal reasoning (especially involving possible worlds semantics) has been used effectively to arrive at philosophical conclusions in such fields as metaphysics, philosophy of language, philosophy of mind, philosophy of mathematics, and philosophy of science. Some contemporary philosophers have gone so far as to claim that possible worlds semantics "ought to be an absolutely vital part of every philosopher's repertoire" (Hintikka and Hintikka, 1989: 73).

It turns out that the following alternative responses to Quine's challenge have involved in some way or another the use of modal notions. In each case, nominalistic versions of mathematics have been proposed which require at some stage the use of modal ideas.[24]

Field's mathematical instrumentalism

As early as 1945, the logical positivist Carl Hempel had argued that mathematics, unlike an empirical science, is completely empty of factual content.[25]

---

[23] However, owing primarily to the work of Solomon Feferman, there is growing evidence that much (if not all) of the mathematics of contemporary science is reproducible within a predicative framework. See Feferman, 1998f: sects. 8 and 9. See also Feferman, 1998g, where the author answers the question 'What do the indispensability arguments amount to?' with the words: "As far as I am concerned, they are completely vitiated" (297). For an overview of the predicative program, see Hellman, 1998: sects. 1 and 2.

[24] This will be brought out, for the case of Field's fictionalism, in Chapter 11 below.

[25] In Hempel, 1964, which was first published in the *American Mathematical Monthly* in 1945.

Such a doctrine raises the question: How is it that a discipline as scientifically useful as mathematics can have no factual content? Here is what Hempel wrote:

[M]athematics (as well as logic) has, so to speak, the function of a theoretical juice extractor; the techniques of mathematical and logical theory can produce no more juice of factual information than is contained in the assumptions to which they are applied; but they may produce a great deal more juice of this kind than might have been anticipated upon first intuitive inspection of those assumptions which form the raw material for the extractor. (Hempel, 1964: 379)

Many years later, Hempel's idea that mathematics is a theoretical juice extractor was rendered more precise by Hartry Field in *Science Without Numbers*. He was convinced that Platonism was false and that a nominalist could account for the usefulness of mathematics without assuming that mathematics was a body of truths. In particular, he thought that scientists could use mathematics to see what is logically implied by their nominalistic scientific theories without fear of presupposing the existence of mathematical objects, because of a distinctive feature that mathematics has: it is "conservative". What it means to say that a theory is conservative will be become clearer when Field's "conservation principle" is better understood.

Here is a first approximation to Field's conservation principle:

Let N be a nominalistic theory, formulated in first-order logic,[26] and let M be a consistent mathematical theory, also expressed in first-order logic. Then, for any nominalistic sentence ϕ,

if $N + M \models \phi$, then $N \models \phi$.

This is only an approximation, because we need to get clearer on just *what a nominalistic theory is supposed to be* and also we need to say more precisely what M and ϕ are.

To be more precise, let us take M to be a specific mathematical theory that is already expressed in first-order logic and that is generally accepted as giving us classical mathematics. For this purpose, let us choose M to be ZFU (Zermelo-Fraenkel set theory with urelements). ZFU is obtained from the

---

[26] I discuss Field's conservation principle in terms of first-order logic despite the fact that Field at times works with second-order logic in Field, 1980, because later, in response to an objection by Shapiro, he opted for restricting his views about mathematics to the first-order case. See Field, 1985: 241.

more standard ZF by altering slightly the axiom of extensionality so that it reads:

$$(\exists z)(z \in x) \rightarrow ((u)(u \in x \leftrightarrow u \in y) \rightarrow x = y)^{27}$$

Let N be a formal theory that does not overlap in vocabulary with ZFU, and let ϕ be a sentence of N. The intuitive idea here is that, following Quine, we can think of set theory as encompassing all of the mathematics used in science, and since the vocabulary of N does not overlap ZFU, N can be taken to make no reference to any mathematical objects and hence to not presuppose the existence of any mathematical objects. Since ϕ is a sentence of N, it makes no reference to any mathematical objects and can be regarded as a nominalistic sentence.

We can now state a better approximation to the conservation principle:

if $N + ZFU \models \phi$,   then $N \models \phi$.

This is still only an approximation because what is said is nothing new—it's something that logic students knew already. Intuitively, since N and ZFU have no vocabulary in common, if the antecedent holds, then ZFU is not contributing to the proof of ϕ, so we get the consequent. Besides, if ZFU is to be useful in scientific and other applications, there must be a connection between ZFU and N.

So Field devises a set theory ZFU*, whose main feature is that the vocabulary of N is allowed to appear in the condition

$$- - - z - - -$$

of the subset axiom[28] of ZFU:

$$(x)(\exists y)(z)(z \in y \leftrightarrow (- - - z - - - \& z \in x))$$

A still better approximation to Field's conservation principle can then be stated:

if $N + ZFU^* \models \phi$,   then $N \models \phi$.

This is a fairly close approximation to the semantical conservation theorem Field proves in his book. It tells us that, if a nominalistic sentence ϕ is a

---

[27] This alteration allows there to be more than one entity in the domain that has no members.

[28] This is how Field explained the conservation principle in Field, 1980. However, in Field, 1989c, he also allows the replacement axiom to have instances in which occurrences of terms from the vocabulary of N appear (56).

consequence of the theory obtained by adding mathematics (ZFU*) to a nominalistic theory N, then ϕ is a consequence of N alone. By making use of the completeness and soundness theorems of first-order logic, one can obtain a syntactical version of the conservation theorem:

if N + ZFU* $\vdash$ ϕ, then $N \vdash \phi$.

This tells us that, if a nominalistic sentence ϕ is derivable from the theory obtained by adding mathematics (ZFU*) to a nominalistic theory N, then ϕ is derivable from N alone. We have then two conservation theorems, one that tells us that mathematics is "semantically conservative over nominalistic theories" and one that tells us that mathematics is "syntactically conservative over nominalistic theories".

The proof of the above semantic conservation principle that Field supplies (Field, 1980) is a model-theoretic one presupposing the existence of inaccessible cardinals. He also gives a rough sketch of a proof of a syntactic conservation principle. What is of special interest here is what Field seems to think he has proven. These arguments show, he says, that "the gap between the claim of consistency and the full claim of conservativeness is, in the case of mathematics, a very tiny one" (1980: 13). He amplifies his claim with more details of his thoughts about what he thinks he has proven:

What the facts about mathematics I have been emphasizing here show is that even someone who doesn't believe in mathematical entities is free to use mathematical existence-assertions in a certain limited context: he can use them freely in deducing nominalistically-stated consequences from nominalistically-stated premises. And he can do this not because he thinks those intervening premises are true, but because he knows that they preserve truth among nominalistically stated claims. (1980: 14)

Thus, we get Field's instrumentalist account of the usefulness of mathematics that Hempel put forward earlier in inchoate form: Mathematics is a theoretical juice extractor.[29]

The above account suggests a way for a nominalist to meet the Quinean challenge: she can attempt to formulate nominalistic versions of scientific theories, from which she could freely use mathematics to draw nominalistic

---

[29] Some readers familiar with Field's views on mathematics may be inclined to take issue with my characterizing his version of nominalism as a form of "instrumentalism", since he quite frequently calls his own view "fictionalism". It should be noted that, in his introduction to Field, 1989d, he writes: "I will use the words 'fictionalism' and 'instrumentalism' interchangeably" (4 n.).

conclusions without having to assume that the mathematics being used was true. Then, if it could be shown that the resulting theories would be empirically equivalent to, or at least not significantly weaker than, our actual scientific theories, Quine's challenge would be met. Thus, suppose a nominalist believed, as Field did when he wrote *Science Without Numbers*, that "there is an alternative formulation of science that does not require the use of any part of mathematics that refers to or quantifies over abstract entities".[30] Such a nominalist would believe that she could legitimately use mathematics in her scientific work without having to believe in mathematical objects, while at the same time believing that her scientific theories were the equal of the Platonist's. For the mathematics used would be functioning merely as a theoretical juice extractor—an instrument that facilitates the drawing of nominalistic conclusions within the theory. Thus, there would be no reason to maintain that the purely mathematical assertions were true and hence that there are mathematical objects. Note: one does not ontologically commit oneself to mathematical objects merely by using a mathematical theory whose quantifiers would have to range over mathematical objects in order that the theory be true. It is only by affirming or believing the truth of such a mathematical theory that the ontological commitment is made.[31] Field, of course, maintains that the theorems and assertions of standard mathematics are, for the most part, false, because he is convinced that no mathematical objects exist.

I should emphasize that I am not claiming that, in *Science Without Numbers*, Field was actually attempting to respond to what I have called "the Quinean challenge to the nominalist". Indeed, Field tells us that he was attempting to refute what he called the Quine–Putnam indispensability argument—an argument that differs in subtle ways from Quine's challenge. I am only claiming here that the various elements of Field's position in that book can be fashioned into a response to the challenge.

---

[30] Field, 1980: 2. I believe that the above quote would be clearer had Field written "does not require the truth of ..." instead of "does not require the use of ...". In a footnote to this page, he writes: "Very little of ordinary mathematics consists merely of the systematic deduction of consequences from such axiom systems: my claim however is that ordinary mathematics can be replaced in applications by a new mathematics which consists only of this." In other words, for any scientific theory T, there is an alternative nominalistic formulation T*, such that a replacement M* of ordinary mathematics could be found which consists merely in the systematic deduction of consequences from the axioms of the nominalistic formulation.

[31] The expression 'ontologically commit' is taken from Quine's criterion of ontological commitment. For a detailed discussion of this criterion, see Chihara, 1973: ch. 3, sect. 3.

I should add, in this connection, that some things Field has written do suggest that he was responding to the challenge. For example, in discussing an objection by Stewart Shapiro, he writes:

An evaluation of the Quine–Putnam argument requires that we evaluate whether ... there are any advantages of platonistic versions of Newtonian physics over nominalistic versions, or vice versa. (Field, 1989b: 141)

He then goes on to suggest that the nominalistic version of a physical theory he has sketched would be as good as the Platonic version because "it is hard to imagine any situation in which the excess content of the platonic version would receive empirical support" (143). In this discussion, there was no attempt to determine which of the alternative versions of Newtonian physics was the one physicists actually used. The only question was: which of the alternatives was preferable? And it is clear from his argumentation that he was intent on showing that the nominalistic version was as good as the Platonic version.

Again, in discussing a Quinean objection to his instrumentalist view of mathematics, he asserts that the objection "can only be undercut by showing that there is an alternative formulation of science that does not require the use of any part of mathematics that refers to or quantifies over abstract entities" (Field, 1980: 2). He tells us that he believes that such a formulation is possible and that he will illustrate a new strategy for achieving such a complete reformulation. His belief in a complete reformulation of science that is free of commitment to mathematical objects is further evidence that, at times, he saw himself as responding to the Quinean challenge.[32]

Hellman's modal structuralism

Hellman's own "modal-structuralist" contribution to the Platonism–nominalism controversy can be regarded as another attempt to meet Quine's challenge.[33] Hellman's general strategy in *Mathematics Without Numbers* is to

---

[32] The attentive reader may wonder just where modality enters Field's account of mathematics. See Chapter 11 below for an answer.

[33] It should be noted, however, that in a work published more recently, he writes of his modal structuralism: "What it does imply is that confirmation of our best contemporary science cannot be expected to confer support on set-theoretic mathematics through the mechanism of indispensability arguments, even if such a mechanism can in principle be integrated within an otherwise satisfactory account of confirmation" (Hellman, 1999: 29). Here he is pointing his account of mathematics directly against the indispensability argument based upon holism (which I take up later in this chapter).

represent statements of pure mathematics as elliptical for modal conditionals of a certain sort. In the case of the arithmetical statement ϕ, the modal-structuralist interpretation of ϕ would be (roughly):

If there were any ω-sequences, ϕ would be true in them.[34]

More specifically, Hellman's "modal-structuralist translate" of ϕ turns out to be of the form:

$$\Box Q(A \rightarrow \phi)$$

where 'Q' is to be replaced by a string of universal quantifiers and 'A' is to be replaced by a conjunction of the axioms of second-order Peano arithmetic.[35]

The perspective from which Hellman makes his "modal-structuralist interpretations" of classical mathematics involves adopting a powerful theory, framed in a second-order monadic quantificational S5 modal language,[36] which requires various non-trivial modal assumptions. Hence, it might appear doubtful that such a position can be classified as "nominalistic". But Hellman claims that "it is possible to read the ms [or modal-structuralist] interpretations given so far *entirely* nominalistically" (Hellman, 1989: 47). The sort of nominalism he envisages is one that postulates only "individuals" and "mereological sums of individuals [to be explained below]". The concept of a mereological sum is something defined in mereology (the logic of part-wholes) that some philosophers have regarded as nominalistically acceptable. Hellman's idea is to regard the first-order variables of his theory as ranging over a totality of concrete objects that are pairwise discrete (that is, have no part in common); the second-order (monadic) variables are to be regarded as ranging over the mereological sums of the concrete objects. In this way, the second-order variables can be regarded as functioning as

[34] As will be indicated below, the specific logical form of the translation is the necessitation of a universally quantified conditional. See Resnik, 1997: part 1, sect. 3, for a more detailed description of Hellman's method of "translating" statements of mathematics into his modal-structuralist version. Resnik also mounts a spirited criticism of Hellman's view in the work cited.

[35] For a more detailed and precise statement of the modal-structuralist interpretation of ϕ, see Hellman, 1989: 24. The semantics of the modal system being presupposed in this analysis is that of Cocchiarella, 1975. It should be noted that my student Randolph Goldman has raised some objections to parts of Cocchiarella's formalization; he provides an alternative formulation of second-order quantificational S5 modal logic in Goldman, 2000, to correct what is perceived to be its flaws. See Hellman, 1989: 36–7 for Hellman's discussion of the semantics of his modal operators.

[36] The reader not wishing to struggle through the complications of Cocchiarella's system might first wish to tackle an easier version of modal logic. A formalization and discussion of a version of first-order quantificational S5 modal logic can be found in Chihara, 1998: ch. 1.

classes of the concrete objects, even though "these [mereological] sums are, ontologically, on a par with the [concrete objects] themselves: they are every bit as concrete" (Hellman, 1989: 49).

Although I have some doubts and worries about various aspects of Hellman's views (see Appendix A below), there is no question that his modal structuralism has opened up a promising and intriguing line of investigation that warrants continued research.[37]

The impact of the nominalistic responses

Each of the above nominalistic responses to Quine's challenge has been criticized in various ways.[38] Still, despite these criticisms, these responses have convinced many philosophers of the dubiousness of Quine's claim that the nominalist *"is going to have to accommodate his natural sciences unaided by mathematics"*. I know of no philosopher who still believes that a nominalist is precluded from using any non-trivial mathematical theory at all. These investigations have shown that there are many promising avenues for the nominalist to pursue.[39]

4. THE DIRECT INDISPENSABILITY ARGUMENT

In more recent years, Quine has been understood to have constructed a more direct argument for belief in the existence of mathematical objects, perhaps because he has claimed that traditional metaphysical assertions such as "Numbers exist" or "There are sets" can be supported by empirical evidence. He writes:

I think the positivists were mistaken when they despaired of evidence in such cases and accordingly tried to draw up boundaries that would exclude such sentences as meaningless. Existence statements in this philosophical vein do admit of evidence, in the sense that we can have reasons, and essentially scientific reasons, for including

---

[37] For example, Hellman might be able to develop the position of Agustin Rayo and Stephen Yablo (as set out in Rayo and Yablo, 2001) into a clearly more acceptable nominalistic version of his modal structuralism.

[38] Some of the criticisms aimed at my constructibility approach will be taken up in later chapters. There have been many critics of Field's approach. See Chihara, 1990: ch. 8, sect. 5 for a sampling of some of the objections raised. I detail a new set of objections to Field's approach later in this work (in Chapter 11). My doubts and worries about Hellman's views are given in Appendix A.

[39] For example, Mark Colyvan has recently written: "Although I am not yet convinced that Field's program will be successful, I have no doubt about the importance of his program. Indeed, I, like Field, believe that the correct philosophical stance with regard to the realism/anti-realism debate in mathematics hangs on the outcome of his program" (Colyvan, 2001: 89–90).

numbers or classes or the like in the range of the values of our variables. Numbers and classes are favored by the power and facility they contribute to theoretical physics and other systematic discourse about nature. (Quine, 1969*b*: 97–8)

Thus, Quine has been taken to have argued as follows:

[T]he use of mathematics in science provides empirical evidence for the truth of mathematics. Moreover, since mathematical claims involve quantification over abstract mathematical entities, accepting the truth of mathematics requires accepting mathematical objects as well. (Vineberg, 1998: 118)

Quine is evidently arguing that, contrary to what the logical positivists thought, contemporary science provides genuine scientific evidence for the existence of mathematical objects. How does "including numbers or classes or the like in the range of the values of our variables" contribute power and facility to theoretical physics? Presumably by allowing the physicist to use mathematics both to express the kind of complex relationships that are needed in high-level physics and also to draw mathematical inferences from the statements of physics so expressed. These contributions to the physics are, one might say, indispensable.

Granted all this, why should these "indispensable benefits" that the postulation of mathematical objects provides count as scientific evidence for the existence of mathematical objects? After all, these benefits do not appear to amount to what is ordinarily counted as evidence: experimental data or observed facts. To understand Quine's reasons for maintaining such a position, we need to turn to an early paper he wrote entitled "Posits and Reality"—a paper concerned with the question: 'What is the nature of evidence?' Quine begins this article with an examination of the "molecular hypothesis": the hypothesis that there are molecules. What supports the hypothesis? In answer, he lists:

(a) Simplicity of theory;
(b) Familiarity of principle;
(c) Wide scope of application;
(d) Fecundity (leads to fruitful developments);
(e) Testable consequences that have been confirmed.

Quine then raises the question of evidence: do the above constitute genuine evidence for the truth of the theory? *Or do they merely give us a reason for using the theory?* He wonders whether "the benefits conferred by the molecular doctrine give the physicist good reason to prize it, but afford no evidence of its truth" (Quine, 1966b: 235). Later, he says:

[T]he molecular physicist is, like us, concerned with common place reality, and merely finds that he can simplify his laws by positing an esoteric supplement to the exoteric universe. ... No matter if physics makes molecules or other insensible particles seem more fundamental than the objects of common sense; the particles are postulated for the sake of a simple physics. (1966b: 236–7)

The suggestion is, then, that we do not have genuine evidence for the existence of molecules—*we only have pragmatic reasons for positing molecules.*

Quine then introduces a surprising new twist to an old story. He asks: *what evidence do we have for the existence of the "bodies" of common sense?* That is, things like tables, chairs, dogs, buildings, even people? Here, we are apt to think that, unlike molecules, these things can be directly seen, felt, smelled, heard, and tasted. Not so, suggests Quine:

What are given in sensations are variformed and varicolored visual patches, varitextured and varitemperatured tactual feels, and an assortment of tones, tastes, smells, and other odds and ends; desks are no more to be found among these data than molecules. (1966b: 237)

Quine concludes:

If we have evidence for the existence of the bodies of common sense, we have it only in the way in which we may be said to have evidence for the existence of molecules. The positing of either sort of body [molecules or commonsense bodies] is good science insofar merely as it helps us formulate our laws. (1966b: 237)

At first sight, it would seem that the positing of the bodies of common sense merely "helps us formulate our laws" and hence does not actually provide us with evidence for the existence for such things as tables and chairs. But Quine wants us to draw a different conclusion. He says:

Unless we change meanings in midstream, the familiar bodies around us are as real as can be; and it smacks of a contradiction in terms to conclude otherwise. Having noted that man has no evidence for the existence of bodies beyond the fact that their assumption helps him organize experience, we should have done well, instead of disclaiming evidence for the existence of bodies, to conclude: such, then, at bottom, is what evidence is, both for ordinary bodies and for molecules. (1966b: 238)

And:

The benefits of the molecular doctrine which so impressed us [earlier], and the manifest benefits of the aboriginal posit of ordinary bodies, are the best evidence of reality we can ask. (1966b: 238–9)

In other words, we should take the "benefits" that are created by an existential postulation to be genuine evidence for the truth of the postulation, on pain of being reduced to an absurdity—the absurdity of having to say that the bodies of common sense are unreal. This is how Quine justifies his pragmatic analysis of evidence.

It is no wonder that, with such a conception of evidence, Quine would go so far as to claim that the question of what exists "is at bottom just as arbitrary or pragmatic a matter as one's adoption of a new brand of set theory or even a new brand of bookkeeping" (Quine, 1966: 125). *If finding evidence for the truth of some theory is simply a matter of finding pragmatic reasons for using a theory (in the way we may find pragmatic reasons for using a new brand of bookkeeping), then we certainly can find evidence for the existence of mathematical objects.* In this light, we can understand the quotation I gave you earlier. We can see why Quine thinks that "the positivists were mistaken when they despaired of evidence" for deciding such questions as 'Do numbers exist?' He requires only that one find pragmatic reasons for using a theory that assumes the existence of numbers—which boils down to finding pragmatic reasons for using number theory—in other words, a mathematical theory that is ontologically committed to numbers. Given Quine's view of evidence, it is an easy step to the position that we have scientific evidence for the existence of mathematical objects.

Quine's indispensability argument, understood in the above way as based upon a pragmatic conception of evidence, has found very little acceptance among philosophers of science or philosophers of mathematics. Quine's analysis of evidence seems to be crude, counterintuitive, even bizarre. And the reasoning in support of Quine's conception of evidence is highly questionable. He argues that we must accept such a pragmatic conception of evidence, since otherwise we would have to allow that we have no evidence for the existence of "the bodies of common sense" and hence that such bodies are "unreal". Supposedly, we have a reductio ad absurdum of the rejection of his pragmatic conception of evidence. But is this so? Must we accept Quine's pragmatic conception of evidence on pain of being reduced to absurdity?

I suggest that we question the cogency of Quine's reasoning on this point. What I claim is that it by no means *follows from* the hypothesis that we have no evidence for the existence of the bodies of common sense that, in that case, such bodies would have to be taken to be unreal. Before accepting Quine's conclusions regarding evidence, we should consider the following possibility. Our belief in the existence of physical objects—the bodies of common sense—may be so fundamental to our thinking, to our theorizing about the world, that all talk of evidence for the existence of such things is simply inappropriate, the reason being that our practices of gathering observational data, performing experiments, assessing data, confirming hypotheses, and testing theories take place within a framework of ideas and beliefs in which physical objects are presupposed. In such a situation, it would make no sense to try to gather evidence for the existence of physical objects. It is not that we would be in a situation in which it would be reasonable to look for evidence for the existence of physical objects but in which, for some reason, we just couldn't find any. Rather, it would make no sense to speak of evidence for something so fundamental to our whole practice of gathering evidence. If something like this is the case, then it would be unreasonable to infer from the absurdity of maintaining that physical objects are unreal that we must adopt the Quinean pragmatic conception of evidence.[40]

As I noted earlier, Quine has won over few adherents to his view of evidence. Nowadays, most philosophers of science attempt to analyze the concept of evidence by studying the scientific practice of gathering and assessing evidence, and by examining what the experts on matters of scientific evidence say; that is, they consult the writings of theoretical scientists. They do not give "philosophical arguments" of the sort Quine gave to determine *what must count as evidence*. Not surprisingly, practically all philosophers who believe that Quine advanced an indispensability argument for the existence of mathematical objects interpret the argument in a way that avoids bringing in Quine's questionable conception of evidence.[41]

5. THE INDISPENSABILITY ARGUMENT BASED ON HOLISM

Most philosophers of mathematics now take Quine's indispensability argument to proceed from his holism: the doctrine he attributes to Pierre Duhem that "our statements about the external world face the tribunal of sense experience not individually but only as a corporate body" (Quine, 1961*b*: 41). For example, here's how Maddy understands Quine's reasoning:

Our best scientific theory of the world makes indispensable use of mathematical things. ... To draw a testable consequence from our theory requires the use of various

---

[40] Susan Vineberg has suggested to me (personal communication) an alternative way of responding to Quine's argument. She believes that, even if it were granted that the "benefits" of assuming the existence of "bodies" do not constitute evidence for their existence, that would not necessarily imply that we could then have no evidence for the existence of "bodies": one could still make a case, on Bayesian grounds, that one has evidence for their existence.

[41] There is more to Quine's direct argument than I have indicated above. For a more detailed discussion of this argument and for additional objections to it, see Chihara, 1990: ch. 1, sect. 2.

far-flung parts of the theory, including much mathematics, so the confirmation resulting from a successful test adheres not to individual statements but to large bodies of theory. (This is holism.) Finally, our theory is committed to those things that it says 'there are'. (This is Quine's criterion of ontological commitment.) It follows that our theory, and we who adopt it, are committed to the existence of mathematical things.[42]

Elliott Sober gives a similar reading to Quine's argumentation:

This indispensability argument for mathematical realism gives voice to an attitude towards confirmation elaborated by Quine. Quine's holism—his interpretation of Duhem's thesis—asserts that theories are confirmed only as totalities. A theory makes contact with experience only as a whole, and so it receives confirmation only as a whole. If mathematics is an inextricable part of a physical theory, then the empirical success of the theory confirms the entire theory—mathematics and all. (Sober, 1993: 35)

Sober's version no doubt draws its inspiration from what Quine has written about verification and Duhemian holism, for example, that "our statements about the external world face the tribunal of sense experience not individually but only as a corporate body".[43]

Resnik sees the indispensability argument, as presented by Maddy and Sober, as being essentially the following argument:

(1) Mathematics is an indispensable part of our best scientific theories.
(2) Mathematics shares whatever confirmation accrues to the theories using it (Quinean holism).
(3) So mathematics shares the confirmation accruing to our best scientific theories.

---

[42] (Maddy, 1997), p. 133. It is interesting to compare this formulation of the argument with one given in her earlier work (Maddy, 1990). There she talks about the Quine/Putnam indispensability argument, which she formulates as follows: "We are committed to the existence of mathematical objects because they are indispensable to our best theory of the world and we accept that theory" (30). Notice that in the earlier version, there is no talk about confirmation. For purposes of the present investigation, I shall take the more recent formulation of Maddy's to be simply a more detailed version of the earlier argument. Susan Vineberg has suggested to me (in correspondence) that Maddy may be simply following Quine in equating acceptance with evidence: this is because, "for Quine, being the best overall theory (where best is understood in his characteristic pragmatic way) is his criterion for acceptance, but it also appears to be constitutive of evidence". However, to the community of philosophers of science, there is a very significant difference between acceptance and evidence, and formulating the indispensability argument in terms of acceptance, as Maddy did in the earlier work, raises theoretical questions that are quite different from those raised by her more recent formulations in terms of evidence and confirmation.

[43] Quine, 1961b: 41. Sober gives a critique of Quine's view on verification in Sober, 2000.

(4) We are committed to the truth of our best scientific theories (naturalism).
(5) So we are also committed to the truth of the mathematics they contain.[44]

So far as I know, Quine never explicitly gave the specific indispensability argument, (1)–(5), being attributed to him above, but the argument is clearly in the spirit of Quine's philosophy, and one can argue that the principles and premises being used in the argument have all been supported, in one way or another, by Quine.[45]

Notice that the above version of the indispensability argument (especially (2)) maintains that essentially every confirmation of a scientific theory confirms mathematics—not just such and such a principle of mathematics or such and such a theorem of mathematics, but *mathematics*. Why should, say, confirmation of Mendel's genetic theory, which involved the use of little more mathematics than arithmetic, be taken to confirm, say, the least upper bound theorem or the power set axiom?

Perhaps the idea is this: Quine had once argued that we should believe in the existence not of all the various sorts of mathematical objects discussed by mathematicians—numbers, functions, ordered pairs, matrices, spaces, and so on—but only of sets, the reason being "ontological economy":

Researches in the foundations of mathematics have made it clear that all of mathematics [needed in science] can be got down to logic and set theory, and that the objects needed for mathematics in the above sense can be got down to a single category, that of *classes*. (Quine, 1966c: 231)

One could, in this way, simplify considerably the ontology of mathematics by proceeding from the axioms of set theory. Since simplicity is taken by Quine to be evidence for the truth of the theory, by taking the mathematics to be used in science to be set theory, we would thus have additional evidence for the truth of this mathematics. Imagining, then, that the mathematics used in science is carried out in set theory, it is easy to see how one might think that every confirmation via science of any part of mathematics amounts to confirmation of the axioms of set theory and hence of set theory itself (which can be said to be, in this setting, *mathematics*).[46]

---

[44] This statement of the argument consisting of the lines (1)–(5) is presented exactly as Resnik gives it on p. 232 of Resnik, 1998.

[45] Commenting on the Maddy–Sober interpretations of Quine's indispensability argument, Resnik has written: "Let me grant immediately that whether or not Quine has offered this argument, it can be constructed from his principles" (Resnik, 1998: 232).

[46] There are passages in Maddy, 1990 that suggest such a way of reading Quine's version of set-theoretical realism: see, for example, 60–2.

There is another way in which one might come to believe that any confirmation of a scientific hypothesis or theory amounts to a confirmation of all the mathematics that is used in science. If one believes, as did Quine, that in confirming a single hypothesis *h*, one confirms not only *h*, but all the auxiliary assumptions one may use in the confirmation—assumptions about the apparatus used, about the electrical power needed to run the instruments, about the states of mind of the technicians running the equipment, about the constancy and variability of sense organs and mental capacities of people ... (and of course all the auxiliary assumptions that were used in the confirmations of the aforementioned auxiliary assumptions, etc.). According to this scenario, practically everything one believes becomes tested by a single confirmational situation. Perhaps this is why Resnik sees the indispensability argument as depending on a principle of holism that he characterizes as follows: "The observational evidence for a scientific theory bears upon the *theoretical apparatus as a whole* rather than upon individual component hypotheses" (Resnik, 1997: 45, italics mine).

This way of understanding the indispensability argument makes each confirmation of any scientific hypothesis into a confirmation of practically everything one believes. It implies that a geneticist's confirmation of a biological hypothesis, say about where in a human's chromosomes a particular gene is to be found, amounts to a confirmation that humans have eyes. I can imagine Wittgenstein saying: If every test of any hypothesis is a test of every hypothesis one believes, then no test of any hypothesis is a test of any hypothesis one believes. Certainly, one could argue that such an extreme holistic view of confirmation results in a notion of confirmation that is stretched beyond recognition.

Notice that neither of the above versions of the indispensability argument utilizes, or even mentions, Quine's pragmatic conception of evidence; it is Quine's holism that provides the basis for this argument. Holism evidently gives rise to a view of confirmation according to which, somehow or other, the verification of a single scientific hypothesis is equally a verification of all of mathematics—at least all the mathematics that is used in science.[47] One can see how Maddy could come to characterize Quine's views of confirmation in her words: "Mathematics is part of the theory we test against experience, and *a successful test supports the mathematics as much as the science*" (Maddy, 1990: 27, italics mine).

---

[47] Cf. Philip Kitcher's objections to the Quine–Duhem version of holism, according to which the doctrine is "a product of the underrepresentation of scientific practice" (Kitcher, 1993: 247).

There is another feature of these versions of the Platonic argument that deserves comment. The above arguments are extremely crude ("mushy") arguments for the existence of mathematical objects in so far as they supply no detailed or finely tuned analysis of the actual uses of mathematics in scientific reasoning. The underlying idea seems to be that *it does not matter how mathematics may be used* in carrying out a specific scientific confirmation, or even what particular branch of mathematics may be involved in the scientific theorizing—any confirmation of the scientific theory thereby confirms *mathematics*. Such an idea is certainly suspicious and deserving of critical investigation.[48]

Suppose, for example, that in testing a certain scientific theory T, a mathematical theory is only used to describe a sort of simplified model of the actual experimental situation, and that it is generally agreed by the experts in the field that this use of mathematics could be avoided by giving a more accurate, but considerably more complicated, description of the actual experimental situation. Would the defenders of the holistic version of the indispensability argument still claim, in this situation, that the mathematics, so used would be confirmed by any confirmatory test of T? Surely, how mathematics is used in some portion of a scientific theory should be taken into account in determining whether some scientific test of this theory confirms mathematics. I shall return to this doubt about the indispensability argument in Chapter 9.

Finally, it should be noted that all the above versions of the indispensability argument based on holism make use of the term 'confirmation', with no suggestion that it is being used in a Pickwickian or specialized sense. Indeed, in discussions of the argument, it is clear that those involved—Maddy, Sober, and Resnik—think of themselves as using the term in the ordinary way in which non-philosophical scientists use it. Of course, they may give radically divergent analyses of confirmation.

6. PUTNAM'S VERSION OF THE INDISPENSABILITY ARGUMENT

Field has maintained that the most thorough presentation of Quine's indispensability argument is to be found not in Quine's works but in the writings of Putnam.[49] Here is how Putnam expressed the argument in his short

---

[48] Anthony Peressini has begun an investigation into this aspect of the indispensability argument in Peressini, 1997.   [49] See Field, 1980: 107 n. 4, where this is claimed.

monograph *Philosophy of Logic*:

[Q]uantification over mathematical entities is indispensable for science, both formal and physical; therefore we should accept such quantification; but this commits us to accepting the existence of the mathematical entities in question. (Putnam, 1971: 57)

This rough statement of the argument cannot legitimately be said to be a version of any of the three types of indispensability arguments I have thus far discussed. However, in some later works, Putnam argued for Platonism in a way that brings to mind the holistic version. The result is an argument that is closely related to, but significantly different from, the holistic argument for the existence of mathematical objects.[50]

There are several features of Putnam's reasoning that are not to be found in any of Quine's works. One novel feature of Putnam's argument is its appeal to the thesis that various principles and theories of mathematics, as well as principles linking mathematical entities with certain theoretical entities of an empirical science, have been given "quasi-empirical" justifications.[51] However, Putnam did not equate "quasi-empirical" justification with empirical confirmation. Here's how he explained his notion:

By 'quasi-empirical' methods I mean methods that are analogous to the methods of the physical sciences except that the singular statements which are 'generalized by induction', used to test 'theories', etc., are themselves the product of proof or calculation rather than being 'observation reports' in the usual sense.[52]

Thus, Putnam maintained that the remarkable fruitfulness of accepting and utilizing the principle that there is a one-one correspondence between the real numbers and the points on a line in physical space constitutes in part a quasi-empirical, and also in part an empirical, justification of that principle.[53]

---

[50] See, for example, Maddy, 1990: 29–30.

[51] See Chihara, 1990: ch. 6, sect. 4 for a discussion of Putnam's argument based upon such "quasi-empirical" justifications.

[52] Putnam, 1979: 62. Unfortunately, Putnam did not go on to explain how this kind of justification is related to the more heavily studied notion of empirical confirmation. Notice that his thesis that quasi-empirical methods may confirm mathematical principles is similar to Gödel's idea that our mathematical intuitions may "confirm" set-theoretical principles in a way analogous to the way sense perceptions may confirm a new scientific law or general hypothesis (which I discussed in Section 1 of this chapter).

[53] Putnam, 1979: 65. For a rebuttal of Putnam's reasoning on this point, see Chihara, 1990: ch. 6, sect. 4.

Notice that Putnam's justification does not appeal to what amounts to Quine's pragmatic conception of evidence; nor does it depend upon the general thesis of holism, at least as Maddy, Sober, and Resnik understand it. He certainly never claimed that any empirical confirmation of any scientific theory amounts to a confirmation of *all the mathematics* used in science. The "quasi-empirical justification" of Putnam's argument is more or less directly tied to the principle that is supposed to be justified. In particular, when Putnam attempted to show, at particular points in his reasoning, that particular mathematical principles had received "quasi-empirical" justifications, he was proceeding in a way that none of the others believed Quine had proceeded.

However, the overarching idea that guides Putnam's reasoning is the thesis that what justifies taking the theorems of mathematics to be true is the "success" of applying these theorems in the various sciences. Thus, he claims that "the real justification of the calculus is it success—its success in mathematics, and its success in physical science" (Putnam, 1979: 66). Indeed, he went on to suggest that "the hypothesis that classical mathematics is largely true accounts for the success of the physical applications of classical mathematics" (1979: 75). In this respect, Putnam is very close to Quine.

Another novel feature of Putnam's argument in favor of mathematical Platonism is its premise that our scientific theories cannot even be expressed without the heavy use of mathematical concepts and mathematical theories. As Putnam put it, "mathematics and physics are integrated in such a way that it is not possible to be a realist with respect to physical theory and a nominalist with respect to mathematical theory" (1979: 74).[54]

It should be mentioned, however, that Putnam has, in his subsequent writings, completely abandoned the kind of mathematical realism which allied him with Gödel and Quine. Putnam no longer defends any version of the indispensability argument and he has even set forth some anti-realist arguments.[55]

---

[54] Susan Vineberg has emphasized this feature of Putnam's reasoning in Vineberg, 1998: 119–20, where she notes that even "if we accept the view that the use of mathematics in formulating scientific theories provides no empirical support for mathematics, Putnam's observations seem to show that the truth of at least some mathematics is presupposed by science, assuming that the scientific laws formulated using mathematics are regarded as true". It might be replied, following Nancy Cartwright's reasoning, that the fundamental explanatory laws of physics are not true (do not state facts). See Cartwright, 1983: Essay 3. For another discussion of Putnam's version of the indispensability argument and its differences from Quine's version, see Chihara, 1990: 164–5.

[55] See, for example, Putnam, 1980.

7. RESNIK'S VERSION OF THE ARGUMENT

Michael Resnik, building on the work of Quine and Putnam, has put forward an argument for the existence of mathematical objects, which he calls "the Holism-Naturalism indispensability argument" (or "H-N argument" for short). This indispensability argument, which is quite similar to the version I presented earlier, is based upon the following three philosophical theses:

> *Indispensability*: Referring to mathematical objects and invoking mathematical principles is indispensable to the practice of natural science.
>
> *Confirmational Holism*: The observational evidence for a scientific theory bears upon the theoretical apparatus as a whole rather than upon individual component hypotheses.
>
> *Naturalism*: Natural science is our ultimate arbiter of truth and existence.[56]

Here is how the reasoning is supposed to go. By Indispensability, mathematical objects and principles are presupposed by the natural sciences. By Holism, whatever empirical evidence we have for a natural science "is just as much evidence for the mathematical objects and principles it presupposes as it is for the rest of its theoretical apparatus" (Resnik, 1997: 45). Then, Resnik concludes by Naturalism, that "this mathematics is true, and the existence of mathematical objects is as well-grounded as the other entities posited by science" (1997: 45).

Notice that this holistic version of the indispensability argument has the weakness of the version mentioned earlier. It is philosophy of science done with a five-foot brush. In addition, Resnik's reasoning is by no means tight. A person who accepted the above three principles would not be logically committed to the conclusion Resnik arrives at. By Indispensability and Holism, one can indeed infer that the observational evidence for a scientific theory "bears upon" the truth of the mathematical principles that are indispensable to the science in question. Furthermore, we can allow by Naturalism that natural science is our ultimate arbiter of the truth of the mathematical principles mentioned above. But all of this does not logically imply that these mathematical principles are true or that mathematical objects exist. At best, we can infer that there is some evidence "bearing upon" the existence of mathematical objects and on the truth of the mathematics presupposed. But the thesis that there is some evidence bearing upon a proposition P is not

---

[56] Resnik, 1997: 45. Resnik gives essentially the same argument in Resnik, 1995, but there he calls the argument the 'Confirmational Indispensability Argument'.

tantamount to the thesis that we have good grounds for believing that P is true. After all, the above three principles do not logically imply that there is empirical evidence actually supporting any mathematical principle, nor do they say anything about the *strength* of any evidence supporting a mathematical principle.

Of course, Resnik could make his argument more convincing by specifying that what he means by 'bears upon' is something like 'is just as much evidence for'. In other words, he could blunt the above objection by giving his second premise as:

Confirmational Holism: The observational evidence for a scientific theory is just as much evidence for the theoretical apparatus used in the theory as it is for the individual component hypotheses.

The "Confirmational Holism" premise would have to assert some such proposition to allow Resnik to conclude that any observational evidence for a scientific theory is just as much evidence for the mathematical principles used in the theory. Now there are grounds for thinking that Resnik had in mind just such a premise for his indispensability argument, since in a more recent article, he presents an indispensability argument (which he allows "can be constructed from [Quine's] principles") with the following premise:

[#] Mathematics shares whatever confirmation accrues to the theories using it (Quinean holism).[57]

Notice that, in this principle, we again have the idea that it is *mathematics* that shares whatever confirmation accrues to the scientific theory using any mathematical theorem or principle of any mathematical theory; there is no suggestion that only the particular mathematical theorem or principle that is used in the confirmation is confirmed. This is a huge claim that certainly deserves careful examination.

8. Sober's Objection to the Indispensability Argument

An objection has been raised to this sort of indispensability argument by Sober, who, in "Mathematics and Indispensability" (Sober, 1993), challenges premise [#]. In other words, Sober takes aim at *Confirmational Holism* as the

[57] Resnik, 1998: 232.

term was specified above to make Resnik's reasoning acceptable. Sober attacks Resnik's holism from the vantage point of a position he calls "contrastive empiricism". Contrastive empiricism holds that the degree of empirical support observational data provides a theory is a relative matter. Empirical evidence, according to this view, should be understood as favoring some theories over their competitors. Thus, observations enable us to discriminate among the alternative theories we can think up (Sober, 1993: 39).

The fundamental principle of contrastive empiricism is the *Likelihood Principle*, which says that:

Observation O favors hypothesis H_1 over hypothesis H_2 iff $P(O/H_1) > P(O/H_2)$, where $P(O/H_1)$ is the conditional probability of O, given H_1.

The contrastive empiricist would use the Likelihood Principle to determine which of the competing hypotheses $H_1, H_2, H_3, \ldots, H_n$ is the most favored by the observation O.

Consider now the indispensability argument from the perspective of contrastive empiricism, and focus, in particular, on [#] or the premise of holism that forms the basis for the claim that we have empirical evidence for the truth of mathematics. The holist maintains that the observations that confirm some empirical theory H_1 also confirms the mathematics used in H_1. Now suppose that $H_2, H_3, \ldots, H_n$ are the competitors to H_1. It would be reasonable to suppose that mathematics would be an essential part of each of the competitors—that any reasonably sophisticated scientific theory would involve mathematics in some essential way. In that case, writes Sober, mathematical propositions do not obtain any evidential support from the observational data:

If the mathematical statements M are part of *every* competing hypothesis, then, no matter which hypothesis comes out best in the light of the observations, M will be part of that best hypothesis. M is not tested by this exercise, but is simply a background assumption common to the hypotheses under test. (Sober, 1993: 45)

Without committing myself to adopting Sober's contrastive empiricism, I would like to support Sober's skepticism about [#]; his doubts are plausible and in agreement with our intuitions about a great many examples. For example, it seems absurd to suppose that Mendel's careful experiments with peas "is just as much evidence for the mathematical objects and principles it presupposes as it is for" the truth of the Mendelian principles of inheritance. The mathematical principles Mendel used in his calculations amount to little more than principles of simple arithmetic, and it is grotesque to claim that Mendel's experiments constitute significant evidence for such principles (whereas it is not grotesque to suppose that these experiments provide important evidence for the truth of Mendel's genetic principles).

Let me make my position clear and unambiguous. I second Sober's skepticism about [#]. I do not think that any genuine and plausible grounds have been given for the truth of [#] by any of the supporters of this version of the indispensability argument. My skepticism is independent of Sober's arguments based upon his contrastive empiricism, but I welcome his supporting arguments, based as they are upon his long study of confirmation and evidence. Furthermore, having a worked-out alternative to the holistic principle [#] is heuristically useful for seeing what supporters of the principle need to do. Now, insofar as the version of the indispensability argument being discussed depends upon [#], this version, in my judgment, is highly questionable.

Resnik's defense of the indispensability argument against Sober's objection is curious. He seems to be quite willing to grant Sober that the observational successes of scientific theories do not provide *evidence*, in any ordinary sense of that word, for the truth of the mathematical principles used in that theory (Resnik, 1997: 124). He even allows that "[n]either scientists nor mathematicians take scientific experiments as providing evidence for the mathematics used in designing them" (Resnik, 1998: 231). But he denies that holism is thereby refuted (Sober, 1993: 119). However, what he means by 'holism' in this context is not what the word had to mean to generate the conclusions of "the Holism-Naturalism indispensability argument". Here, he is using the term 'Holism' to be "the thesis that no claim of theoretical science can be confirmed or refuted in isolation but only as a part of a system of hypotheses" (Resnik, 1997: 114). This thesis of holism is based upon the logical point that "if a hypothesis H only implies an observational claim O when conjoined with auxiliary assumption A, then we cannot deductively infer the falsity of H from that of O but only that of the conjunction of H and A" (Resnik, 1997: 115). Of course, nothing Sober has written refutes this logical point.[58]

---

[58] Indeed, Sober does not contest this logical point, writing: "I do not doubt the logical point with which Duhem and Quine begin" (Sober, 2000: 267). Elsewhere, he writes: "The simple logical point is that hypotheses rarely make observational predictions on their own; they require supplementation by auxiliary assumptions if they are to be tested ... The controversial, and I think mistaken, empistemological point ... is that what gets confirmed and disconfirmed by observations is not H taken by itself, but the conjunction H&A" (Sober, 1999: 54).

But this logical point supporting Resnik's thesis of holism does not justify [#], which after all is a premise about how evidence is shared. Clearly, logic alone cannot dictate how evidential support is to be apportioned to the parts of a theory. I wish to emphasize here that [#] is the crucial point of contention and that what Sober's attack makes evident is the weakness of the defense of this principle that defenders of the holistic version of the indispensability argument have provided.

One could, of course, question the view of confirmation from which Sober mounts his objection. In particular, one might dispute Sober's claim that mathematical statements are not tested by empirical evidence. Susan Vineberg notes that a version of Bayesianism that does not accord all true mathematical statements probability 1 could admit the empirical confirmation of mathematical statements.[59] Since, for Bayesians, probabilities are subjective degrees of belief, if M is a mathematical statement such that $P(M)$ is less than 1, then, using the Bayesian criterion of confirmation according to which evidential statement E confirms hypothesis H iff $P(H/E) > P(H)$, one could conceivably come up with an appropriate E which confirms M.[60]

Hellman raises a similar objection in arguing that "alternatives to standard mathematical assumptions have been proposed and can even be plugged into suitable instances of the likelihood principle" (Hellman, 1999: 33). He then describes a possible situation in which the finitistic hypothesis that there are no infinite wholes or sets competes with the infinitistic mathematical hypothesis that such infinite things exist. Hellman then envisages amassing evidence for the hypothesis that space-time is infinite in extent, thereby confirming the latter of the two competing hypotheses.

Notice, however, that neither of the above two objections hits the main point of Sober's attack on the indispensability argument, namely, its

---

[59] Personal Communication. Generally, Bayesians accord true logical and mathematical statements probability 1. But the "Old Evidence Problem" has prompted some to attempt to find a more realistic way of assigning probabilities to logical and mathematical truths. See, for a statement of the problem, Glymour, 1980: 85–93. For a detailed discussion of the various attempts to revise Bayesian confirmation theory so as to obviate the problem, see Earman, 1992: ch. 5. I discuss one attempt to defend Bayesian confirmation theory against this problem in Chihara, 1994: 168–71.

[60] Vineberg, 1998: 121–2. Those unfamiliar with the basics of Bayesianism can find an excellent introduction to the foundations of Bayesianism in Eells, 1982: chs. 1–3. For a judicious evaluation of the difficulties facing Bayesianism, see Earman, 1992. For a discussion of some specific problems with Bayesian confirmation theory, see Chihara, 1987.

questioning of [#]. Remember that the version of the indispensability argument that Sober is criticizing maintains that, if "mathematics is an inextricable part of a physical theory, then the empirical success of the theory confirms the entire theory—mathematics and all". The fact that one could conceivably come up with experiments or observations that confirm some mathematical proposition or theory in no way supports [#]—the position being attacked by Sober. Furthermore, and this point should be emphasized since it seems to be so frequently missed, *Sober nowhere denies that one could conceivably come up with the sort of confirmation of a mathematical proposition or theory described by Vineberg and Hellman*. What he does deny is that any experiment or observation that confirms a scientific theory expressed in mathematical terms *thereby* confirms mathematics. Thus, the Vineberg–Hellman points do not support what Sober denies.

Colyvan's defense of the Hellman objection

Mark Colyvan has tried to defend and expand Hellman's objection by attacking the following passage from Sober's paper:

Formulating the indispensability argument in the format specified by the Likelihood Principle shows how unrealistic that argument is. For example, do we really have alternative hypotheses to the hypothesis of arithmetic? If we could make sense of such alternatives, could they be said to confer probabilities on observations that differ from the probabilities entailed by the propositions of arithmetic themselves? I suggest that both these questions deserve negative answers. (Sober, 1993: 45–6)

Colyvan has replied to these claims of Sober's by agreeing with Hellman that we do have alternatives to "the hypothesis of arithmetic":

Frege showed us how to express most numerical statements required by empirical science without recourse to quantifying over numbers. Furthermore, depending upon how much analysis you think Field has successfully nominalized, there are alternatives to that also. (At the very least he has shown that there are alternatives to differential calculus). (Colyvan, 1999: 325)

Colyvan agrees with Sober, however, that neither of the above alternatives would confer probabilities on observations that differ from those conferred by standard classical mathematical theories. But this, he thinks, just shows the weakness of Sober's contrastive empiricism.

"The question of which is the better theory will be decided on the grounds of simplicity, elegance and so on—*grounds explicitly ruled out by contrastive empiricism.*"[61]

It is hard for me to see just why Colyvan thinks the above is a serious defense of the indispensability argument. Theory choice, involving as it may pragmatic considerations (such as simplicity and elegance), is one thing; confirmation and evidence, directed at the probability of the truth of theories, is something else. Sober's contrastive empiricism was not put forward as an account of theory choice—of what ought to go into deciding whether one scientific theory or hypothesis is preferable to its competitors. Clearly, Sober's articles on this topic should be grouped with the huge literature on confirmation—not theory choice.

Supporters of the indispensability argument, who do not distinguish con-firmational grounds from grounds for theory choice, tend to accept Quine's pragmatic conception of evidence, according to which grounds of simplicity, elegance, and so on count as evidence. But unless they come up with better arguments for such a view than Quine has supplied, I cannot see them con-verting the majority of philosophers of science to their view. The following quotation from Bas van Fraassen's writings expresses, I believe, a more pre-valent view:

Values of this sort [such as simplicity and elegance] ... provide reasons for using a theory, or contemplating it, whether or not we think it true, and cannot rationally guide our epistemic attitudes and decisions. (Van Fraassen, 1980: 87)

[T]hey do not concern the relation between the theory and the world, but rather the use and usefulness of the theory: they provide reasons to prefer the theory inde-pendently of questions of truth. (Van Fraassen, 1980: 88)

So I judge Colyvan to be simply off-base in suggesting that contrastive empiricism explicitly rules out deciding theory preference on grounds of simplicity, elegance, etc. More important, it can be seen that nothing Colyvan has brought up above constitutes a defense of [#]—a principle about

---

[61] Colyvan, 1999: 326, italics mine. Colyvan then adds: "Indispensability theory does not propose to settle all discrimination problems by purely empirical means, so of course it flounders when forced into the strait-jacket of contrastive empiricism." But what is "indispensability theory"? Are all defenders of the indispensability argument supposed to accept the same theory of confirmation and theory preference? Should one suppose that Quine, Putnam, Hellman, and Resnik (to mention just a few supporters of the argument) all espouse the same theory of confirmation and theory preference? That would clearly be absurd.

confirmation. As I noted, Colyvan may be lumping together theory choice and confirmation because he has accepted Quine's extreme pragmatic view of evidence which was discussed in detail earlier, but Sober (and indeed most philosophers of science and probability theory) would simply reject that view of evidence.

I wish to emphasize that I am not endorsing here Sober's view of confirmation. I do not presuppose or endorse any particular brand of confirmation theory or evidence—orthodox or radical. However, I do from time to time make appeal to more or less orthodox views about confirmation or theory choice in order to show where legs of the indispensability argument appear to be shaky. If supporters of the argument wish to flout orthodox views or principles, that is where they need to shore up their reasoning. We should not lose sight of the fact that the burden of justification rests with those who are putting forward an argument supposedly showing that we have overwhelming evidence for the existence of mathematical objects (as is supposedly shown by the holistic version).

Is mathematics disconfirmed?

I wonder if supporters of the version of the indispensability argument under discussion have ever attempted to sort out the implications of taking their holistic view of confirmation seriously. If, as [#] says, mathematics shares whatever confirmation accrues to the theories using it, then would it not be reasonable to suppose that mathematics shares whatever *disconfirmation* accrues to the theories using it? If so, then we can infer that, over the many hundreds of years in which mathematics has been used in the empirical sciences, mathematics has received a great deal of disconfirmation. Should we then reject classical mathematics as scientifically discredited?

Essentially the above objection to the holistic view of confirmation was raised by Sober (in Sober, 1993: 53). Hellman has advanced a defense of the holistic view (and by implication [#]) against this objection by arguing as follows. Suppose some scientific hypothesis H which, when coupled with auxiliary statements A and mathematical theory M, entails observational statement O. What gets confirmed by an observation that established the truth of O is, according to the holist, the whole conjunction (H & A & M). Similarly, if −O is observed to be the case, then what gets disconfirmed is that same conjunction.

Hellman believes, however, that there is an asymmetry at the next step in assigning "credit" or "blame" to the individual conjuncts. According to Hellman, if the conjunction is confirmed, then M receives confirmation automatically (Hellman, 1999: 30). On the other hand, he goes on to argue, if a conjunction is false, it does not follow that any specific one of the conjuncts is false. Thus, for the specific case at hand, he may have concluded that if (H & A & M) is disconfirmed, M does not automatically receive disconfirmation: "independent grounds are needed" (Hellman, 1999: 34). Hellman then explains why M would not be blamed for such a disconfirmation. In this way, Hellman implies, the holist can defend her view that mathematics would always receive confirmation from any observations that confirm science but never any disconfirmation.

Let us now examine the steps in Hellman's reasoning. Why should one believe that, if a conjunction is confirmed, then each of the conjuncts is thereby confirmed "automatically"? This is something, evidently, that the holist believes, but why should we? Hellman does not provide any rationale for accepting the principle, but he should, since it has serious counterintuitive consequences and most researchers in confirmation theory reject the principle.

Consider the following example. Suppose that D is a scientific hypothesis concerning the genetics of some species of Drosophila, and suppose that D is confirmable by some observation O made in some specific experimental situation. Let L be the hypothesis that there is some kind of life on Mars. It is easy to see how O, by confirming D, can also confirm the conjunction (D & L). But it would be extremely counterintuitive to conclude that O thereby confirms L. How could a typical experiment dealing with the genes of Drosophila here on earth confirm the hypothesis that there is some kind of life on Mars? It is easy to see why the general rule of confirmation mentioned above is not a widely accepted rule of confirmation.

The conjunction rule of confirmation specified above is a special case of the following principle called the "special consequence condition of confirmation":

If α confirms ϕ, ϕ entails θ, and this entailment is known, then α confirms θ.

It is quite possible that holists accept the conjunction rule of confirmation because they accept this principle. Sober suggests (2000: 264) that it is because the holist accepts this special consequence condition that she espouses the principle that mathematics is confirmed by any observation that confirms a scientific hypothesis using mathematics. But the special consequence condition of confirmation is a principle that most philosophers of science have rejected; and the reasons for this rejection are not hard to find.[62] Here's a

---

[62] Sober, 2000: 265. For a detailed discussion of the special consequence condition, see Hesse, 1974: 141–52.

simple example Sober gives to show the implausibility of this principle.[63] Suppose that one draws, at random and without looking at its value, a card from a standard deck of cards (with no jokers). Let H be the hypothesis that this card is the seven of hearts; let R be the hypothesis that the suit of this card is red; and let S be the hypothesis that this card is a seven. Suppose that we were to learn by some means or other that the suit of this card is red— suppose, that is, that we were to learn R. This would clearly confirm H, since the probability of H would increase significantly as a result of learning R. Now H implies S, and we all know that it does. So the question is: would R confirm S as well, as the special consequence condition requires? Clearly not. The probability that the card drawn is a seven remains 1/13, despite the added information provided by R.[64]

From these considerations, it can be concluded that there are serious grounds for questioning the case for the asymmetry between confirmation and disconfirmation that Hellman describes. Are there grounds for holding that, in fact, there is a symmetry between confirmation and disconfirmation? At several points, Hellmen appeals to agreement with Bayesianism to support his own account of holistic confirmation, writing: "without claiming that Bayesianism provides an adequate confirmation theory, it *is* a useful framework" (Hellman, 1999: 30). However, he does not attempt to show that the *asymmetry* he attributes to confirmation and disconfirmation follows from Bayesianism. Had he done so, he probably would have come to a very different conclusion about the confirmation and disconfirmation of mathematics. For Sober has shown, contrary to the asymmetry inferred in Hellman's account, that Bayesianism yields just the opposite result, namely that Bayesianism implies a *symmetry* between confirmation and disconfirmation. That is, Bayesianism implies that a hypothesis H is confirmed by O iff H is disconfirmed by −O.[65]

The gulf between [#] and the Duhemian point

What the Sober objection makes abundantly clear is that there is a huge logical gap between the logical point emphasized by Resnik (that "if a hypothesis H only implies an observational claim O when conjoined with auxiliary assumption A, then we cannot deductively infer the falsity of H from

[63] Sober, 2000: 264.

[64] One can even show that there are conceivable situations in which α confirms ϕ, ϕ entails θ, this entailment is known, but α disconfirms θ. See Eells, 1982: 56–7.

[65] Sober, 2000: 265.

that of O but only that of the conjunction of H and A") and the principle of confirmation [#] that is used to generate the indispensability argument.

The principle [#] is extremely far-reaching and has many very counter-intuitive consequences. Recall the earlier quote from Quine with which I began Section 2 of this chapter:

[M]athematics—not uninterpreted mathematics, but genuine set theory, logic, number theory, algebra of real and complex numbers, differential and integral calculus, and so on—is best looked upon as an integral part of science, on a par with the physics, economics, etc., in which mathematics is said to receive its applications. (Quine, 1966c: 231)

Notice that, for Quine, both set theory and logic are to be included as integral parts of science, so that we can infer from [#] that both set theory and logic will also share whatever confirmation accrues to science in general. Furthermore, we can infer from other things that Quine has written about analyticity that even definitional statements such as 'all bachelors are unmarried' can be inferred to share in the confirmation. Thus, if modus ponens is used in the development of some empirical theory, then whatever confirms the theory is also supposed to confirm modus ponens. We can infer that an observation confirming a theory about the incidence of colon cancer among bachelors over the age of sixty will confirm 'all bachelors are unmarried' as well as such axioms of set theory as the power set axiom. These very striking counterintuitive consequences should be weighed in evaluating the cogency of [#].

It can now be seen that the gulf between the Duhemian logical point emphasized by Resnik and [#] is huge. This is a gulf that Resnik has simply failed to bridge, and until he does so, his version of the indispensability argument will remain unconvincing. Indeed, Resnik seems, in the end, to abandon all of the above versions of the indispensability argument. Instead, he proposes a new version of the argument that "separates questions of indispensability from questions of confirmation" (Resnik, 1998: 233). However, before taking up this new version, let us consider another series of objections aimed at the above versions.

9. MADDY'S OBJECTIONS

Having for many years championed an extreme version of Platonism, Penelope Maddy has recently undergone a surprising conversion to the anti-Platonic camp, due primarily to the discovery of what she considers convincing objections to the indispensability argument. Maddy's objections, although not in my opinion nearly as compelling as she seems to believe, do add further grounds for questioning the holistic version of the indispensability argument.

Making extensive use of historical material relevant to the early twentieth-century controversy over the reality of atoms, Maddy notes (in Maddy, 1997) that by 1900, the atomic theory enjoyed all five of the virtues that Quine attributed to the "molecular hypothesis" in "Posits and Reality" (Quine, 1966b). However, despite these virtues, scientists did not agree that atoms were real—many theoreticians were skeptical of the existence of such things. Thus, Maddy concludes that "the actual behavior of the scientific community in this case does not square with the Quinean [holistic] account of confirmation" (Maddy, 1997: 142). She infers from these studies that "the case of atoms makes it clear that the indispensable appearance of an entity in our best scientific theory is not generally enough to convince scientists that it is real" (1997: 143).

Susan Vineberg, commenting on Maddy's historical case studies, agrees that to be convinced of the reality of some new theoretical entity scientists require more than its merely being an entity postulated by our best available scientific theory; she suggests that a more direct test of its existence is needed. Such a direct test typically involves laboratory manipulations which make it likely that entities of this sort exist in the event of a positive result and make it unlikely that entities of this sort exist if the test yields a negative result.[66] It was just such a direct test that Jean Perrin devised with his painstaking Brownian motion experiments and it was due largely to Perrin's work that the scientific community became convinced of the reality of atoms.[67] Of course, given the causal inertness of mathematical entities, it is hard to see how scientists can so directly test the reality of the mathematical entities that are supposedly indispensable for science. Vineberg concludes with Maddy that "scientific practice suggests that at best the empirical evidence for mathematical entities would be so weak that it would be rational to suspend belief in them, just as I assume it was rational to doubt the existence of atoms prior to Perrin's experiments" (Vineberg, 1998: 123).

---

[66] Vineberg, 1998: 123. A similar sort of view was effectively argued by Ian Hacking in Hacking, 1985.

[67] I discussed the significance of Perrin's experiments for the case for belief in mathematical objects in an early article (Chihara, 1982: 214–15). Maddy discusses Perrin's experiments in much greater detail (in Maddy, 1995: ch. 6).

Colyvan's rebuttal

Colyvan has attempted to rebut Maddy's claim that the indispensable appearance of an entity in our best scientific theory is not generally sufficient to convince scientists that it is real—that something more by way of some sort of empirical test is required. Maddy's claim was based to a great degree on the behavior of scientists during the controversy over the existence of atoms and molecules. Colyvan's response is to dig in his heels and to accuse the skeptical scientists of being wrong in their assessment, suggesting that they were indulging in dishonest behavior by "using atoms, say, in our best chemical theories, then denying the existence of these very same atoms" (Colyvan, 2001: 99, 100). Colyvan admits that this rebuttal rests only upon an intuition and he appreciates that many would not share his intuition.

What I find striking is that the "intuition" that grounds this rebuttal does not seem to be the result of a careful study of the actual thoughts and arguments of the skeptical scientists. Colyvan does not cite the writings of even one of these scientists to indicate why he thinks they are being intellectually dishonest. Yet an investigation of the details of the controversy over molecular reality is surely called for, if one is to make a reasonable case that these skeptics were not being honest.

The controversy in question (about the existence of atoms) began with the publication of John Dalton's *System of Chemical Philosophy* in 1808, in which it was claimed that there are different particles of matter—indivisible atoms—for each element, the postulation of which was thought to explain a number of accepted regularities and laws of chemistry. Dalton's hypothesis, however, was complicated by a number of related but incompatible hypotheses. For example, Prout had postulated that the hydrogen atom was a "basic unit of matter contained in different proportion in each chemical element" (Nye, 1972: 2)—a hypothesis that was debated for some time. And Humphry Davy insisted that "the 'atom' was a unit of chemical reaction, rather than a material entity" (Nye, 1972: 3). Even supporters of Dalton's hypothesis questioned certain features of Dalton's atomic theory. For example, J. J. Berzelius "considered Dalton's conception of diverse geometrical configurations for different atoms a flight of fancy", and he "regarded the basic hypothesis as unsatisfactory, for it simply could not explain enough" (Nye, 1972: 4).

In this unsettled atmosphere, James Clerk Maxwell's work on the electromagnetic theory of light made its appearance. Maxwell produced detailed mechanical analogies for an electromagnetic field, on the basis of which he came up with a series of equations that could be empirically verified. The mechanical analogies were not intended to be real in any way, and Maxwell explicitly freed his equations from the analogical models (Nye, 1972: 15). Mary Jo Nye describes what happened next:

[T]he British school in general began formulating differential equations to demonstrate that the most varied phenomena—such as transmission of heat, propagation of electricity in conductors, and hydrodynamical problems—could be explained in the very same way. Soon the older theories themselves, such as the wave theory of light, theories of gases, and the geometrical schemes of the chemists, were considered simple mechanical analogies, and the philosophy of Maxwell was generalized into the dictum that knowledge is nothing other than the discovery of analogies. (Nye, 1972: 15)

Not surprisingly some scientists began to suspect that the theory of atoms of Dalton, involving as it does the postulation of invisible and indivisible particles of matter, was basically nothing more than another mechanical analogy, to be discarded as a sort of fiction when the verifiable equations had been constructed. Wilhelm Ostwald, in particular, "became more and more convinced that molecules, atoms, and ions were only mathematical and *a priori* fictions and that the real underlying component of the universe was energy in its various arrays".[68] Other influential scientists who studied, as did Ostwald, the remarkable scientific advances being made in thermodynamics were Ernst Mach and Pierre Duhem, who also became strong opponents of the atomic theory (Nye, 1972: 16).

Without going into more of the details of this controversy—I have only touched on a few elements of the dispute—it should be clear to the reader that the controversy over the reality of atoms was scientifically sophisticated and theoretically complicated.[69] Despite what Colyvan has suggested, there were grounds for being suspicious of the atomic theory, and one can see how, in this situation, reasonable scientists might be reluctant to assent to a theory that postulated entities of a kind that had never been detected. Colyvan's charge of intellectual dishonesty, in the absence of a thorough

---

[68] Nye, 1972: 17. Nye goes on to tell us that "Ostwald's purge of mechanical hypotheses from physical theories was greeted sympathetically by a large number of figures in chemistry and physics" (17).

[69] To readers wishing to study the scientific controversy over the reality of atoms in more detail, I can recommend Nye, 1972.

study of the reasoning of these scientists, seems to me to be highly questionable.[70]

Maddy's second objection

Maddy has a second objection to the indispensability argument based on her examination of scientific practices. The idea is that a great deal of scientific theorizing involves reasoning about highly idealized situations (for example, the earth's surface is taken to be flat, the ocean is treated as infinitely deep, etc,) and greatly idealized entities (such as rigid bodies, frictionless planes, and incompressible spheres)—situations and entities that clearly are not to be found in physical reality. Thus, the many scientific theories that are ostensibly theories about such idealized situations and entities are strictly speaking not even true. In that case, even if

(1) mathematics is indispensable for a scientific theory of some type of phenomenon,

(2) this scientific theory is acknowledged to be the best theory of this phenomenon we now have,

and

(3) any mathematical theory, which is indispensable for some scientific theory, shares whatever evidential support there is for the scientific theory,

we cannot infer that the empirical evidence supporting this scientific theory provides us with sufficient evidence for acceptance of the truth of the mathematical principles utilized in that theory. For if the theory is itself false and acknowledged to be false by its scientific practitioners, then the empirical evidence in support of it must not be sufficient to warrant belief in the truth of the theory, and hence can hardly be sufficient to warrant belief in the truth of the mathematical principles utilized in the theory.

---

[70] Colyvan has sketched (2001: 100–1) an alternative way of trying to rebut Maddy's objection to the indispensability argument, but this alternative seems directed at defending a somewhat different version of the indispensability argument than the one I have been discussing, and also it seems to require a substantial revision of the Quinean position on confirmation of theories. For example, he suggests that the Quinean could espouse the view that the scientists involved in the controversy over molecular reality should have some degree of belief in the hypothesis somewhere between zero and one. But how the Quinean could fit a theory of degrees of belief into the framework of Quine's view of pragmatic evidence and confirmation is not at all obvious. (See in this regard Chihara, 1981.) Given the sketchiness of Colyvan's rebuttal, I am convinced that an adequate discussion of it would take up more space than is appropriate here.

Of course, it can be argued in reply that not all scientific theories involve idealizations of the sort discussed above, and that the non-idealizing theories are sufficient to generate the evidence required to justify belief in mathematical entities. But such defenders of the argument presumably would then have to sort out those parts of science that are composed of the non-idealizing theories and then specify the parts of mathematics that are thereby empirically justified—no small feat. One wonders whether belief in sets (the type of mathematical object that was favored by Maddy during her set-theoretical realism period) would in this way be plausibly justified, for Maddy would agree with Vineberg that, "there is a possibility that all cases of continuum mathematics involve idealization, in which case even standard real analysis would fail to receive support through its indispensable use in mathematics" (Vineberg, 1998: 126; cf. Maddy, 1997: 152–3).

It may be useful at this point to give some indications of the case Maddy musters for her contention that physicists have raised doubts about the assumption that physical reality (in particular, physical space) is continuous. Maddy provides her readers with a fund of quotations from theoretical physicists that clearly express such doubts. For example, Chris Isham is quoted as writing:

In a gross extrapolation from daily experience, both special and general relativity use a model for spacetime that is based on the idea of a continuum, i.e. the position of a spacetime point is uniquely specified by the values of four real numbers (the three space, and one time, coordinates in some convenient coordinate system). ... [F]rom the viewpoint of quantum theory, the idea of a spacetime point seems singularly inappropriate: by virtue of the Heisenberg uncertainty principle, an *infinite* amount of energy would be required to localize a particle at a true point; and it is therefore more than a little odd that modern quantum field theory still employs fields that are functions of such points. It has often been conjectured that the almost unavoidable mathematical problems arising in such theories (the prediction of infinite values for the probabilities of physical processes occurring, and the associated need to 'renormalize' the theory ...) are a direct result of ignoring this internal inconsistency.[71]

Richard Feynman is quoted as writing: "I believe that the theory that space is continuous is wrong, because we get these infinities and other difficulties ... I rather suspect that the simple ideas of geometry, extended down into

---

[71] Maddy, 1997: 151. The reader can find there additional portions of the passage from which the above quote was taken, as well as further discussions of the point.

infinitely small space, are wrong."[72] Maddy concludes that the continuity of physical space must be considered open.

Hellman's reply

Hellman has produced an interesting reply to the above objection of Maddy's that warrants investigation. He takes Maddy to be providing us with a sort of "meta-induction": we have a huge number of cases in which literally false assumptions were used (indispensably) to generate empirical predications that were subsequently confirmed, so in the case in which mathematics was used (indispensably) to generate empirical predictions that were subsequently confirmed, with what confidence can we move from the indispensability of mathematics to its truth? He replies to this objection from the perspective of his own modal structuralism, claiming that the problem does not even arise for this view of mathematics. Idealized scientific statements of the sort noted by Maddy are, from this perspective, really statements interpreted over "idealized modal or possible situations" (Hellman, 1999: 31). Either they are statements about an ideal type of system or they are consequences of such statements describing "how such a system *would* behave under the assumed conditions" (1999: 32). Thus, confirmation of scientific hypotheses are to be understood as confirming counterfactual conditionals of the following sort: if the physical system under study were reasonably well modeled by such and such, then it would behave with high probability thus and so. And these counterfactuals are not known to be false. Thus, from this perspective, the meta-induction does not even get off the ground.

One problem with this defense of the indispensability argument is that it provides very little comfort to the mathematical realists who use the argument to support their belief in mathematical objects. Resnik, for example, explicitly rejects the modal-structuralist position from which this defense is launched.[73] He could hardly adopt the position at one moment to refute Maddy's objection, and in the next, attack this same position, with the aim of bolstering his thesis that the various nominalistic alternatives to classical Platonic mathematics are unsatisfactory.

Besides, how does Hellman's point support the version of the indispensability argument Maddy is attacking? Remember, this version is an argument that attempts to show that our *actual* mathematical theories have been abundantly confirmed by empirical evidence amassed over many years

---

[72] Maddy, 1997: 149. For amplifications and references, see 143–54.
[73] See Resnik, 1997: ch. 4, sect. 3.2, 4, and 5.

of scientific work. And the meta-induction attributed to Maddy is supposedly aimed at undermining that argument by casting doubt on the move from the indispensability of the actual mathematics used in science to its truth.

Now no one—certainly not Hellman—identifies the actual mathematics used in science with the highly philosophically motivated nominalistic-mereological version of the modal-structuralist formal theories presented in Hellman's writings. I think it is safe to say that Hellman's modal-structuralist system has not been used in the verification of even a single substantial scientific theory. So let us grant, for the sake of argument, that the problem posed for the indispensability argument by the Maddy meta-induction does not get off the ground when the mathematics in question is Hellman's philosophically motivated mathematical theory. Since the indispensability argument is concerned with actual mathematics and not Hellman's version, we do not thereby absolve the version of the argument Maddy is attacking from the meta-induction.

Maddy's third way

Maddy has a third way of attacking the indispensability argument which should also be noted. On the basis of her studies of the writings of scientists, she says: "science seems not to be done the way it would be done if the interrelations of mathematics and science were as the indispensability argument requires" (Maddy, 1997: 154). Focusing on Feynman's introductory physics lectures, Maddy infers that, as a rule, "physicists seem happy to use any mathematics that is convenient and effective, without concern for the mathematical existence assumptions involved".[74] This lack of concern is shown in two ways:

(1) The existential and structural assumptions that are required by the mathematical theories being used by the physicist "are not held to the same epistemic standards as ordinary physical assumptions"— physicists do not demand some sort of direct test of the existence of the mathematical entities assumed by the mathematical theory or of the structural assumptions made in using the mathematical theory (e.g. that time and space are continuous) as was demanded for the case of atoms (Maddy, 1997: 156).

---

[74] Maddy, 1997: 155. Feynman, a renowned physicist who was awarded the Nobel Prize in 1965, made important contributions to the theories of superfluidity and quarks, and proposed with Murray Gell-Mann the theory of weak nuclear force. He is known for the "Feynman diagram": a method of describing particle interactions.

(2) The confirmation of a scientific theory making use of a mathematical theory containing such existential and structural assumptions "is not regarded as confirming evidence for those assumptions" (Maddy, 1997: 156).

Thus, she is led to doubt that the natural sciences are concerned with assessing mathematical ontology. "If it were in that business, it would treat mathematical entities on an epistemic par with the rest, but our observations clearly suggest that it does not" (1997: 157). All of this implies for Maddy that the confirmation of any scientific theory involving the use of mathematics does not, as is claimed in the indispensability argument, confirm mathematics and all its ontological commitments, for if confirmation worked in the way supporters of the indispensability argument believe—so that any confirmation of a scientific theory confirmed mathematics—then (Maddy seems to be arguing) scientists would surely realize that mathematics is confirmed in this way, and they would have to be concerned about assessing the ontological commitments of mathematics.

Colyvan's rebuttal

Colyvan takes aim at Maddy's claim that "physicists seem happy to use any mathematics that is convenient and effective, without concern for the mathematical existence assumptions involved". Here again, Colyvan suggests that he would not "endorse such apparently dishonest behaviour" (Colyvan, 2001: 102). Furthermore, he argues, contrary to Maddy's claim, that scientists do worry about the mathematical existence assumptions of the mathematics they use, and he cites historical examples to support his case. The cases he cites are the earliest uses of the calculus, Dirac's delta function, and the first use of complex numbers by Cardano to solve quadratic equations.

I do not find these cases very supportive of Colyvan's thesis. As he himself notes, each of the first two cases could very well be understood to be a worry not about the ontological commitments of the new mathematical theory being used, but rather about whether this mathematics was consistent, coherent, or mathematically acceptable. In the case of the calculus, Bishop Berkeley pointed out in "The Analyst" (Berkeley, 1956) the absurdities underlying some of the fundamental ideas of the Newtonian calculus. For example, an infinitesimal was supposed to be a quantity greater than zero, while being so small that no multiple of it, however great, attains a measurable size. Surely it would be rash to infer that the reluctance of a scientist to use a mathematical theory resting on such seemingly incoherent notions is due to the scientist's worries about the ontological commitments of the theory.

What about Dirac's delta function? Here's is what Colyvan writes:

If, as Maddy writes, physicists are inclined to simply use whatever mathematics is required to get the job done, without regard for ontological commitments, why was Dirac so intent on dispelling doubts about the use of this new "function"? (Colyvan, 2001: 103)

But the very quotation that Colyvan gives to support his case provides an answer to his question. Dirac is quoted as writing:

The use of improper functions thus does not involve any *lack of rigour* in the theory, but is merely a convenient notation, enabling us to express in a concise form certain relations which we could, if necessary, rewrite in a form not involving improper functions, but only in a cumbersome way which would tend to obscure the argument. (Colyvan, 2001: 103, italics mine)

Evidently, Dirac was concerned to show that his use of the delta function does not involve a "lack of rigour". There is nothing to suggest that he was concerned about the unwanted ontological commitments of his mathematics.

What about the worries that were engendered by the introduction of complex numbers into the theory of equations in the sixteenth century? The first thing that strikes me about this example is the fact that all the researchers mentioned by Colyvan in supporting his case (Cardano, Descartes, Newton, and Euler) are mathematicians. Of course, some of those (e.g. Newton) are also empirical scientists. But again, I am inclined to classify their worries as primarily mathematical. Descartes was worried that no *number* could be the square root of -1. This seems to me to be a worry about what is mathematically permissible and not a worry about the acceptability of using in physics a mathematical theory that has unwanted ontological commitments. It should be remembered that Descartes was very much involved in the seventeenth-century controversy over what constructions are to be allowed in geometry. The worry over introducing complex numbers into algebra strikes me as being similar to the worry about introducing all sorts of new methods of constructing points and curves into geometry.[75] Besides, Maddy is putting forward a view about the practice of contemporary or twentieth-century scientists. The fact that Colyvan has to reach way back to the sixteenth-century

---

[75] See Mancosu, 1996: 72–81 for a discussion of Descartes's views on these matters.

introduction of complex number to try to justify his case seems to me, if anything, to support Maddy's position.

In any case, I shall be arguing later in this work that contemporary scientists are right to behave as Maddy describes them as behaving. Contrary to what Colyvan believes, I shall argue, the behavior of physicists can be analyzed in a way that shows it to be perfectly reasonable.

Colyvan does raise an objection to one aspect of Maddy's reasoning that I do find convincing: he questions a principle of her naturalism according to which, if philosophy conflicts with some scientific practice, it is philosophy that *must always give* (Colyvan, 2001: sect. 5.2). I agree with Colyvan that such a principle is questionable. However, I have presented Maddy's objections in a way that avoids the need for such a principle, even though I suspect that, in places, she was indeed relying upon such a principle.

10. RESNIK'S NEW INDISPENSABILITY ARGUMENT

As I said earlier, Resnik does not attempt to rebut Maddy's historical points or to defend his H-N argument against the Sober or Maddy objections to [#].[76] Instead, he responds by producing a completely new form of the indispensability argument. Here's how it reads:

(1) We are justified in engaging in scientific activity, because it is the best available means of achieving our predictive, explanatory, and technological goals.
(2) Our success in achieving these goals depends on drawing conclusions within and from science.
(3) We are justified in drawing conclusions from and within science only if we are justified in taking the mathematics used in science to be true.
(4) So, we are justified in taking the mathematics used in science to be true.
(5) So, mathematics is true.[77]

---

[76] See Resnik, 1998: 233. There, Resnik admits "to Maddy's worry that if our grounds for accepting mathematics are limited to those we have for accepting our best scientific theories, then we might be justified in accepting very little mathematics".

[77] I am here using the formulation of Resnik's argument given in Vineberg, 1998: 129–30. Resnik's own formulations in Resnik, 1997 is very close to the above, but the version expressed in Resnik, 1998 is an enthymeme.

Vineberg's evaluation

Vineberg raises a number of preliminary worries about several steps in Resnik's argument, but her main objection boils down to the charge that the reasoning is infected with a kind of fallacy of ambiguity. We must, she claims, distinguish two sorts of justification: *pragmatic* justification and *epistemic* justification. Using a fundamental notion of decision theory, Vineberg gives the following decision-theoretic definition of pragmatic justification: "One is pragmatically justified in believing or accepting φ if the expected value of doing so is greater than the expected value of not doing so".[78] On the other hand, Vineberg does not attempt a definition of epistemic justification, being content to give the following rough characterization: "One is epistemically justified in believing or accepting φ if one has sufficient evidence supporting belief in φ" (Vineberg, 1998: 131). Armed with these distinctions, Vineberg examines the above form of the indispensability argument. In particular, she focuses on premise (3). The term 'justified' occurs in both antecedent and consequent of that premise. So the question arises: What sort of justification is intended? Clearly, both occurrences are intended to have the same sense of 'justification', but which one? Well, suppose that Resnik is talking about pragmatic justification. Then, (4) would be the conclusion:

(4*) We are pragmatically justified in taking the mathematics used in science to be true.

Since Resnik thinks we can conclude (5) from (4), if Resnik has in mind pragmatic justification, so that (4) is really (4*), Resnik would be advocating inferring the truth of the mathematics used in science from the mere fact that we are pragmatically justified in taking it to be true. But such an inference would then be very questionable indeed.

To see the dubiousness of such an inference, imagine a criminal lawyer who finds that he does a much better job of defending his clients if he believes that they are innocent of the crimes of which they are accused. It is not hard to fill out this example in such a way that the lawyer is pragmatically justified in believing that a particular client is innocent of the crime for which he is on trial. Only the most extreme Quinean would believe that this pragmatic justification for believing in the innocence of his client constitutes genuine evidence of the client's innocence and hence amounts to epistemic

---

[78] Vineberg, 1998: 131. Vineberg does not define 'expected utility', but I assume that she has in mind what Beyesians call 'subjective expected utility'. The reader can find a useful discussion of this notion in Eells, 1982.

justification for believing in his innocence. We would not infer the truth of the Christian doctrines about the afterlife from the premise that the expected utility of accepting these doctrines is greater than the expected utility of not accepting these doctrines.

It would be reasonable to conclude, then, that the term 'justification' has the sense of 'epistemic justification' in (3). In other words, Resnik is talking about epistemic justification in (3) and (4). Vineberg then turns her attention to (1), and asks: What kind of justification is being asserted in that premise? Here, she claims, it would have to be pragmatic justification that is intended:

> To see this, observe that Resnik defends premise (1), that we are justified in engaging in scientific activity, by appealing to the fact that it is the best available means for achieving our predictive, explanatory and technological goals. What this means is that according to our best judgment, scientific activity either will, or is expected to, be more effective for achieving these goals than any alternative activity. ... As it stands, this justification is merely pragmatic, in that it only appeals to the expected consequences of doing science as being more desirable than the consequences of other present alternatives. (Vineberg, 1998: 133)

Thus, Vineberg concludes, this new form of the indispensability argument trades fallaciously on an ambiguity, substituting epistemic justification for pragmatic justification to obtain the needed conclusion.

Resnik's alternative argument

Resnik has also supplied another argument in support of his thesis that *we are justified in taking mathematics to be true*. This argument is based upon the following three premises:

(i) We are justified in using science to explain and predict.
(ii) The only way we know of using science to explain and predict involves drawing conclusions from and within science.
(iii) We are justified in drawing conclusions from and within science only if we are justified in taking the mathematics used in science to be true.[79]

This argument falls victim to a similar objection. Consider the justification mentioned in premises (i) and (iii). Is this justification pragmatic or epistemological? Evidently, it is pragmatic, since he writes:

> It is likely that our ultimate justification for using science to explain and predict is that doing so appears to promote our theoretical and practical interests better than

---

[79] The premises are given in Resnik, 1997: 48.

any other method we know. If so, then our justification for accepting the mathematics used in science also has this character. But isn't this what one would expect from a pragmatic argument? (Resnik, 1997: 48 n.)

Then, if the argument is not to rest upon a fallacy of ambiguity, the conclusion must be that we are *pragmatically* justified in taking mathematics to be true. But this conclusion provides a rather flimsy basis for accepting Resnik's Platonic ontology of abstract mathematical objects—especially for Resnik, since (as we saw in the previous chapter) his realism has a number of very counterintuitive consequences that conflict with common-sense principles.

Other objections to both arguments

There is another point at which Resnik's pragmatic version of the indispensability argument, as well as his holistic version I discussed earlier, may be questioned. Premise (3) of the former and premise (iii) of the latter state that *we are justified in drawing conclusions from and within science only if we are justified in taking the mathematics used in science to be true*. Might we not be justified in drawing inferences within and from science even though we are not justified in taking some mathematical system used in the reasoning to be true? I shall take up this question in Chapter 9.

Finally, we should reconsider Resnik's belief in the existence of positions in structures that are not the structures studied in mathematics and applied in the empirical sciences—structures such as S* of the previous chapter. None of the indispensability arguments—even Resnik's most recent one—give any reasons for accepting the existence of the positions in these structures. The vast majority of structures we can think up do not have positions whose existence can be given the kind of indispensability justification, weak though it may be, advanced in this chapter. So it can be seen that such an existential posit rests unsupported—not what rational philosophers should accept, given the highly paradoxical and counterintuitive nature of the hypothesis.

11. CONCLUDING ASSESSMENT OF THE VARIOUS ARGUMENTS

Having examined, in this chapter, various versions of the indispensability argument that have been proposed and defended by twentieth-century mathematical realists—Quine's challenge, the direct argument based upon Quine's pragmatic conception of evidence, the direct argument based upon holism, the pragmatic version propounded by Resnik—my conclusion is that none of these arguments should convince a knowledgeable and reasonably cautious contemporary philosopher that belief in the existence of mathematical objects is warranted. As I evaluate the argumentation, the reasoning is simply not compelling.[80]

In Chapter 9, I shall give additional reasons, from the perspective of the structural view of mathematics developed in this work, for questioning these arguments. I shall argue that certain basic assumptions of these indispensability arguments are simply false.

---

[80] It has been suggested to me that some philosophers supporting a holistic version of the indispensability argument make use of a radical non-standard theory of confirmation, so that appeals to what is implied by orthodox or standard accounts of confirmation in order to raise doubts about the argument would be fruitless. However, so far as I know, none of these philosophers have spelled out the details of their unorthodox theory of confirmation so that its reasonableness could be assessed by philosophers and historians of science. Thus, these unorthodox confirmation theorists can hardly be judged to have produced a convincing indispensability argument showing that we ought to believe in the existence of mathematical objects.

6

Minimal Anti-Nominalism

John Burgess and Gideon Rosen (in Burgess and Rosen, 1997) have advanced an overall general evaluation of the various "nominalistic reconstructions" of science and mathematics that have been proposed in recent years. Judging from the many remarks found in their evaluations and from comments made by other philosophers of mathematics, it would seem that among the target "reconstructions" of their evaluations are the three views described in the previous chapter as responses to the Quinean challenge.[1] The Burgess–Rosen criticisms of the nominalist's reconstructions of mathematics occur as part of their overall account of contemporary nominalism, and their evaluations cannot be easily grasped without some understanding of this overall account. For this reason, I shall give a brief description of this account.

1. THE BURGESS–ROSEN ACCOUNT OF NOMINALISM

According to Burgess and Rosen, nominalism is the philosophical view that everything that exists is concrete and not abstract (Burgess and Rosen, 1997: 3). Thus, according to them, one can also characterize contemporary nominalism as the view that *there are no abstract entities*. They view the contemporary nominalist, then, as someone who "denies that abstract entities exist, as an atheist is one who denies that God exists" (1997: 11).

These characterizations raise a number of tricky questions: (1) What makes something abstract and not concrete? (2) Why should one not believe in the existence of any abstract entities? (3) Why should philosophers who do not believe in abstract entities try to produce nominalistic interpretations or constructions of mathematics? That is, why should nominalists attempt to reconstrue or "reconstruct mathematics" in such a way that its assertions can be seen not to assert the existence of abstract entities?[2]

---

[1] Thus, in objecting to the reconstructive theories of Field and Hellman, Resnik asserts (in Resnik, 1997: 80 n.) that Burgess and Rosen "make related criticisms of the scientific merits of nominalistic systems" in their book. [2] Burgess and Rosen, 1997: 12.

Burgess and Rosen devote a whole section to question (1), but for reasons that will become clear later, I shall not go over their worries about the distinction at issue. I turn instead to (2) and (3). The latter question needs some preliminary discussion. What is a *nominalistic* interpretation or construction of mathematics? The nominalist attempts to construct a system of mathematics and/or physics that is consistent with his/her skepticism about abstract entities. It is an attempt to reproduce what ordinary mathematics accomplishes without presupposing the existence of such things as numbers and sets.

Let us now reconsider question (2): Why should one not believe in the existence of any abstract entities? In his article "Epistemology and Nominalism" (Burgess, 1990), Burgess quotes a number of nominalistic reconstructivists, Daniel Bonevac, Dale Gottlieb, and Hartry Field, who cite an argument of Paul Benacerraf's published in his article "Mathematical Truth" (Benacerraf, 1973) as providing motivation for undertaking their reconstructions. This argument, based upon a "causal account of knowledge", is supposed to show that "belief in an assertion or theory implying or presupposing that there are numbers or objects of some similar sort cannot be knowledge" (Burgess, 1990: 1). Burgess then spends most of the article blunting this argument,[3] arriving in the end with the rather awkward double negative conclusion that "it is not known that it cannot be known that there are numbers" (12).

In their book (Burgess and Rosen, 1997), Burgess and Rosen take up, with great attention to detail and to subtleties, not only Benacerraf's original "causal theory" argument, but also refinements of that argument which nominalists have put forward to motivate their reconstructive programs. In particular, they take up Field's argument based upon a "reliability thesis" (41–9), as well as an epistemological argument involving the theory of reference (49–60). Burgess and Rosen provide assessments of these arguments, concluding, in each case, that a kind of stalemate between the nominalist and the anti-nominalist results, where no decisive conclusion can be drawn.

That there is such a stalemate between the two antagonists in this dispute is taken by these authors to be very significant. For Burgess and Rosen are clearly most sympathetic with a position that may be called "moderate realism" or

---

[3] I do not agree with some of Burgess's arguments against the causal account of knowledge. Later, in Part 2 of this chapter, I shall give a specific objection to one of Burgess's refutations of the causal account of knowledge.

"minimal anti-nominalism".[4] Minimal anti-nominalists do not present arguments for their belief in mathematical objects as does the typical realist. Nor do they attempt to explain how set theorists have come to know that, say the null set exists or that the pair set axiom is true. They seem completely unconcerned with the sort of questions that Platonic philosophers of mathematics typically worry about, such as why knowledge of these undetectable non-physical mathematical entities is needed to discover empirical facts about the physical world. Instead, the minimal anti-nominalist starts with a "fairly uncritical attitude towards, for instance, standard results of mathematics".

Having studied Euclid's Theorem, we are prepared to say that there exist infinitely many prime numbers. Moreover, when we say so, we say so without conscious mental reservations or purpose of evasion. ...

For those of us for whom something like this is the starting-point, any form of nominalism will have to be revisionary, and any revision demands motivation. (Burgess and Rosen, 1997: 10–11)

In other words, minimalists start with their "fairly uncritical attitude" about mathematical existence, and demand of the nominalist a proof or convincing argument that their starting point cannot be maintained. It is a strategy that consists in throwing the onus upon the nominalist to show that the minimal anti-nominalist's position is untenable. Thus, having arrived at the stalemates described above, they can feel justified in maintaining their minimal anti-nominalistic stance: in the absence of any compelling arguments for the revisions demanded by nominalism, the burden of proof required of the nominalist, it can be argued, has not been met.[5]

---

[4] Burgess characterized his own philosophical position (in Burgess, 1983) as a "moderate version of realism": it was a moderate kind of realism insofar as it held merely that our current scientific theories seem to assert the existence of mathematical entities and that we do not have good reason to abandon these theories. (Burgess's moderate realism will be discussed in more detail later in Part 2 of this chapter). The position that I am describing here as "minimal anti-nominalism" is essentially Burgess's moderate kind of realism.

[5] I am inclined to call the Burgess–Rosen brand of realism 'aphilosophical realism' instead of 'moderate realism' or 'minimal anti-nominalism', since it does not subject the doctrine being advanced to philosophical analysis or development, being content to assert merely the existence of mathematical objects, without attempting to make sense of the doctrine, to reconcile the doctrine with the various epistemological views that are widely accepted, or to explain a number of very puzzling questions that arise in connection with what mathematicians and scientists do and say. If risk aversion is one's chief concern, then minimal anti-nominalism may be a good strategy, but from the perspective of those sympathetic with the view of philosophy described in the Introduction, minimal anti-nominalism is simply a "cop-out".

Additional reasons for rejecting the nominalist's position

In the last section of their book (Burgess and Rosen, 1997), a section entitled 'Conclusion', Burgess and Rosen question the philosophical motivation for developing the kinds of nominalistic reconstructions of mathematics discussed above. "What good are these reconstructions?" they can be understood to be asking. The authors confine themselves only to two possible rationales for such reconstructions: *hermeneutic* and *revolutionary*. On the one hand, the nominalist's reconstructions of mathematics might be part of an attempt to give analyses of the mathematical propositions being asserted by actual practicing mathematicians. In other words, the nominalist might be claiming that the propositions of his or her reconstructions are what practicing mathematicians are actually asserting. A philosopher who held such a view would be classified by these authors as a *hermeneutic nominalist*. On the other hand, the nominalist might be proposing a new kind of mathematics that scientists ought to accept in place of their current mathematical theories. On this view, there is no suggestion that our actual mathematical theories are nominalistic; instead, it is argued that we ought to replace our current mathematical theories with the nominalistic reconstructions being developed. A philosopher who took this route would be classified as a "revolutionary nominalist".

To understand adequately the reasoning underlying the Burgess–Rosen evaluations, it is necessary to review a paper cited earlier in which Burgess attacks reconstructive nominalism and defends his "Moderate Realism". What follows is a sketch of the argument.

We first need some definitions that Burgess gives:

An *instrumentalist* maintains that "science is just useful mythology, and no sort of approximation to or idealization of the truth".

A *hermeneutic nominalist* holds that when the language of mathematics is properly analyzed, one will see that the scientist in asserting mathematical propositions is not really asserting the existence of any abstract mathematical objects.

A *revolutionary nominalist* is a nominalist who proposes a new version of science—a nominalistic version—in which there are no assertions of the existence of abstract mathematical objects.

The argument proceeds from just one premise:

[%] A nominalist is either an instrumentalist, or a hermeneutic nominalist, or a revolutionary nominalist.

Burgess's strategy is to argue that none of the above three positions is really tenable. So let's turn to the details of his reasoning.

Few philosophers are "instrumentalists" in the extreme form given by the definition above. And I agree with Burgess that this sort of instrumentalism is not plausible. Thus, if one accepts Burgess's premise, one is faced with choosing between being a *hermeneutic nominalist* or being a *revolutionary nominalist*. Hermeneutic nominalism is a thesis about the proper analysis of sentences of mathematics—and since ordinary mathematics is expressed in English or some such natural language, it is a thesis about the proper analysis of sentences of ordinary languages. Now, Burgess can find no linguistic evidence supporting the hermeneutic nominalist's thesis. And no philosopher has adduced any evidence supporting such a thesis (that is, evidence that a linguist would take seriously).

The revolutionary nominalist, on the other hand, is proposing a new version of science. So the question is: why should we adopt the revolutionary version? What grounds are we given for accepting such a new version of science? There are, he claims, no good scientific grounds for such a revolution, and there are strong practical reasons for not proceeding with such a revolution. He says:

[A]ny major revolution involves transition costs: the rewriting of text books, redesign of programs of instructions and so forth. [I]t would involve reworking the physics curriculum [to allow the student to take courses to learn the logical or philosophical concepts required to understand the nominalist's reconstructions]. (Burgess, 1983: 98)

Thus, Burgess concludes:

Unless he is content to lapse into a mere instrumentalist or "as if" philosophy of science, the philosopher who wishes to argue for nominalism faces a dilemma. He must search for evidence for an implausible hypothesis in linguistics, or else for motivation for a costly revolution in physics. Neither horn seems very promising, and that is why I am not a nominalist. (Burgess, 1983: 101)

In a more recent work, Burgess argues that the hermeneutic and the revolutionary nominalists both make weighty claims the burdens of proof of which have "not yet been fully met" (Burgess, 1990: 7). Returning to their even more recent book, Burgess and Rosen concentrate, for the most part, on the revolutionary nominalist's rationale for their reconstructions. Specifically, they question "the scientific merits of a nominalistic reconstruction as an alternative to or emendation of current physical or mathematical theory" (Burgess and Rosen, 1997: 205), and they argue that the ultimate judgment of the scientific merits of the nominalist's reconstructions should be made by the scientific community and not by the philosophical community. They go so far as to claim that "the true test would be to send in the nominalistic reconstruction to a mathematics or physics journal, and see whether it is published, and if so how it is received" (206). They then write:

If nominalistic reconstruals are not plausible as analyses of the ordinary meaning of scientific language, and if nominalistic reconstructions are not attractive by our scientific standards as alternatives to current physical or mathematical theories—if nominalism makes no contribution to linguistic science, nor to physical or mathematical science—then must the programme of nominalistic reconstrual be judged . . . an intellectual entertainment addressed to no serious purpose? (208)

This last question is addressed at the very end of their book, where they ask: "What is accomplished by producing a series of such distinct and *inferior* theories?" (1997: 238, italics mine).

Notice that the nominalist's reconstructions are now judged to be "inferior". It is not clear to this writer why Burgess and Rosen feel justified in using the term 'inferior' in their question.[6] Perhaps it is because these authors are sure that the nominalist's reconstructions would be judged inferior to current theories, using the test they espouse of submitting the reconstructions to physics journals. Another possibility is suggested by the criteria they advance for theory evaluation; these are similar to those put forward by Quine in (Quine, 1966b): accuracy, precision and breadth of predictions generated by the theory, explanatory range, consistency and coherence, economy of assumptions, familiarity with established theory, perspicuity of its basic notions and assumptions, overall fruitfulness in solving problems, and capacity for being extended to answer new questions.[7] The nominalistic reconstructions, it could plausibly be argued, all fall short by the criterion of "familiarity with established theory", and this alone might lead one to suppose that they are all inferior to the standard Platonic versions of mathematics used in science. Or perhaps they judge the reconstructions to be inferior because of the pragmatic considerations Burgess put forward in his paper (Burgess, 1983). My own view is that their judgment of inferiority is

---

[6] They write: "Since anti-nominalists reject all hermeneutic and revolutionary claims, from their viewpoint the various reconstruals or reconstructions are all distinct from and inferior to current theories" (1997: 238). But I fail to see what grounds they have for concluding that such reconstructions should be judged inferior to current theories.

[7] See Burgess and Rosen, 1997: 209.

based upon a mistaken understanding of what the reconstructions were devised to accomplish. If one thinks tweezers are nail pullers, then one would undoubtedly regard tweezers as inferior tools indeed.

In any case, they answer their question (the one that ends the paragraph preceding the last one) with the words: "No advancement of science proper, certainly; but perhaps a contribution to the philosophical understanding of the character of science." Here is how they imagine such a contribution being made:

> Devising alternatives distinct from and inferior by our standards to our actual theories, but in principle possible to use in their place, is a way of imagining what the science of alien intelligences might be like, and as such a way of advancing the philosophical understanding of the character of science.
>
> It is just such an advance, we want to suggest, that is accomplished by the various reconstructive nominalistic strategies surveyed in this book. (Burgess and Rosen, 1997: 243)

Thus, the work of the nominalist is likened to the work of science fiction writers, who help us to imagine what the intellectual products of alien intelligences might be like—quite a damning simile.[8]

Naturalized epistemology

Another Quinean view that has been extremely influential is his doctrine of naturalized epistemology. Quine rejected the traditional a priori way in which philosophers tended to practice epistemology and advocated instead taking epistemology to be a sub-area of psychology. He suggested that epistemology should study a "natural phenomenon, viz. a physical human subject":

> This human subject is accorded a certain experimentally controlled input—certain patterns of irradiation in assorted frequencies, for instance—and in the fullness of time the subject delivers as output a description of the three-dimensional external world and its history. The relation between the meager input and the torrential output is a relation that we are prompted to study for somewhat the same reasons that always prompted epistemology; namely, in order to see how evidence relates to theory, and in what ways one's theory of nature transcends any available evidence. (Quine, 1969a: 82–3)

Quine's naturalized view of epistemology was adopted by Burgess and Rosen. Since these authors regard the typical nominalist as attempting to

---

[8] See Burgess, 1990: 13–14 for amplifications of Burgess's view of how reconstructive nominalism might be regarded as suggesting what the science of aliens might be like.

justify his (or her) position on epistemological grounds, it was reasonable for them to pay special attention to the field of epistemology. Thus, they contrast "the traditional alienated conception of epistemology"—the advocates of which they describe with the words: "the epistemologist remains a foreigner to the scientific community, seeking to evaluate its methods and standards"— with the Quinean "naturalized conception of epistemology", according to which the epistemologist "becomes a citizen of the scientific community, seeking only to describe its methods and standards, even while adhering to them". They argue that, since the alienated epistemologist attempts to evaluate the methods and standards of the scientific community, it needs to use methods and standards of evaluation that are "outside and above and beyond those of science" (Burgess and Rosen, 1997: 33).

It is clear where the sympathies of these two authors lie when they describe the "pretensions of philosophy to judge common sense and science from some higher and better and further standpoint", and then go on to characterize the following words of David Lewis as giving an especially forceful expression of the rejection of such "pretensions":

Renouncing classes means rejecting mathematics. That will not do. Mathematics is an established, going concern. Philosophy is a shaky as can be. To reject mathematics on philosophical grounds would be absurd. ... I laugh to think how presumptuous it would be to reject mathematics for philosophical reasons. How would you like to go and tell the mathematicians that they must change their ways, and abjure countless errors, now that philosophy has discovered that there are no classes? ... Not me![9]

All in all, Burgess and Rosen have assembled quite a case against the nominalist. As I noted in the Introduction, the title of their book, "A Subject with No Object", suggests that the Burgess–Rosen criticisms are aimed at undermining the very object of nominalism. Shapiro, for one, thinks that they have to a large extent succeeded, writing that Burgess and Rosen have

---

[9] Lewis, 1991: 58, quoted in Burgess and Rosen, 1997: 34. Burgess and Rosen do not tell the reader, nor note the irony, of the fact that this same philosopher is willing to proclaim to the whole world (and presumably to astronomers) that he has made the remarkable discovery that there are planets in existence, distinct from those in our solar system, on which there are intelligent beings, who speak English, philosophize as we do, who have developed mathematics and science identical to our own, and have a political system that is indistinguishable from the American system, with a President who looks, talks, thinks, and acts exactly like G. W. Bush—all this on the basis of philosophical reasoning! For a discussion of how Lewis thinks he has achieved such a discovery, see Chihara, 1998: ch. 3.

raised "sharp and penetrating criticisms of the nominalistic projects and of the whole point of nominalism" (Shapiro, 1998: 600).

2. NOMINALISTIC RECONSTRUCTIONS OF MATHEMATICS REEXAMINED

I shall now give reasons for holding that Burgess and Rosen have not succeeded in undermining nominalism, at least when it is directed at the sort of nominalism I advocate. To do this, I shall present a rationale for the nominalist's reconstructions of mathematics that is very different from the epistemological one described by Burgess and Rosen. First of all, unlike the nominalist of the Burgess–Rosen book, the kind of nominalist I have in mind does not begin with the thesis that there are no abstract entities. I see these nominalists, rather, as anti-realists. They start with the idea that mathematics should not be understood as the kind of Platonic theory described by such realists as Gödel and Quine.

Why some nominalistic reconstructivists should not be classified as alien espistemologists

The nominalistic reconstructivists of the sort I have in mind do not attempt to judge common sense and science from some higher, better, and further standpoint. They seek to piece together their account of mathematics in a way that is compatible with both what science teaches us about how we humans obtain knowledge and also what we already know about how humans learn and develop mathematical theories. Furthermore, these nominalists do not reject mathematics—a fortiori, they do not reject mathematics on the basis of "some higher and better and further standpoint". On the contrary, their goal is to understand the nature of mathematics in a way that is compatible with the other features of the Big Picture they are attempting to construct.

It is true that, as nominalists, they do not believe that sets, classes, or extensions of concepts actually exist. But their skepticism about the existence of such things is not based upon the conviction that they have some decisive knock-down a priori argument showing that mathematical objects cannot (or do not) exist. Rather, it is based upon a lot of different considerations, such as:

(a) The fact that set theorists, sitting in their offices and merely thinking about sets, were able to discover somehow the truth of the axioms of set theory (taken to be assertions about mathematical objects) is something that nominalists typically find either utterly mysterious or unintelligible.[10]

(b) Nominalists have serious doubts about the arguments (such as the various versions of the indispensability argument discussed earlier) that Platonists have given to support their belief in such things. All such arguments have seemed to them to be highly questionable.[11]

(c) They find implausible and unscientific the philosophical theories that Platonists have advanced to account for their supposed knowledge of the things they postulate.[12]

(d) The Platonist's position that empirical scientists need to discover complex and complicated relationships between entities that do not exist in the physical world[13] in order to develop their empirical theories has seemed bizarre and counterintuitive to nominalists. They question the explanations that Platonists have given of why knowledge of mathematical objects is required to obtain sophisticated scientific knowledge of the physical universe.[14]

(e) Their assessment of the many doctrines, advanced by philosophers over the ages, that postulate some type of non-physical, undetectable substance or object in order to account for some feature of our language or our beliefs has taught them to be skeptical of all such doctrines.

(f) They are convinced that the descriptions Burgess and Rosen give of the nominalist ("the pretensions of philosophy to judge common sense and science"), as well as Lewis's diatribe against the presumptuousness of philosophers, are fundamentally misleading. First of all, contrary to what Lewis has suggested, there are many nominalists who maintain neither that our mathematical theories consist of false assertions nor that

---

[10] Penelope Maddy has attempted to give explanations, based upon the work of the neurophysiologist Donald Hebb, as well as her theory of perception, for how the truth of some of the axioms of set theory (understood realistically) have been discovered in Maddy, 1990: ch. 2. Needless to say, I have not found her explanations believable.

[11] Of course, the major arguments of this sort put forward by contemporary Platonists were discussed and evaluated in the previous chapter.

[12] For example, Gödel's mathematical intuition (discussed in Chapter 1) was supposed to be something like a faculty of perceptions. (See Chihara, 1982 for details.) Maddy's view (also discussed in Chapter 1) that humans are able to literally see sets is another example. Cf. also Brown's view sketched in Chapter 1.

[13] Gödel explicitly asserted that the objects of transfinite set theory do not exist in the physical world. See Gödel, 1964b: 271.   [14] See, for example, Brown, 1999: ch. 4, esp. 47–9.

mathematicians have to change their ways because of what philosophers have discovered. Secondly, one could easily get the impression from the above-mentioned Platonists that it is only the presumptuous philosopher (with pretensions of having some special means of judging the affirmations of science) who has doubts about the existence of mathematical objects. In fact, many (if not most) of the outstanding researchers in mathematics have serious doubts about the Platonist's account of mathematical theorems and knowledge.[15] And it has certainly been my experience that many empirical scientists find the Platonic view of mathematics quite fantastic. Also, anyone who has taught philosophy of mathematics (as I have for a great many years) will no doubt remember the look of disbelief, and sometimes of amazement, on the faces of some students on being told that highly respected philosophers and mathematicians at major universities believe that, in addition to planets, stars, galaxies, atoms, electrons, molecules, and photons, such things as numbers, sets, and functions also exist. These reconstructivists feel that any account of mathematics should take account of the skeptical attitudes of non-philosophers towards the Platonic account. It suggests to them that the belief that Platonists have in mathematical objects is not at all like the belief that, essentially, everyone has in ordinary material objects like tables and chairs, despite what some Platonists have argued.[16]

Such considerations as the above lead these nominalists to doubt the appropriateness of the Platonist's conception of mathematical theories as descriptions of a realm of objects that do not exist in the physical world. These nominalists are skeptical of the Platonic doctrine that to achieve genuine mathematical knowledge, one needs to know that such objects truly

---

[15] Recall the judgment of Paul Cohen, quoted in Chapter 1, that "probably most of the famous mathematicians who have expressed themselves on the question have in one form or another rejected the Realist position" (Cohen, 1971: 13). Recall also Soloman Feferman's anti-realist views quoted earlier: "Briefly, according to the platonist philosophy, the objects of mathematics such as numbers, sets, functions, and spaces are supposed to exist independently of human thoughts and constructions, and statements concerning these abstract entities are supposed to have a truth value independent of our ability to determine them. Though this accords with the mental practice of the working mathematician, I find the viewpoint philosophically preposterous" (Feferman, 1998c: p. ix).

[16] Penelope Maddy at one time argued that Gödel is not required to supply a theoretical justification for his belief in sets since we humans have perceived and believed in sets since prehistoric times, suggesting that our belief in sets is very much like our belief in material objects. See Chihara, 1982: Part 2, esp. 226, where I raise objections to such a position.

exist and are related to one another as the theory in question affirms. As a result, these nominalists search for an alternative way of understanding the nature of mathematics, and *their reconstructions are seen as an aid to achieving their overall goal of arriving at such an understanding.* Furthermore, most reconstructivists believe that their skepticism about mathematical objects does not require of them the rejection of set theory. They are convinced that, with a correct understanding of the nature of set theory, one will see not only how set theory can be a coherent and useful mathematical theory even though no such entity as a set actually exists, but also how it can be reasonable for outstanding researchers in set theory to continue to do work in the field even though they do not believe that there are such things as sets.

I would like to emphasize that my own reconstructive works were never motivated by Benacerraf-type epistemological arguments. My first reconstructive book, *Ontology and the Vicious-Circle Principle*, was published in 1973—the year Benacerraf's paper "Mathematical Truth" was published—and almost all of the main reconstructive ideas had been developed many years earlier. I certainly based nothing in my book on the Benacerraf paper. Indeed, I began the book with the quotation from *Word and Object* in which Quine declares that the nominalist would have to "accommodate his natural sciences unaided by mathematics", and then wrote:

When I first read these words in *Word and Object* several years ago, I wrote in the margin: "This philosophical doctrine should be soundly refuted." It was only much later, while I was working on an essay on the vicious-circle principle, that an idea came to me as to how one might construct such a refutation.[17]

What I felt should be refuted was Quine's claim that the nominalist would be unable to develop a non-trivial system of mathematics for the natural sciences that would not be "ontologically committed" to mathematical objects. I had this conviction long before any talk of causal theories of knowledge or Benacerraf-type anti-realist arguments were heard in philosophical circles. The aim of my reconstruction, then, was to refute the Quinean claim by producing a non-trivial system of mathematics that does not require quantification over mathematical objects. But the nominalistic reconstruction was not supposed to be the end of the story: it was to be an aid toward gaining an understanding of the functioning of the mathematical theories mathematicians actually use and develop.

---

[17] Chihara, 1973: p. xiii. The essay mentioned was written in 1964–5.

Should nominalists submit their papers to a physics journal?

Let us now take up the Burgess–Rosen suggestion that the "true test" of the nominalist's view is to submit it to a physics or mathematics journal to see if it is published. As I argued earlier, the three "nominalistic reconstructions" described above can be construed as responses to the Quinean indispensability challenge—a philosophical argument proposed by a philosopher in support of a traditional philosophical position in metaphysics. Are empirical scientists or mathematicians best equipped intellectually and by training to deal with the many and complex philosophical issues that such a dispute may involve? Should they be entrusted with the final word on the merits (scientific, logical, and philosophical) of such responses? Would sending papers detailing these reconstructions to physics journals constitute "true tests" of the merits of these works?

Take, for example, Hellman's modal structuralist system of mathematics. How many physicists have a sufficiently deep understanding of second-order S5 modal quantificational logic and mereology (not to mention the immense literature concerning epistemology and metaphysics, especially those focusing on possible worlds semantics, modality, and mathematics) to give a truly competent assessment of the merits and demerits of modal structuralism? Very few, I should think, if any. Given the paucity of physicists competent to evaluate such matters, why should we expect their assessments of Hellman's work to provide the final word on the merits of such systems of mathematics?

Suppose that one of the papers proposing a nominalist reconstruction of mathematics were to be deemed, by an editor, to be unsuitable for publication in some physics journal. Why should we assume (as Burgess and Rosen evidently do) that such an editorial decision would have to be based upon the judgment that the sort of ontological economy proposed by the nominalist has no scientific merit? Might not such a decision be based, for example, upon the judgment that the journal's readership would not be an appropriate audience for a philosophical paper of that sort?

In *Constructibility and Mathematical Existence* (Chihara, 1990), I proposed another system of mathematics—"the Constructibility Theory"—to meet the Quinean challenge. Like my earlier predicative system, this new theory makes use of constructibility quantifiers to avoid existentially quantifying over mathematical objects. However, it abandons the predicative restriction of the vicious-circle principle in order to develop the mathematics of simple type theory. Because of these changes, this new system seems to overcome the kind of doubt that plagued the earlier system—the kind of doubt that questions the reproducibility of the mathematics needed for our contemporary scientific theories (including quantum physics) within its logical framework. The mathematics of simple type theory seems to be more than adequate for the needs of the empirical scientist, capturing as it does all the mathematics developable within *Principia Mathematica*. Additional details of this "nominalistic reconstruction" will be given in Chapter 7. Here, let me respond to Burgess's argument for his "moderate realism" from my own constructibility perspective.

A response to Burgess's argument for moderate realism

According to Burgess's principal premise, the reconstructive nominalist must espouse one of three positions: scientific instrumentalism, hermeneutic nominalism, or revolutionary nominalism. Since none of the nominalists I described above espoused scientific instrumentalism (certainly in the form characterized by Burgess), Burgess's argument can be regarded as resting on the view that restricts the nominalist's choices to either hermeneutic nominalism or revolutionary nominalism.[18]

Now why does Burgess think that these are the only choices open to the nominalist? (I am reminded, here, of Hume's view that all of human knowledge must be knowledge either of relations of ideas or of matters of fact.) Burgess's disjunction suggests that there are only two alternative views that the nominalist can be advancing: a view about relations of ideas (an analysis of the meaning of mathematical statements) or a view about a matter of fact—"Such and such scientific theory is most probably true" (a revolutionary view).

I do not know if Burgess was influenced by empiricist views that go all the way back to Hume, but in any case, the position I developed in my book does not fit comfortably any of the above descriptions of the possible types of

---

[18] Susan Vineberg has suggested to me that there are many alternatives to the tripartite premise of Burgess's argument that potentially undermine the argument but that are not even considered by Burgess. For example, there are philosophically defended alternatives to the extreme version of instrumentalism that figures in Burgess's premise that should be examined. Nancy Cartwright, for one, has articulated a view of science according to which its laws are, typically, not true. She argues:

Most scientific explanations are *ceteris paribus* laws. These laws, read literally as descriptive statements, are false, not only false but deemed false even in the context of use. This is no surprise: we want laws that unify; but what happens may well be varied and diverse. We are lucky that we can organize phenomena at all. There is no reason to think that the principles that best organize will be true, nor that the principles that are true will organize much. (Cartwright, 1983: 52–3).

nominalism that Burgess provides. My view was not a form of instrumentalism; nor was it a version of either hermeneutic or revolutionary nominalism. I was not espousing an instrumentalist view of science, since I in no way considered science to be a myth. My view was that a great many (if not most) of the assertions of scientists are, if not literally true, at least close to being true, given that the proper conditions are expressed.

I also made it clear in my book that my Constructibility Theory was not meant to be an analysis of the mathematical statements asserted by practicing mathematicians. Here's what I wrote in the first chapter:

If we are puzzled about certain aspects of classical mathematics, why not construct *another kind of mathematics* that will avoid those features of the original system that gave rise to the puzzles? A study of these alternative mathematical theories will give us a new perspective from which to view classical mathematics, which could prove to be extremely enlightening. (Chihara, 1990: 23, italics not in the original)

Clearly, my position was not that of the hermeneutic nominalist.

So was I proposing a revolutionary nominalism? Not at all. I would have regarded it as absurd to make a general proposal which would involve answering such practical questions as:

'How should scientific theories be taught in our universities?' 'How should practicing scientists formulate their theories?' 'How much mathematical logic, modal logic, and metaphysics should practicing scientist have to learn?'

I certainly was not proposing that scientists should actually adopt my Constructibility Theory in place of classical mathematics for the purposes of formulating their laws and expressing their scientific theories, any more than the mathematical foundationalist, who shows how all the mathematics the scientist uses could be done in set theory, is advocating the revolutionary thesis that all the mathematics that is now being taught in the universities and that is in actual use in scientific research should be replaced by set theory.

In summary, my position in that book was not that of instrumentalism, nor of hermeneutic or revolutionary nominalism. I was proposing that the Constructibility Theory provides the nominalist with a response to the Quinean challenge. Thus, I was proposing an answer to a highly theoretical and deeply philosophical question: can our contemporary scientific theories be reformulated or reconstructed in a way that will not require the assertion or the presupposition of abstract mathematical objects? This is a modal question. It is not a practical question of how best to teach physics in our secondary schools, colleges, and universities. Answering this question in the affirmative does not require committing oneself to the proposition that scientists should actually reformulate all their scientific theories using the Constructibility Theory instead of classical mathematics.

It can be seen that the constructibility theory does provide us with grounds for questioning the Quinean idea that contemporary science and logical consistency requires us to believe in the existence of mathematical objects. As I wrote:

[T]he point of showing . . . how mathematics can be done in terms of constructibility quantifiers was not to convert scientists to using a new system of mathematics, but rather to show that the undeniable usefulness of mathematics in science did not require that one believe in the [things apparently] talked about in mathematics. (Chihara, 1990: 188)

This point can be illustrated by taking up again Steiner's version of Quine's argument. Recall that he had suggested that one cannot describe "any [genuine non-trivial] thinkable experience" without presupposing numbers, since any such description would involve rate of change: "Imagine defining *rate of change* without the resources of analysis", he had challenged (Steiner, 1978: 19–20), confident that such a definition was impossible. William Goodwin has responded to Steiner's argument (in a personal communication to me) by wondering if Steiner thinks that before the nineteenth-century arithmetization of analysis, and certainly before Newton and Leibniz came up with their versions of the calculus, humans were unable to describe any genuine non-trivial thinkable experience? In any case, one can define rate of change using the Constructibility Theory, and this without presupposing numbers. So Steiner's challenge can be met by the nominalist.

These developments also raise the question as to whether the use of even standard classical mathematics in science genuinely commits one to accepting the existence of abstract mathematical entities. If, as I have been arguing for many years, science can be done using the Constructibility Theory instead of standard versions of classical mathematics, without serious theoretical loss, this suggests that the role that mathematics is required to play by our contemporary scientific theories is not that of *referring to* and *providing information about* esoteric objects that do not exist in physical space (since the constructibility theory does not do these things). It encourages the thought that our contemporary scientific theories do not truly require belief in mathematical objects, as Quine and other Platonists have thought. In other words, since the Constructibility Theory does not make reference to mathematical objects or provide any information about such things, it would seem that the scientist does not have to refer to mathematical objects in order to develop her scientific theories. After all, no Platonist has adequately explained why information about things that do not exist in the physical world is *necessary* for us to gain knowledge about events and processes that take place in the physical world. Perhaps, the use of even standard versions of classical mathematics in science does not truly ontologically commit one to a Platonic ontology of abstract mathematical objects. As we shall see shortly, the Constructibility Theory will play an important role in answering the above questions.

These last points link up nicely with the case Susan Vineberg has marshalled against the Burgess–Rosen objection and in support of the relevance of these nominalistic reconstructions for deciding whether belief in mathematical objects is required by science. Vineberg investigates various views of compelling evidence that do not require a causal connection between the scientist and what she has evidence for. Of the above sort, Vineberg finds three: the well-known Bayesian view, the eliminativist view, advocated by Philip Kitcher,[19] and the contrastive view, proposed by both Elliot Sober and Larry Laudan.[20] Omitting here the details of her reasoning, Vineberg concludes that, regardless of which of these theories of evidence one may choose, "substantial confirmation requires eliminating, or revealing as improbable, alternative theories". Not surprisingly, it is common scientific practice to look for alternative theories that fit or explain the relevant experimental data and observations that have been gathered and to attempt to find evidence that allows us to eliminate all but one of these theories. Within this context of scientific theory evaluation, the various nominalistic views described above can be seen to be fitting and reasonable: in the absence of genuine evidence (as opposed to mere pragmatic considerations such as "familiarity with established theory") that allows scientists to eliminate these alternatives, the nominalistic reconstruction provides us with rational grounds for being skeptical about the existence of mathematical objects.

[19] See Kitcher, 1993: esp. 237–47.

[20] Sober's eliminativist view was discussed in the previous chapter. Laudan's view is presented in Laudan, 1997. The relevant idea, for our discussion above, is given by the following: "[T]he evaluation of a theory or hypothesis is relative to its *extant* rivals. To accept H is to hold that it is more reliable than its known rivals; to reject H implies that it is worse than at least one of its known rivals" (114). As Vineberg has noted, James Hawthorne has analyzed Bayesian induction in such a way that it can be seen to be a probabilistic form of eliminative induction, i.e. "a method for finding the truth by using evidence to eliminate false competitors" Hawthorne, 1993: 99.

Vineberg's point can be illustrated by the following. It is thought by Platonists that the existence of mathematical objects is required to solve certain problems or puzzles and that this fact alone counts strongly in favor of belief in mathematical objects. For example, according to Steiner, Frege completely solved what Steiner calls "the metaphysical problem of applicability" (this problem and the Fregean solution will be given in Chapter 9). Since Frege's solution presupposes the existence of mathematical objects, it may be thought that the very success of Frege's solution amounts to a kind of evidence for the existence of mathematical objects.[21]

The existence of the above nominalistic alternatives, however, allows one to devise alternative solutions to "the metaphysical problem of applicability" (how the Constructibility Theory can be used to resolve this problem will be shown in detail in Chapter 9)—solutions that do not presuppose the existence of mathematical objects—thus undercutting the idea that the postulation of mathematical objects is required to solve the problem. We can thus see how these nominalistic alternatives can serve to undermine the claims of the Platonist that we have compelling scientific grounds for postulating the existence of mathematical objects.

---

[21] Frege's solution is supposed to consist in showing how mathematical entities relate, not directly to physical objects, but rather through concepts. Commenting on this solution, Steiner writes: "That physical objects may fall under concepts and be members of sets is a problem only for those who do not believe in the existence of sets" (Steiner, 1995: 138). To this, I reply: what about those who do not believe in the existence of concepts?

7

The Constructibility Theory

In investigating the reasons Shapiro and Resnik have proposed for being realists (with respect to the "parts" or "positions" in structures), I have concentrated thus far on the positive reasons these philosophers have given for believing in mathematical objects. I have not yet investigated Shapiro's principal reason for not being an eliminative structuralist—what he called the "main stumbling block of the eliminative program" (Shapiro, 1997: 86). As a useful preliminary to such an investigation, I shall begin a characterization of the Constructibility Theory I discussed earlier in Chapter 6. Since this theory has been described in detail in my *Constructibility and Mathematical Existence* (Chihara, 1990), I shall give here only a brief description of the theory, setting out its main philosophically salient features, before giving a presentation of how (finite) cardinality theory is developed within it—a cardinality theory to be used in later sections of this work. This short exposition will be given in the early sections of this chapter. In Sections 4, 5, and 6, I shall take up the objections to the Constructibility Theory that Shapiro and Resnik have detailed in their respective books on structuralism. By means of these discussions, I hope to clear up some widespread misconceptions and confusions about the Constructibility Theory and to provide the reader with some substantial insights into its finer points.

However, before getting involved with the details of this theory, it needs to be emphasized that the term 'Constructibility Theory' is used in this work ambiguously: sometimes to denote my general theory of constructibility put forward in my book, and sometimes to denote the formal theory of constructibility of that work.[1] (Both are theories of constructibility.) When I feel

---

[1] My general theory of constructibility contains many theses, arguments, and positions that are not found in the formal theory. The ambiguity of the term 'constructibility theory' is similar to an ambiguity found in the use of the expression 'Russell's theory of types': sometimes the expression refers to Russell's general theory about the types of propositional functions there supposedly are, and sometimes it refers to the specific formal theory of propositional functions given in *Principia Mathematica*.

that it is important to indicate that the formal theory is being cited, I shall use the expression 'Ct' instead of 'the Constructibility Theory'.

1. A Brief Exposition of the Constructibility Theory

Set theory is a theory about sets. It tells us what sets exist and how these sets are related to one another by the membership relationship. The Constructibility Theory is a theory about open-sentences: it tells what open-sentences (of a certain sort) are constructible and how these constructible open-sentences would be related to one another by the *satisfaction relation*. (Thus, in this theory, one open-sentence can satisfy another.) The theory is formalized in what is basically a many-sorted first-order logical language[2] that utilizes, in addition to the existential and universal quantifiers of standard first-order logic, *constructibility quantifiers*.

Constructibility quantifiers are sequences of primitive symbols: either (C—) or (A—), where '—' is to be replaced by a variable of the appropriate sort. Using '$\Psi\phi$' to be short for 'ϕ satisfies Ψ', '$(C\phi)\Psi\phi$' can be understood to say:

It is possible to construct an open-sentence ϕ such that ϕ satisfies Ψ,

whereas '$(A\phi)\Psi\phi$' can be understood to say:

Every open-sentence ϕ that it is possible to construct is such that ϕ satisfies Ψ.

And just as '$(\exists x)Fx$' is equivalent to '$-(x)-Fx$', and '$(x)Fx$' is equivalent to '$-(\exists x)-Fx$', we have '$(C\phi)\Psi\phi$' is equivalent to '$-(A\phi)-\Psi\phi$', and '$(A\phi)\Psi\phi$' is equivalent to '$-(C\phi)-\Psi\phi$'.

The Constructibility Theory asserts the constructibility of various sorts of open-sentences. It needs to be emphasized that what are said to be constructible by means of the constructibility quantifiers are open-sentence *tokens* as opposed to open-sentence types. Open-sentence types are classified by Parsons as "quasi-concrete" objects (in contrast to "purely mathematical objects"), because "they are directly 'represented' or 'instantiated' in the concrete" (Parsons, 1996: 273). Despite the "quasi" qualification, it can be argued that these quasi-concrete objects are abstract Platonic objects as epistemologically disreputable as numbers and sets. But open-sentence tokens

---

[2] For a clear and rigorous discussion of such languages, see Enderton, 1972: sect. 4.3.

are not open to the same objection: typically, an open-sentence token consists of particular marks on paper, which exist at a particular place in the universe and at a particular time. Furthermore, to say that an open-sentence of a particular sort is constructible is not to imply or presuppose that any such open-sentence token actually exists or, indeed, that anything exists. Constructibility quantifiers do not carry ontological commitments as do the quantifiers of standard extensional logic.

What does 'possible' mean?

In the phrase 'it is possible to construct', the term 'possible' needs some explanation. There are, of course, many different kinds of possibility: logical, metaphysical, physical, epistemological, technological, to name just a few. Epistemological possibility is concerned with what is known. Thus, to say that ϕ is epistemologically possible for agent X is to say that ϕ is not logically precluded by what X knows, that is, ϕ is logically compatible with everything X knows. To say that ϕ is physically possible is to say something like: ϕ is logically compatible with all the physical laws of our universe. The possibility talked about in the Constructibility Theory is what is called 'conceptual' or 'broadly logical' possibility—a kind of metaphysical possibility, in so far as it is concerned with *how the world could have been*. Every purely logical truth is necessary in this sense, but the set of conceptual necessary truths will include much more. What are called "analytic truths" (such truths as 'All bachelors are unmarried') are also held to be necessary in this broadly logical sense.[3] As a rough guide, Graeme Forbes suggests that 'it is possible that P' should be taken to mean: "There are ways things might have gone, no matter how improbable they may be, as a result of which it would have come about that P" (Forbes, 1985: 2). Another way of expressing this sort of possibility is to take it to mean: the world could have been such that, had it been this way, P would have been the case. There are many different systems of modal logic, but the type of system that is generally believed to correctly formalize the logical features of this broadly logical sense of necessity is S5.[4]

---

[3] See Plantinga, 1974: ch. 1, sect. 1 and Forbes, 1985: ch. 1, sect. 1, for more examples and discussion.

[4] Thus, Kit Fine writes "S5 provides the correct logic for necessity in the broadly logical sense": Fine, 1978: 151. For an interesting discussion of the development of S5 modal logic, see Kneale and Kneale, 1962: ch. 9, sect. 4. See Chihara, 1998 for a discussion of a variety of S5-type systems of modal logic.

What '(Cφ)Ψφ' does not mean

'(Cφ)Ψφ' ('It is possible to construct an open sentence φ such that φ satisfies Ψ') does not mean that *one knows how* to construct such an open-sentence or that one has a method for constructing such an open-sentence. Hence, the constructibility quantifier is not at all like the intuitionist's existential quantifier. Furthermore, it does not mean that one can always, or even for the most part, determine what particular objects would satisfy such an open-sentence or how one would determine what objects would satisfy φ. Nor does it mean that one can determine whether a series of marks, sounds, hand signals, or what have you is or is not such an open-sentence.

It may strike the reader as strange, even objectionable, that I would put forward a theory that asserts the constructibility of open-sentences of various sorts, when it cannot provide the sort of information indicated above. If so, consider the example of Euclid's geometry, which I described earlier: for well over two thousand years, this modal theory was a sort of paradigmatic mathematical system, showing scholars how mathematics should be developed and applied. But notice, Euclid's geometry does not tell us how to recognize straight lines, how to tell if a line is really straight, or if a line really intersects a point. It does not tell us how to construct points, straight lines, or arcs. The important point is this: it doesn't matter that Euclid's geometry does not tell us these things. The usefulness of that kind of modal theory does not depend on its giving us that kind of information. *That's not the way we use that geometry.*[5] Similarly, my Constructibility Theory is not designed to give us information about how to tell what is an open-sentence or what things satisfy any given open-sentence. It is just not that sort theory. It has been designed to provide us with a way of understanding and analyzing mathematical reasoning, in a way that does not presuppose the existence of mathematical entities, as will become apparent in the sections to follow.

The many levels of the Constructibility Theory

The Constructibility Theory is similar to Frege's theory of concepts: just as Frege's hierarchy of concepts is stratified into levels, the open-sentences that are talked about in the Constructibility Theory are of different levels. Thus,

---

[5] Of course, we do have ways of telling when lines drawn on a sheet of paper are approximately straight, roughly intersect a curve at a point, etc., and one might say that the usefulness of Euclid's geometry theory depends upon our having such ways of telling. But it should be noted that we also have ways of constructing open-sentences, and we have ways of telling, for many actual cases, that something satisfies an open-sentence we have constructed.

consider the following situation. On the desk in my office, there are two pieces of fruit, which I have named 'Tom' and 'Sue'. On the blackboard in my office, I write the open-sentence

x is a piece of fruit on the desk in my office.

Both Tom and Sue satisfy this open-sentence. The desk does not. This open-sentence token is of level 1. Now suppose that I write the open-sentence

There is at least one object that satisfies F

in the bottom left corner of my blackboard, where 'F' is being used as a variable of level 1. This open-sentence token is satisfied by the open-sentence token I constructed earlier—the level 1 open-sentence I first wrote on the blackboard. The second open-sentence I constructed is a second-level open-sentence. Clearly, then, we can go on to construct open-sentences of level 3, 4, 5, and so on.

An objection considered

It might strike a critical reader that, in speaking of level 3, 4, 5, ... open-sentences, I am, in effect, presupposing the natural numbers (or the finite ordinal numbers) and in this way presupposing mathematical objects, thus undermining my attempt to provide an analysis of mathematical reasoning that does not assume the existence of such things. Actually, what I do assume here is the rule we all learn as children for constructing and ordering the arabic numerals—numerals which are being used to talk and to theorize about different levels of open-sentences in a way that enables me to order the things I wish to discuss. Such a use does not require genuine quantification over and reference to mathematical entities.

In the next section, I shall sketch the development of finite cardinality theory within the framework discussed above.

2. Preliminaries of the Constructibility Theory of Natural Number Attributes

I shall use the following to refer to the entities of different levels:

Level 0: *Objects x, y, z, ...*

Level 1: *Properties F, G, H, ...*

Level 2: *Attributes $\mathscr{F}$, $\mathscr{G}$, $\mathscr{H}$, ...*

Level 3: *Qualities* **F, G, H**, ...

It needs to be emphasized that what I am calling 'properties', 'attributes', and 'qualities' are just open-sentences. Thus, the open-sentences of level 1 that will be talked about in this theory are to be called 'properties' and the upper-case letters '*F*', '*G*', '*H*', etc. are to be used as variables to refer to these open-sentences. Extrapolating from the level 1 case, one can see that the level 2 open-sentences to be talked about will be called 'attributes' and the script upper-case letters '$\mathscr{F}$', '$\mathscr{G}$', '$\mathscr{H}$', etc. will be used as variables to refer to these second-level open-sentences. Thus, properties, attributes, and qualities are not being used to refer to universals or abstract entities of some sort, as is generally the case in philosophical works. I use this terminology simply to facilitate our keeping distinct and ordered the open-sentences of different levels I shall be talking about.

Quantifiers

(a) Quantifiers containing occurrences of level 0 variables are just the standard quantifiers of first-order logic.
(b) Quantifiers containing occurrences of level 1 or higher variables are constructibility quantifiers.

Relations

All the open-sentences to be talked about in this theory will be monadic open-sentences, that is, such open-sentences as '*x* is a human' that contain occurrences of only one variable. The reason I restrict the theory to just monadic open-sentences is simplicity: it makes the task of formalizing the theory so much easier.

Of course, we will need relations if we are to develop arithmetic in the Fregean way. So how am I to get relations in a system that deals only with monadic open-sentences? There is a similar problem in set theory. How does one get relations in set theory? As far as mathematics is concerned, a relation can be taken to be a set of ordered pairs. That is, everything one needs to do with relations in mathematics can be done by taking a relation to be a set of ordered pairs.[6] In mathematics, the taller-than relation among the set of human beings can be taken to be just the set of all ordered pairs of humans

---

[6] Some might think that only binary relations can be defined to be sets of ordered pairs. Isn't a ternary relation to be defined to be a set of ordered triples? But the ordered triple $<x, y, z>$ can be defined to be the ordered pair $<x, <y, z>>$. Clearly, an ordered *n*-tuple can be defined, in that way, to be an ordered pair.

$<x, y>$ such that x is taller than y. So the problem of defining relations in set theory reduces to the problem of defining ordered pairs in terms of sets.

Kuratowski's solution to this problem is now widely used. He proposed that we take the ordered pair $<x, y>$ to be the set $\{\{x, x\}, \{x, y\}\}$. When $<x, y>$ is so defined, then it can be proved in set theory that $<x, y> = <z, w>$ iff $x = z$ and $y = w$. And that is the crucial condition we want ordered pairs to satisfy.

In what follows, I shall follow the above set-theoretical practice, by defining relations à la Kuratowski.

Couples

A couple $\{x, y\}$ is a property that is satisfied by only the objects x and y.

Example: '$x = $ Tom v $x = $ Sue' is a couple $\{$Tom, Sue$\}$.

Notice that I have given here an example of a monadic open-sentence: only one variable (i.e. 'x') occurs in the open-sentence—of course, there are two *occurrences* of this one variable. Notice also that I say 'a couple' (instead of 'the couple'), because it is possible to construct indefinitely many different such couples. Thus, 'x is the very same person as Tom or x is the very same person as Sue' is also a couple $\{$Tom, Sue$\}$.

Ordered pairs

An ordered pair $<x, y>$ is an attribute that is satisfiable by all and only couples $\{x, x\}$ and $\{x, y\}$ that could be constructed.

Note: An ordered pair is an open-sentence satisfied by other open-sentences—it is not satisfied by the objects x and y.

Example: The open-sentence

F is a couple $\{$Tom, Tom$\}$ or F is a couple $\{$Tom, Sue$\}$ is an ordered pair $<$Tom, Sue$>$.

Relations

A relation **R** is a *quality* that is satisfiable only by ordered pairs such that if an ordered pair $<x, y>$ could be constructed that satisfied **R**, then **R** is satisfiable by every ordered pair $<x, y>$ that could be constructed.

It can be seen that there is nothing mysterious about relations. They are things we can literally construct, say by writing them down on paper or on the blackboard.

Example: '$\mathcal{H}$ is an ordered pair $<x, y>$ such that x is married to y' is a quality which is a relation corresponding to the intuitive relation of *being married to*. Thus, any ordered pair $<$ Hillary Clinton, Bill Clinton $>$ that one constructed would satisfy the above-mentioned relation.

A notational definition

If **R** is a relation, then

'xRy' means$_{def}$ 'an ordered pair $<x, y>$ satisfies **R**'.

Notice that, as the term 'relation' has been defined, if an ordered pair $<x, y>$ satisfies some relation **R**, so that xRy, then every ordered pair $<x, y>$ that is constructible would also satisfy **R**.

In the next section, many of the definitions will look exactly like those given in Frege's theory of cardinal numbers.

Relation **R** correlates F to G

iff, for every x that satisfies F, there is a y that satisfies G such that

xRy;

and for every y that satisfies G, there is an x that satisfies F such that

xRy.

Relation **R** is a one-one relation

iff, for every x, y, and z,
  if xRy and xRz, then $y = z$;
  if xRy and zRy, then $x = z$.

F is equinumerous with G

iff, it is possible to construct a one-one relation that correlates F to G.

Example: Let us suppose that the objects under consideration here are people. Then consider the open-sentences:

  J: 'x is the junior senator from some state.'
  S: 'x is the senior senator from some state.'

and let **R** be the open-sentence:

> $\mathscr{H}$ is an ordered pair $<x, y>$ such that x is the junior senator from a state in which y is the senior senator.

It can be seen that **R** is a one-one relation correlating property J with property S. Hence, J is equinumerous with S.

Compare, in the context of this example, the Fregean truth conditions for the assertion that J is equinumerous with S with the constructibility truth conditions for that same assertion. In the Fregean case, in order that the assertion be true, there must exist some abstract entity—a one-one relation—something that does not exist in physical space, which cannot be detected by any scientific instruments. In the constructibility case, in order that the assertion be true, it must be possible to construct an open-sentence of a certain sort—and in this case, I can (indeed, I have) constructed an open-sentence of the required sort.

The developments in the following section deviate somewhat from the Fregean developments. This is because we don't have objects to serve as the cardinal numbers. In this theory, instead of cardinal numbers, we have cardinal number attributes.

3. CARDINAL NUMBER ATTRIBUTES

Cardinal number attribute of a property

> An attribute $\mathscr{N}$ is a *cardinal number attribute of property* F iff $\mathscr{N}$ is satisfiable by all and only those properties equinumerous with F.

Example of a cardinal number attribute of a property:

> G is equinumerous with 'x is the senior senator from some state'.

Hume's Principle

It is a straightforward matter of proving from the above a constructibility version of what has come to be known as "Hume's Principle":

> F is equinumerous with G iff it is possible to construct cardinal number attributes of both $\mathscr{F}$ and $\mathscr{G}$, and for any $\mathscr{F}$ that it is possible to construct which is a cardinal number attribute of F, and for any $\mathscr{G}$ that it is possible to

construct which is a cardinal number attribute of G, $\mathscr{F}$ and $\mathscr{G}$ are satisfiable by exactly the same properties.[7]

Cardinal number attribute

An attribute $\mathscr{N}$ is a *cardinal number attribute* (or cardinality attribute) iff it is possible to construct some property F such that $\mathscr{N}$ is a cardinal number attribute of property F.

The next step is to define the *natural number attributes*. Thus, I first define:

A *zero attribute* is any cardinal number attribute that is satisfied by '$x \neq x$'.

Then the following five definitions yield what is desired:

(1) *The immediate predecessor relation* P

$\mathscr{M}\,\mathbf{P}\mathscr{N}$ iff it is possible to construct a property F which is such that some object x satisfies it & $\mathscr{N}$ is a cardinality attribute of F & it is possible to construct a property G such that $\mathscr{M}$ is a cardinality attribute of G and G is satisfiable by all (and only those) objects different from x which satisfy F.

(2) P-*hereditary qualities*

A quality $\mathbf{Q}$ is **P**-hereditary iff, for all cardinality attributes $\mathscr{M}$, $\mathscr{N}$, if $\mathscr{M}$ satisfies $\mathbf{Q}$ and $\mathscr{M}\mathbf{P}\mathscr{N}$, then $\mathscr{N}$ satisfies $\mathbf{Q}$.

(3) P-*descendants*

A cardinality attribute $\mathscr{N}$ is a **P**-descendant of cardinality attribute $\mathscr{M}$ iff $\mathscr{N}$ satisfies every **P**-hereditary quality that $\mathscr{M}$ satisfies.

(4) *Natural number attributes*

A natural number attribute is any **P**-*descendant of a zero attribute*.

Comment: the above definition of the natural number attributes takes as its model not Frege's definition of 'natural number' given in *The Foundations of Arithmetic* (Frege, 1959), but rather Russell's definition (see, for example,

---

[7] There is a great deal of discussion in the recent Fregean literature of this principle. For an enlightening overview of these discussions, see William Demopoulos's introduction to Demopoulos, 1995. There is the question as to whether Frege thought that arithmetic might be founded on Hume's Principle instead of his Axiom V. See, in this regard, Heck, 1995 and Heck, 1997. For criticisms of this view, see Nomoto, 2000. It should also be noted that Michael Dummett has strenuously argued against the use of the term 'Hume's Principle' to apply to principles of the sort given above. In Dummett, 1998, he writes: "The term 'Hume's Principle' is productive of intellectual as well as historical confusion, and its use should be resolutely avoided" (387).

chapter 3 of *Introduction to Mathematical Philosophy* (Russell, 1920)). The principal difference is in the definition of 'P-descendant'. Russell's definition of the ϕ-ancestral relation for parameter ϕ (the converse relation of the ϕ-descendant relation) implies that every object x is a ϕ-ancestor of itself—something that Frege wished to avoid since his aim was to capture the intuitive idea of "following in the ϕ-series". Frege did not want his analysis to require that every object follow itself in such a ϕ-series, so his definition of the ϕ-ancestral has an extra clause in the antecedent that is not found in Russell's definition.[8] Russell chose an analysans that gave a less accurate fit with the analysandum in order to have a simpler definition of 'natural number'. The difference is only minor, but I believe that it results in a more easily grasped definition.

Specific natural number attributes

We cannot define the specific natural number attributes in the way Frege defined the specific natural numbers. Recall how Frege proceeded. From the point of view of set theory, Frege proceeded essentially as follows:

For any set α, $C\alpha$ is to be the set of all sets equinumerous with α.[9] Then,

$$1 = C\{x \mid x = 0\}$$
$$2 = C\{x \mid x = 0 \ \text{v} \ x = 1\}$$
$$3 = C\{x \mid x = 0 \ \text{v} \ x = 1 \ \text{v} \ x = 2\}$$

etc.

We cannot take this route since we don't have any objects to function as the "numbers" in such a definition. So this is how I define the specific natural numbers:

A natural number attribute is:

a *one attribute* iff it is possible to construct a zero attribute which immediately precedes it;

a *two attribute* iff it is possible to construct a one attribute which immediately precedes it;

a *three attribute* iff it is possible to construct a two attribute which immediately precedes it,

etc.

---

[8] See Frege, 1959: 92 for Frege's definition.

[9] For purposes of simplicity of exposition, I overlook the fact that in most standard set theories, there is no such set as $C\alpha$. In simple type theory, there is such a set of the appropriate type.

Thus, instead of defining specific natural numbers as Frege did, here we define what attributes would be specific natural number attributes.

Compare how simple cardinality statements would be analyzed in the present system in contrast to the way such statements were analyzed by Frege. The statement

> The number of moons of the earth is one

was analyzed by Frege to be:

> The cardinal number of the concept *is a moon of the earth* = one

whereas, in the present system, we get:

> It is possible to construct a one attribute which 'x is a moon of the earth' satisfies

Addition

What we define here are the "addition attributes": suppose that $\mathcal{M}$ and $\mathcal{N}$ are natural number attributes and that it is possible to construct properties F and G such that F satisfies $\mathcal{M}$, G satisfies $\mathcal{N}$, and no object satisfies both F and G. Then an $\mathcal{M} + \mathcal{N}$ attribute is any attribute satisfied by all and only those properties satisfied by any object that satisfies either F or G.

Contrast the above definition of addition with the usual recursive definition found in typical first-order developments of Peano arithmetic:

> For any natural number x, $x + 0 = x$;
>
> For all natural numbers x and y, $x + y' = (x + y)'$

(where $x' = $ the successor of x).[10]

The above recursive definition is perfectly adequate in the context of attempting to study the mathematical features of a type of structure, since it yields all the usual theorems regarding addition. But this sort of definition tells us nothing about how addition is connected to our use of cardinality theory in common everyday situations in which addition plays a role. It does not aid us in assessing the often-heard philosophical view (to be discussed in Chapter 9) that seven plus five equals twelve is refuted by the fact that seven gallons of alcohol mixed with five gallons of water does not yield twelve gallons of mixture. The above definitions are directed at quite another project: picking out the type of structure with which arithmetic deals. And it is

---

[10] See Russell, 1920: 6 for a brief discussion of this definition in the context of characterizing the Peano system.

perfectly adequate for giving us all the usual theorems regarding addition. But it is not the sort of definition that would be satisfactory for the project I have in mind, which is to gain insights into the way arithmetic is applied in science and everyday life.

Analogues of the above recursive definition of addition are among the following theorems, which can easily be proved:

(1) If $\mathcal{M}$ and $\mathcal{N}$ are natural number attributes, then it is possible to construct an $\mathcal{M} + \mathcal{N}$ attribute and every such attribute is a natural number attribute.

(2) If $\mathcal{N}$ is a natural number attribute and $\mathcal{M}$ is a zero attribute, then every $\mathcal{N} + \mathcal{M}$ attribute is an $\mathcal{N}$ attribute.

(3) If $\mathcal{N}$ is a natural number attribute, and $\mathcal{M}$ is a one attribute, then $\mathcal{N} \mathbf{P}(\mathcal{N} + \mathcal{M})$.

(4) If $\mathcal{M}$ and $\mathcal{N}$ are natural number attributes, $\mathcal{O}$ is an $\mathcal{M} + \mathcal{N}$ attribute, $\mathcal{T}$ is a one attribute, $\mathcal{S}$ is an $\mathcal{N} + \mathcal{T}$ attribute, and $\mathcal{F}$ is an $\mathcal{M} + \mathcal{S}$ attribute, then $\mathcal{O} \mathbf{P} \mathcal{F}$.

Specific sums

Specific sums corresponding to such theorems of arithmetic as '$7 + 5 = 12$' can be easily proved. Thus, it can be proved that:

If $\mathcal{M}$ is a seven attribute, $\mathcal{N}$ is a five attribute, $\mathcal{O}$ is a twelve attribute, then any $\mathcal{M} + \mathcal{N}$ attribute is an $\mathcal{O}$ attribute.

I shall not go on to develop the usual arithmetical theory of the "multiplication attributes", since this can be done in a straightforward way. Let us turn instead to a constructibility version of the theory of natural numbers, which I shall call 'Peano's theory', whose axioms consist of constructibility versions of the following.

Peano's axioms[11]

(P1) Zero is a natural number.

(P2) The successor of any natural number is a natural number.

(P3) If the successors of two natural numbers are equal, then the numbers themselves are equal.

---

[11] When Peano introduced his axioms in 1889, he included among them identity axioms. It is common today to give only the five axioms that concern the natural numbers. See in this regard Gillies, 1982: 67.

(P4) Zero is not the successor of any natural number.

(P5) The Principle of Mathematical Induction.

A constructibility version of (P1) can be given as:

(P1*) Any zero attribute is a natural number attribute.

Given the above definitions of the relevant terms, this assertion is obviously true. So let us turn to the following constructibility version of (P2):

(P2*) If $\mathcal{M}$ is a natural number attribute and $\mathcal{M}$**P**$\mathcal{N}$, then $\mathcal{N}$ is a natural number attribute.

To see that this assertion is true, notice that, since $\mathcal{M}$ is a natural number attribute, it must satisfy every **P**-hereditary quality that a zero attribute satisfies. Since $\mathcal{M}$**P**$\mathcal{N}$, $\mathcal{N}$ must also satisfy every **P**-hereditary quality that a zero attribute satisfies.

(P3*) If $\mathcal{M}$**P**$\mathcal{U}$, $\mathcal{N}$**P**$\mathcal{V}$, and, further, any property satisfies $\mathcal{U}$ iff it satisfies $\mathcal{V}$, then any property satisfies $\mathcal{M}$ iff it satisfies $\mathcal{N}$.

To prove this, I assume the following:

(1) $\mathcal{M}$**P**$\mathcal{U}$ and $\mathcal{N}$**P**$\mathcal{V}$.

(2) Any property satisfies $\mathcal{U}$ iff it satisfies $\mathcal{V}$.

It follows from (1) that

(3) it is possible to construct a property F & there is an object x such that x satisfies F & $\mathcal{U}$ is a cardinality attribute of F & it is possible to construct a property G such that $\mathcal{M}$ is a cardinality attribute of G and G is satisfied by all (and only those) objects different from x which satisfy F

and

(4) it is possible to construct a property F & there is an object x such that x satisfies F & $\mathcal{V}$ is a cardinality attribute of F & it is possible to construct a property G such that $\mathcal{N}$ is a cardinality attribute of G and G is satisfied by all (and only those) objects different from x which satisfy F.

Since it is possible to construct a property F with the features described in (3), let us suppose that:

(5) property F^* has been constructed which is such that there is an object x such that x satisfies F^* & $\mathcal{U}$ is a cardinality attribute of F^* & it is possible

to construct a property G such that $\mathcal{M}$ is a cardinality attribute of G and G is satisfied by all (and only those) objects different from x which satisfy F^*.

We can assume that:

(6) x^* is an object such that x^* satisfies F^* & $\mathcal{U}$ is a cardinality attribute of F^* & it is possible to construct a property G such that $\mathcal{M}$ is a cardinality attribute of G and G is satisfied by all (and only those) objects different from x^* which satisfy F^*.

Since we know that it is possible to construct a property G that has the features described in (6), let us assume that:

(7) property G^* has been constructed which is such that $\mathcal{M}$ is a cardinality attribute of G^* and G^* is satisfied by all (and only those) objects different from x^* which satisfy F^*.

Similarly, applying the same three steps to (4) *mutatis mutandis*, we obtain:

(8) property $G^{\#}$ has been constructed which is such that $\mathcal{N}$ is a cardinality attribute of $G^{\#}$ and $G^{\#}$ is satisfied by all (and only those) objects different from $x^{\#}$ which satisfy $F^{\#}$.

From (2) and (5), we can infer that:

(9) $\mathcal{V}$ is a cardinality attribute of F^* and that F^* is equinumerous with $F^{\#}$.

From (7), (8), and (9), we can infer that:

(10) G^* is equinumerous with $G^{\#}$.

Hence,

(11) any property that would satisfy $\mathcal{M}$ would also satisfy $\mathcal{N}$ and conversely.

From this, we can infer (P3*).

(P4*) It is not possible to construct an attribute $\mathcal{M}$ and a zero attribute $\mathcal{N}$ which is such that $\mathcal{M}\mathbf{P}\mathcal{N}$.

A proof can take the form of a *reductio ad absurdum* as follows:

Suppose that attributes $\mathcal{M}$ and $\mathcal{N}$ have been constructed such that:

(1) $\mathcal{N}$ is a zero attribute and $\mathcal{M}\mathbf{P}\mathcal{N}$.

Then, we can infer:

(2) it is possible to construct a property F & there is an object x such that x satisfies F & $\mathcal{N}$ is a cardinality attribute of F & it is possible to construct a

property G such that $\mathcal{M}$ is a cardinality attribute of G and G is satisfied by all (and only those) objects different from x which satisfy F.

But it is absurd to suppose that it is possible to construct a property F which is such that, for some object x, x satisfies F & $\mathcal{N}$ is a cardinality attribute of F.

(P5*) is the following constructibility version of the Principle of Mathematical Induction:

If

(i) **Q** is a quality satisfied by some zero attribute,

and

(ii) for all cardinality attributes $\mathcal{M}$, $\mathcal{N}$ that it is possible to construct, if $\mathcal{M}$ satisfied **Q** and $\mathcal{M}$**P**$\mathcal{N}$, then $\mathcal{N}$ would also satisfy **Q**,

then

every natural number attribute would satisfy **Q**

Proof: Suppose that $\mathscr{C}$ is a natural number attribute. Then $\mathscr{C}$ must satisfy every quality that is **P**-hereditary and that is satisfied by a zero attribute. Since from (ii) we can infer that **Q** must be **P**-hereditary, we can conclude that $\mathscr{C}$ must satisfy **Q**.

Instead of continuing an exposition of the Constructibility Theory, I shall now take up the objections to my Constructibility Theory advanced by Shapiro and Resnik in their respective books on structuralism.

4. Shapiro's Objections to the Constructibility Theory

Recall that Shapiro's grounds for adopting his *ante rem* structuralism were twofold: on the one hand, he maintained that his realistic version of structuralism was the simplest and most straightforward ontological position to take. On the other hand, he raised objections to his nominalistic rivals, claiming among other things that the nominalistic views were, in some sense, *equivalent* to the realistic positions, and hence did not have any epistemological advantages. In this section, I shall evaluate Shapiro's objections to his nominalistic rivals, concentrating for the most part on his objections to my Constructibility Theory.

Let us consider an objection to my views that Shapiro first raised in his paper "Modality and Ontology" and that he repeated with some minor changes in his book on mathematical structuralism.[12] Shapiro begins his objection by providing his readers with a way of translating the sentences of the Constructibility Theory into the sentences of a set-theoretical version of simple type theory.[13] His instructions are: replace all the variables of the Constructibility Theory that range over level n open-sentences with variables of simple type theory that range over type n sets ($n = 1, 2, 3, \ldots$); then replace the symbol for satisfaction with the symbol for membership; and replace the constructibility quantifiers with ordinary existential quantifiers. (Shapiro, 1997: 231). He sums up as follows:

[T]o translate from Chihara's language to the standard one, just undo Chihara's own translation. *The two systems are definitionally equivalent.* From this perspective, Chihara's system is a notational variant of simple type theory. An advocate of one of the systems cannot claim an epistemological advantage over an advocate of the other. (Shapiro, 1997: 231, italics mine)

In the article, Shapiro is even more blunt and pointed in stating his objection, writing:

From this perspective, Chihara's system is a notational variant of simple type theory (or non-cumulative set theory, with finite ranks). To be crude: here, it seems, *ontology is reduced by envisioning that we have changed the shape of a few symbols of the regimented language.* (Shapiro, 1993: 468, italics mine)

My Constructibility Theory is thus made to appear as if were an utterly trivial rewriting of the set-theoretical version of simple type theory—a mathematical innovation which has no more value than a new printing (in a slightly different font) of an old book developing simple type theory. These are very serious charges that Shapiro has leveled against my views—charges which, if sustained, would be a devastating refutation of my philosophical position.

But before attempting any sort of evaluation of Shapiro's charges, we should investigate in more depth the logic of Shapiro's reasoning. What grounds does he have for concluding that my system is a mere notional variant of simple type theory? To see the underlying rationale for his inferences, we need to see what his general strategy is for rejecting all the anti-realist views that he takes up in his works.[14] Basically, his overall goal is to show that

---

[12] The paper cited is Shapiro, 1993. The book is Shapiro, 1997.

[13] The version of simple type theory Shapiro has in mind is, I believe, essentially the set theory described by Quine in Quine, 1963: sec. 36.

[14] The other anti-realists attacked in this way by Shapiro are Hartry Field, Geoffrey Hellman, and George Boolos.

[t]he epistemological problems facing the anti-realist programmes are just as serious and troublesome as those facing realism. Moreover, the problems are, in a sense, *equivalent* to those of realism. (Shapiro, 1993: 456).

Here, then, is how Shapiro claims to establish these conclusions:

In each case, the structure of the argument is the same. I show that there are straightforward, often trivial, translations from the set-theoretical language of the realist to the proposed modal language, and vice-versa. The translations preserve warranted belief, at least, and probably truth ... The contention is that, because of these translations, neither system can claim a major epistemological advantage over the other. (Shapiro, 1993: 457).

Notice that, according to this scenario, one must come up with a translation scheme *tr* that preserves warranted belief; that is, for any sentence ϕ of the anti-realist language, belief in ϕ is warranted iff belief in $tr(\phi)$ is warranted.

In the quote from Shapiro's book, it was claimed that the two theories in question are "definitionally equivalent". What is definitional equivalence and how is this condition relevant to the question as to whether the Constructibility Theory is a mere notational variant of simple type theory? Here's the definition of the technical term 'definitional equivalence' that Shapiro gives in his article:

Two theories T, T' are said to be *definitionally equivalent* if there is a function f_1 from the class of sentences of T into the class of sentences of T' and a function f_2 from the class of sentences of T' into the class of sentences of T, such that (1) f_1 and f_2 both preserve truth (or theoremhood if the theories do not have intended interpretations) and (2) for any sentence ϕ of T, $f_2 f_1(\phi)$ is equivalent in T to ϕ, and for any sentence ψ of T', $f_1 f_2(\psi)$ is equivalent in T' to ψ. (Shapiro, 1993: 479)

He then proposes to use definitional equivalence "as an indication that the intended structures, and thus the ontologies, of different theories are the same" (Shapiro, 1993: 479).

Before subjecting Shapiro's reasoning to critical examination, it should be noted that some logicians have questioned the appropriateness of Shapiro's conditions for *definitional equivalence*.[15] In order to avoid a lengthy investigation of the concept of definitional equivalence here, I shall regard Shapiro

---

[15] This was done by Albert Visser at my presentation of an earlier version of this material at the *One Hundred Years of Russell's Paradox* conference help in Munich, June 2001. And others there seemed to agree.

as simply putting forward a stipulative definition of the term 'Shapiro equivalence', with no suggestion that he is attempting to capture some pre-existing notion of definitional equivalence. Thus, we can take Shapiro to be claiming that if two theories A and B are shown to be Shapiro equivalent, then we have an "indication" that A and B have one and the same ontology.

Surprisingly, Shapiro claims to have "shown" that the Constructibility Theory, under certain (plausible) conditions, "is [Shapiro equivalent] to a standard "realist" theory", and he concludes that "the intended structure—and the ontology—of [the Constructibility Theory] is the same as that of the corresponding realist theory, and is not to be preferred on ontological grounds" (Shapiro, 1993: 479). I say 'surprisingly', because nowhere either in his article or in his book is there anything like a proof of this Shapiro equivalence. As I noted earlier, he does provide a translation scheme for translating each sentence of the Constructibility Theory into a sentence of simple type theory; and he supplies us with a method of translating back again. But he must do more than that to establish Shapiro equivalence. To justify this conclusion, one must show that the method of translation given in the earlier quotation preserves truth (or at least theoremhood). Not only has Shapiro failed to provide a proof of this crucial element, he has not even given a plausibility argument supporting the belief that truth or theoremhood is preserved by the translation procedures. What has happened here? Has Shapiro simply forgotten that this extra condition needs to be fulfilled? That is hard to believe.

Recall that Shapiro also claimed that his method of translating from the language of the Constructibility Theory to the language of simple type theory preserves warranted assertability. Has he provided a proof of this claim? Again, no such proof is to be found in either Shapiro's article or his book.

Why is Shapiro so sure that the Constructibility Theory is a mere notational variant of simple type theory? Since there is nothing remotely like a proof of this doctrine, I can only wonder at his confidence at affirming such a position. Perhaps he believes that if he can prove that the Constructibility Theory is Shapiro equivalent to simple type theory, then that would be tantamount to a proof that the Constructibility Theory is a notational variant of simple type theory. And perhaps he believes that he has good grounds for thinking the Constructibility Theory is Shapiro equivalent to simple type theory.

If theory A is Shapiro equivalent to theory B, does it follow that A and B are notational variants of one another? Not at all. Suppose theory A is a first-order theory about Mr Jones's cats and theory B is a first-order theory about

Mr Smith's dogs. And suppose that A and B are Shapiro equivalent (the domain of A consists of the cats of Jones, and the domain of B consists of the dogs of Smith; what A says about the number of males Jones has, B says about the number of males Smith has—and it just happens to be the case that both theories are true). Theory A is no mere notational variant of theory B, for B is a theory about dogs, not cats. One could not explain the barking emanating from Smith's house by appealing to theory A.[16]

I suggest that we examine Bertrand Russell's Theory of Types, as it was developed in *Principia Mathematica*,[17] for some insights into the plausibility of Shapiro's claims.

Russell's theory of types

Principia Mathematica was the culmination of over thirteen years of very intensive research by Russell and his colleague Alfred North Whitehead. It was supposed to provide the world with Russell's solution not only to the paradox that bears his name, but also to a whole class of similar "vicious-circle" paradoxes containing such antinomies as the Cantor and the Burali-Forti.[18] The authors claimed that "the theory of types, as set forth in [*Principia Mathematica*], leads both to the avoidance of contradictions, and to the detection of the precise fallacy which has given rise to them" (Russell and Whitehead, 1927: vol. 1, p. 1). It was hoped that *Principia Mathematica* would provide mathematicians with a solid paradox-free foundation for all their theories by showing how classical mathematics could be derived within its formalized logical system.

The Constructibility Theory is similar to Russell's theory of types in two important respects: (1) it is a "no-class" theory just as Russell's theory is; and (2) the mathematical developments in its formal theory are carried out in a kind of simple type theory, in the way that the mathematical part of Russell's formal theory is.[19] So it is not surprising that much of what Shapiro says about the Constructibility Theory can be also said of Russell's theory of types. Here is how this might be done.

---

[16] David Kaplan suggested an example such as the above at the *One Hundred Years of Russell's Paradox* conference.   [17] Russell and Whitehead, 1927.

[18] See Chihara, 1973: ch. 1 for more details.

[19] Of course, there are significant differences. For one thing, there is not even an apparent reference to sets in the Constructibility Theory. Talk instead is of the constructibility of open-sentences that function like sets.

As is well known, set theory is developed in *Principia Mathematica* even though no sets are supposed to be within the range of any of its quantifiers. Mathematics is developed within *Principia Mathematica* in a theory that is only about propositional functions and individuals—these are the only kinds of things that are talked about in the theory and this is why it is called a "no-class" theory. However, there are abbreviating definitions by means of which certain sentences of the theory get transformed into sentences that look like sentences of set theory.[20] Mathematics is developed in *Principia Mathematica* essentially using only those sentences that, by means of these abbreviations, look and function like sentences of set theory. As Quine has remarked: "Once classes have been introduced, [propositional functions] are scarcely mentioned again in the course of the three volumes" (Quine, 1966d: 22).

Now the actual type theory of *Principia Mathematica* is that horror of students: the very complicated ramified theory of types.[21] However, once the Axiom of Reducibility is accepted, the mathematical part of the system—the set-theoretical part—gets simplified, so that it operates very much like the simple theory of types.[22] Indeed, we can distinguish a sort of sub-theory of *Principia Mathematica*, which formalizes the mathematics of this theory. For purposes of reference, let us call this theory 'RST' (for 'Russell's simple theory of types'). RST has as its sentences the sentences of *Principia Mathematica* that get transformed by its abbreviating definitions into straightforward sentences of the standard set-theoretical version of simple type theory (henceforth to be called 'ST'). We have in effect, then, a translation function f_1 that takes each sentence of RST into a sentence of ST, and it would be a simple exercise to come up with a translation function f_2 that would take each sentence of ST into a sentence of RST.[23] The assertions of RST are just those sentences of the theory that are theorems of *Principia Mathematica*. I suggest we investigate whether RST is a mere notational variant of ST.

The fact that one can translate from RST to ST and back again is no guarantee that the two are notational variants of one another. If RST is just a notational variant of ST, then each sentence of RST should have the same

---

[20] See Chihara, 1973: ch. 1, sect. 4.

[21] See Chihara, 1973: ch. 1 for a characterization of Russell's ramified theory of types.

[22] How this comes about is explained in Chihara, 1973: ch. 1, sect. 9.

[23] For purposes of simplicity of exposition, I have skipped over the fact that the rules of translation cannot be simply read off the abbreviating definitions of *Principia*, for as Gödel has noted (in Godel, 1964a: 212), the authors of that logical work were sloppy about the syntax of their system and a certain amount of work would be required to make the translation rules logically adequate.

meaning as its set-theoretical correlate. After all, it is only the notation that is supposed to be different. So consider the following sentence of RST:[24]

[1] $(\exists F)((x)(F!x \leftrightarrow x = x)$ & $(\exists y)F!y)$

This sentence translates into the set-theoretical statement:

[2] $(\exists y)y \in V_1$

where V_1 is the universal set of individuals. But it is highly implausible to suppose that the former has the same meaning as the latter, for the "no-class" sentence logically implies a whole host of other sentences of *Principia Mathematica* (many of which are not sentences of ST) that are not implied by the corresponding set-theoretical sentence. For example, the former sentence of RST logically implies:

[3] $(\exists F)(x)(F!x \leftrightarrow x = x)$

which is not implied by [2]. The set-theoretical sentence says that the universal set of type 1 has a member, and how can that sentence imply that there is a predicative propositional function of order 1 that is satisfied by every individual that is identical to itself?

Thus, the mere fact that one can translate the sentences of one theory into the sentences of another so as to preserve certain mathematically relevant relationships between the sentences does not give one the right to say that the two theories are mere notational variants of one another. Much more needs to be shown.

Consider, then, the question as to whether the above two theories are Shapiro equivalent. To be able to answer this question in the affirmative, one would have to have convincing grounds for thinking that the translation functions preserve at the very least theoremhood (since it is questionable that *Principia Mathematica* has a standard interpretation).[25] So is the translation of

---

[24] For purposes of perspicuity and simplicity of exposition, I have used a simplified and more contemporary notation. The diligent student can reconstruct an accurate version. The exclamation mark in the above formula [1] is used to indicated that the variable 'F' ranges over *predicative* propositional functions of order 1. For details, see Chihara, 1973: 21.

[25] In his 1918 lectures (Russell, 1956), Russell described the language of *Principia Mathematica* as "a language which has only syntax and no vocabulary whatsoever. ... [it] aims at being that sort of language that, if you add a vocabulary, would be a logically perfect language" (198).

every theorem of ST a theorem of RST? Take the Axiom of Infinity and the Axiom of Choice (which Russell called "the multiplicative axiom"). These are fundamental axioms of simple type theory. Their status is no different from all the other axioms. But the corresponding statements of *Principia Mathematica* are treated very differently. The authors do not attempt to justify the acceptance of these statements, as they do for the case of the Axiom of Reducibility.[26] Instead, they assert that the Axiom of Infinity, like the Axiom of Choice, "is an arithmetic hypothesis which some will consider self-evident, but which we prefer to keep as a hypothesis, and to adduce in that form whenever it is relevant".[27] And in a later work, Russell wrote: "As to the truth or falsity of the axiom [of choice] in any of its forms, nothing is known at present" (Russell, 1920: 124). Thus, the Axiom of Infinity and the Axiom of Choice are not treated as genuine axioms of the theory. Not surprisingly, we find the authors of *Principia Mathematica* asserting such things as "if the Axiom of Infinity is true", such and such follows.

Making such conditional assertions is perfectly consistent with the authors' practice of proving theorems having these "axioms" as antecedents. Thus, when Russell and Whitehead encountered a theorem of standard mathematics requiring the Axiom of Choice for its proof—say Zermelo's Well Ordering Theorem—what they proved in *Principia Mathematica* was a conditional sentence of the form

$$AC^* \rightarrow WO^*$$

where AC^* is the sentence of RST that is abbreviated by the Axiom of Choice of ST and WO^* is the sentence of RST that is abbreviated by the Well Ordering Theorem of ST.[28] In short, the Well Ordering Theorem is not really a theorem of RST.[29] Of course, the Well Ordering Theorem is a theorem of ST, so it can be seen that theoremhood is not preserved by the translation scheme described above and that the two theories are not Shapiro equivalent after all.

---

[26] In a section entitled "Reasons for Accepting the Axiom of Reducibility", they claim that "the inductive evidence in its favor is very strong" (Russell and Whitehead, 1927: vol. I, p. 59). No such reasons are produced in favor of accepting the other two axioms.

[27] Russell and Whitehead, 1927: vol. II, p. 203. In Russell, 1920: 117, Russell describes the axiom as "convenient, in the sense that many interesting propositions, which it seems natural to suppose true, cannot be proved without its help; but it is not indispensable, because even without those propositions the subjects in which they occur still exist, though in a somewhat mutilated form". Later in this work, Russell writes: "It is conceivable that the multiplicative axiom in its general form may be shown to be false" (130).

[28] Russell and Whitehead, 1927: vol. I, pp. 536–42. See also Carnap, 1964: 34–5.

[29] It would be grotesque to claim to have proved Fermat's Last Theorem (FLT) in first-order Peano arithmetic on the basis of a proof of the theorem: FLT $\rightarrow$ FLT.

Is the Constructibility Theory Shapiro equivalent to simple type theory?

Consider now Shapiro's claim that my Constructibility Theory is Shapiro equivalent to simple type theory. As I said earlier, the Constructibility Theory is much closer to *Principia Mathematica* than it is to the set-theoretical version of simple type theory. Like *Principia Mathematica*, the Constructibility Theory has no axiom that corresponds to the Axiom of Choice. So we have some reason for thinking that the Constructibility Theory is not Shapiro equivalent to simple type theory, just as we had strong grounds for concluding that RST is not Shapiro equivalent to ST. Furthermore, if the Constructibility Theory is not Shapiro equivalent to simple type theory, then it certainly cannot be a mere notational variant of that theory.

It should now be obvious that the case Shapiro has assembled in support of his claim that the Constructibility Theory is a mere notational variant of simple type theory is very weak indeed. All that he has done is to provide us with a method for translating sentences of the Constructibility Theory into sentences of simple type theory and a method for translating sentences of simple type theory into sentences of the Constructibility Theory, these translations preserving certain mathematically significant relationships. As the above discussion of Russell's Theory of Types shows, there is much more that must be established in order to conclude that the two theories are mere notational variants of one another. We can have such methods of translation, even when a sentence in one theory is quite different in meaning and logical significance from the sentence of the other theory into which it is translated.

I shall now provide additional grounds for concluding that the Constructibility Theory is no mere notational variant of the standard set-theoretical version of simple type theory.

The mystery of infinitely many different empty sets

Let us begin our evaluation of Shapiro's objection by comparing the Constructibility Theory with the simple theory of types regarding the existence of null sets. Some of you may wonder at my use of the plural of the word 'set' in the previous sentence. Is there not at most one null set? Can there be more than one? But in simple type theory, there is a null set of type 1, a null set of type 2, a null set of type 3, ... Quine has expressed his distaste for this feature of the set theory, writing:

One especially unnatural and awkward effect of the type theory is the infinite reduplication of each logically definable class. ... This reduplication is particularly strange in the case of the null class. One feels that classes should differ only with respect to their members, and this is obviously not true of the various null classes. (Quine, 1938: 131)

What distinguishes the null set of type 1 from the null set of type 2? They have exactly the same members, namely, none at all. So how can they be different?

Now the Constructibility Theory is not subject to this paradoxical feature of simple type theory. What corresponds in the Constructibility Theory to the statement of simple type theory, 'There is a null set of type 1', is the statement, 'It is possible to construct an open-sentence of level 1 that is true of no objects'. The latter statement can be seen to be true, since one can construct, for example, the open-sentence

$$x \neq x$$

which is an open-sentence of level 1 that is true of no object. The statement of the Constructibility Theory that corresponds to the set theoretical statement that there is a null set of type 2 is the statement that it is possible to construct an open-sentence of level 2 that is not true of any open-sentence of level 1. This statement can be seen to be true, since one can construct, for example, the level 2 open-sentence

> F is a level 1 open-sentence that is satisfied by an object that does not satisfy F.

There is no mystery about how there can be such distinct open-sentences. The open-sentence of level 1 constructed above is easily distinguished from the open-sentence of level 2 that was constructed. The mystery of infinitely many distinct null sets simply does not arise for the Constructibility Theory.[30] The fact that the Constructibility Theory does not give rise to a mystery that bedevils the simple theory of types is evidence that these two mathematical theories are not mere notational variants of each other.

Differences in how the two theories are confirmed

Another reason for questioning Shapiro's thesis that the Constructibility Theory is a notational variant of simple type theory can be found by comparing the grounds we have that support the truth of the two theories. If the two theories are mere notational variants of one another, then any

---

[30] Cf. Chihara, 1980: 27.

confirmation of one would automatically be a confirmation of the other. Now one consequence of the axioms of the Constructibility Theory is this:

(4) It is possible to construct an open-sentence of level 1 that is not satisfied by any object.

By Shapiro's translation scheme, the corresponding sentence of simple type theory is the sentence:

(5) There exists a set of type 1 of which nothing is a member.

If the two theories in question are mere notational variants of one another, then the grounds supporting the one belief should be the same as the grounds supporting the other. But among the grounds I have for my belief in the former is the fact that I have literally constructed an open-sentence of the kind in question. By a law of modal logic ("actuality implies possibility"), this implies the truth of (4). But that law of modal logic does not allow me to infer (5) from my construction: the fact that I have constructed some open-sentence does not imply, in modal logic, that the null set of type 1 exists. Thus, we get a kind of direct verification of the constructibility statement that one cannot get from the set-theoretical case. After all, how does one justify the belief that there exists a null set of type 1? If one justifies the belief at all, it will be on very general grounds—no doubt involving Quinean considerations about what is our best scientific theory of the world. Nothing like that is needed in the open-sentence case.

Of course, if the two theories in question were indeed mere notational variants of one another, then one would expect all the justifications of the axioms of one theory to be translatable into justifications of the translated versions of that axiom. But it is hard to see how the kind of modal justifications of the axioms of the Constructibility Theory I give in my book can be translated into justifications of the axioms of the set-theoretical version of simple type theory. Thus, consider the Abstraction Axiom of the lowest level. It tells us that for any condition on x expressible in the language of **Ct**,

$$- - - x - - -$$

it is possible to construct a level 1 open-sentence that is satisfied by all and only those objects that satisfy the condition.[31] But any condition on x expressible in the language of **Ct** is an open-sentence of the sort required that

---

[31] For purposes of simplicity here, I am considering only the case in which the condition in question contains no occurrences of any free variable other than x.

it is possible to construct. So it obviously is possible to construct an open-sentence that is satisfied by all and only those objects that satisfy the condition. But this modal justification is not directly translatable (certainly not by the simple method of translation that Shapiro gives) into a set-theoretical justification of (this case of) the set-theoretical version of the Abstraction Axiom.

I conclude that Shapiro's thesis that the two theories in question are mere notational variants of one another is not only unfounded, but its credibility is overwhelmingly undermined by the considerations cited above.

5. OTHER OBJECTIONS SHAPIRO HAS RAISED

Shapiro's objection to my use of possible worlds semantics

As I noted in the beginning section of this chapter, the kind of possibility with which I am concerned in the Constructibility Theory is that of the "conceptual" or "broadly logical" sort, the system of derivations for which is widely believed to be formalized in S5 modal logic. The semantics of an S5 system is generally given in terms of possible worlds. The overarching idea behind such a semantical system is this. Imagine that there is a huge totality of universes (or "possible worlds") such that anything that could happen does happen in at least one of these universes. Then, to say that $\diamond\phi$ (it is possible that ϕ) is to say that there is a possible world w such that ϕ is true in w. To say that $\square\phi$ (it is necessary that ϕ) is to say that, for every possible world w, ϕ is true in w. Using these ideas, one can devise model-theoretical definitions in terms of which the fundamental logical notions of validity, semantical consistency, soundness, and completeness can be defined. One can define, within this framework, what an "interpretation" of the language is, the idea being that the interpretation will supply:

  a non-empty set **W** to represent the set of possible worlds;

  a member **a** of **W** to represent the actual world;

  a function U to assign to each member **w** of **W** a set of elements to represent "the universe of **w**" (the set of things that exist in **w**);

  a function to assign to every pair, whose first element is a unary predicate F of the language and whose second element is a member **w** of **W**, a subset of what U assigns to **w** to represent the extension of F in the world represented by **w**;

(and similarly for the case of binary, ternary, ... predicates).

One can then define what it is for a sentence of the language to be true under an interpretation, and in terms of this relative notion of truth, one can define the above-mentioned logical notions of validity, semantical consistency, soundness, and completeness. Standard versions of derivational systems of S5 quantificational modal logic can then be formulated and proved to be sound and complete.[32]

Consider now the constructibility quantifier. This functions as a kind of logical constant and is taken to be a primitive of my system in much the way 'is a member of' is a primitive of set theory. However, it was important to provide readers of *The Worlds of Possibility* (Chihara, 1998) with explanations of how this kind of quantifier functions and also of how inferences involving this primitive are to be made. Since the semantics of S5 type modal systems was already well known, I chose to explain the basic features of the constructibility quantifier in terms of the possible worlds idea, using the kind of set-theoretic semantical system that was already in use in modal logic. However, as Shapiro notes explicitly, I warned the reader that my possible worlds semantical explanations should not be taken to be literal descriptions of how constructibility quantifiers function. I said: "[T]his whole possible worlds structure is an elaborate myth, useful for clarifying and explaining the modal notions, but a myth just the same" (Chihara, 1990: 60).

It is on this point that Shapiro jumps in with an objection, writing:

If the structure really is a myth, then I do not see how it explains anything. One cannot, for example, cite a story about Zeus *to explain a perplexing feature of the natural world such as the weather.* ... In everyday life, a purported explanation must usually be true, or approximately true, in order to successfully explain. (Shapiro, 1997: 232–3, italics mine)

Here, Shapiro seems to be confusing two very different sorts of explanations:

(A) scientific explanations of natural phenomena,

and

(B) explanations of ideas or the meaning and use of expressions.

Now Shapiro can certainly make a plausible case that the former, type (A) explanations, must be true, or approximately true, in order to successfully explain some natural phenomenon, although, as Bas van Fraassen has emphasized, "There are many examples, taken from actual usage, which show

---

[32] See Chihara, 1998: ch. 1 to see how this is done.

that truth is not presupposed by the assertion that a theory explains something" (van Fraassen, 1980: 98), and I can think of cases in which one makes use of imaginary states of affairs in order to explain some complicated natural process. However, I shall not contest Shapiro's case as it applies to type (A) explanations, since the objection only goes through if it also applies to type (B) explanations. Instead, I shall argue that, first, my explanations of how the constructibility quantifiers function, using possible worlds semantical ideas, were type (B) explanations; second, that type (B) explanations are not precluded from using myths or imaginary things and states of affairs.

Clearly, my explanations of how the constructiblity quantifier functions was a type (B) explanation. I was introducing, in my logical system, a novel logical constant. So a type (B) explanation was called for. Besides, philosophers only rarely put forward scientific explanations of natural phenomena. So why should one suppose that I was offering a type (A) explanation?

Now if one is explaining or clarifying some idea or one's use of a term or phrase, what is to prevent one from making use of a myth or purely imaginary things and situations? Why could one not use a blatantly false theory in explaining an idea, if the logical features of the false theory are well understood? Let me give you an example. For many years, intuitionism was a very puzzling and confusing theory to many classical mathematicians: the intuitionistic restrictions of mathematical operations seemed arbitrary or simply unintelligible. Then a number of interpretations of intuitionistic ideas were given within the framework (and using the concepts) of classical mathematics. These interpretations made intuitionism intelligible to many classical mathematicians, even though the classical theory in terms of which intuitionism was being explained was, to many intuitionists, false. From the point of view of the intuitionist, a false theory that was well understood was being used to explain a true theory that was not understood at all. Similarly, I wanted to make the logic of my Constructibility Theory intelligible to mainstream philosophers and logicians, who were familiar with possible worlds semantics, even though I was convinced that all this talk of possible worlds was fictional.[33] By proceeding in the above way, I expected those familiar with modal logic to learn how to use my constructibility quantifiers.

Take the imaginary world Flatland of Edwin Abbott.[34] This is a world in which all things in it are portrayed as existing and moving about as flat

---

[33] See Chihara, 1998 for a detailed expression of my attitude towards possible worlds semantics.
[34] The reader can find a short description of Abbott's work, as well as some selections from *Flatland*, in Newman, 1956: 2383–96.

objects on a two-dimensional plane. The inhabitants of Flatland turn out to be such geometric objects as triangles. Trying to imagine how the inhabitants of Flatland would view a three-dimensional object, say a sphere, entering their world and passing through it has been thought to shed light on certain aspects of relativity theory.[35] Ian Stewart has suggested that the scientific purpose of the work was "serious and substantial": "Abbott's sights were focused not on the Third Dimension—familiar enough to his readers—but on the Fourth Dimension. Could a space of more than three dimensions exist?" (Stewart, 2001: vii).

Is one to react to Flatland by claiming, as Shapiro has, that no myth can be used to explain anything? Surely the philosophical literature is filled with examples in which imaginary situations and worlds are used to explain various ideas and conceptual possibilities. Anyone who has worked in modal logic and the philosophy of necessity can come up with countless examples of the use of such imaginary situations and states of affairs to explain some principle, concept, or doctrine.[36] And surely those who have studied Wittgenstein's writing can cite abundant examples in which imaginary societies or tribes (i.e., "myths") are effectively used to explain the author's ideas or his use of some term.[37] Indeed, such devices are so common in the writings of analytic philosophy that few contemporary philosophers, I dare say, would accept Shapiro's thesis on explanations.

The constructibility quantifiers

Closely related to the "notational variant" objection I discussed earlier is another that Shapiro raises against my constructibility quantifiers. By reasoning that brings to mind this "notational variant" objection, Shapiro argues that my constructibility quantifiers are no different in logical meaning from standard existential quantifiers. More specifically, Shapiro's contention is that "the constructibility quantifier has virtually the same semantics as the ordinary existential quantifier." Here is how Shapiro reasoned:

As he [Chihara] describes the system, in a given model, the variable ranges are dutifully distributed into different possible worlds, but this fact plays *no role* in the definition of satisfaction in the modal system he develops. Every object (and every predicate) is rigid and world-bound, and each constructibility quantifier ranges over all objects (of appropriate type) in all worlds. Thus, the worlds do not get used

---

[35] This was proposed in an anonymous letter published in the 12 February 1920 issue of *Nature*.
[36] Even a cursory perusal of my book (Chihara, 1998) will yield numerous examples of the sort being described.   [37] See, for example, Wittgenstein, 1958 and Wittgenstein, 1953.

anywhere—just the objects in them. In short, the (mythical) semantics that Chihara develops is just ordinary *model theory*, with some irrelevant structure thrown in. ... Thus, my contention is that the constructibility quantifier has virtually the same semantics as the ordinary existential quantifier. (Shapiro, 1997: 233).

As in the previous cases, this is a very harsh criticism. But is it accurate? First of all, is it true that all the objects in the modal system are "world-bound"? Where did Shapiro get that idea? I make it clear that an object, thing, or person in one world can appear in many possible worlds. For example, I write: "If, in one possible world, someone makes certain speech sounds, and if in another possible world, *this same person* makes the very same sounds, we are not forced to conclude that the very same open-sentence token was produced" (Chihara, 1990: 60, italics added). Furthermore, when one examines the semantics of the language in which Ct is formulated, one finds no requirement, explicit or implicit, that the objects in each world must occur in only one world. So Shapiro's claim that the objects in the modal system are all world-bound is simply false.[38]

In describing the formal language in which Ct is formulated, I make it clear that I have omitted certain logical features of the language for purposes of facilitating the exposition of the system. For example, I make it clear that the vocabulary of the language in question has been simplified so as to contain only the non-logical constants that express satisfaction and identity. Of course, to use the theory to validate inferences involving the use of mathematics that we routinely make in science and everyday life, one would need predicates other than satisfaction and identity. But since these other predicates would not be needed to describe the axioms of the theory, I dropped them from consideration in the exposition. This is why I wrote: "In a fuller exposition, I would include an infinite number of predicates" (Chihara, 1990: 56).

For similar reasons, I dropped from consideration all existential quantification over the type n entities ($n > 0$). But in a fuller exposition of the language, I would certainly include a discussion of existential quantification of that sort. I would definitely want to express in the language, such thoughts as:

[1*] An open-sentence ϕ exists which is such that exactly two objects satisfy ϕ.

---

[38] I think Shapiro was misled by a suggestion I made that one could avoid a certain problem by treating the open-sentence tokens in each possible world as "world-bound". See Chihara, 1990: 60–1. Of course, this suggestion was never intended to apply to all the objects in each possible world, but only to the open-sentence tokens.

Thus, in a fuller exposition, the language would contain the following sentence:

[1] $(\exists x_1)(\exists x)(\exists y)(Sxx_1 \;\&\; Syx_1 \;\&\; (z)(Szx_1 \rightarrow (z = x \;\text{v}\; z = y)))$.

Now as the semantics of the language makes clear, from this sentence, one could infer that it is possible to construct an open-sentence which is such that exactly two objects satisfy it. That is, one could infer:

[2] $(Cx_1)(\exists x)(\exists y)(Sxx_1 \;\&\; Syx_1 \;\&\; (z)(Szx_1 \rightarrow (z = x \;\text{v}\; z = y)))$.

But from [2], one cannot infer [1]—which shows clearly that constructibility quantifiers are not just existential quantifiers. These logical differences should make evident, even to those unfamiliar with possible worlds semantics, that the worlds do play an important role in the semantics of the constructibility quantifiers.

In other words, the apparent similarities that led Shapiro to his conclusions are due to the fact that the constructibility quantifiers were being explained and discussed in a very simplified and impoverished context—this for didactic and heuristic purposes.

In claiming that "the constructibility quantifier has virtually the same semantics as the ordinary existential quantifier", Shapiro seems to have overlooked or ignored completely a whole chapter of the book (chapter 2) which is devoted entirely to the possible worlds semantics of the constructibility quantifier. Now, a constructibility statement does not, strictly speaking, entail that an *entity* that satisfies some condition can be brought into existence: the construction of an open-sentence may consist simply in the waving of a flag in various ways. However, within the set-theoretical setting in which possible worlds semantics is developed, the constructibility of something is *represented* by there being some element of the appropriate type in a possible world. For example, I give an example of a language with constructibility quantifiers for which possible worlds interpretations (called 'K*-interpretations') are defined. In this language, there are two kinds of variables: starred variables to be used in constructibility quantifiers and unstarred variables to be used in ordinary existential quantifiers. There are two kinds of things talked about in the language, when it is given a K*-interpretation: 0-things, which are the kind of things that concern the existential quantifier, and the 1-things, which are the kind of things that are said to be constructible. Truth under a K*-interpretation is defined in such a way that:

(1) the truth (under a K*-interpretation) of an existential statement depends only upon what 0-things exist in the actual world;

(2) the truth (under a K*-interpretation) of a constructibility statement depends upon what 1-things exist in all the possible worlds.

Thus, in discussing the truth conditions for existential statements, I wrote:

[A]s far as what is relevant to truth under a K*-interpretation, the 0-things from all possible worlds other than the actual world are not significant: the domain of the standard [existential] quantifiers can be regarded as the set of things in the actual world....[On the other hand] the constructibility quantifier can be regarded as ranging over the totality of 1-things from all the possible worlds. (Chihara, 1990: 35).

This semantical definition reflects the fact that existential statements (using the standard existential quantifier) assert the actual existence of something, that is, the existence of something in the actual world, whereas constructibility statements do not—they only assert that something (an open-sentence, say) could exist, that is, that something of the appropriate sort exists in some possible world (not necessarily the actual world). Clearly, the distinction between the actual world and the other worlds in **W** is used, and plays a significant role, in the semantics of constructibility quantifiers. So it is hard to see how the differences (between how truth under an interpretation is defined for the existential and the constructibility statements respectively) are compatible with maintaining, as Shapiro does, that "the constructibility quantifier has virtually the same semantics as the ordinary existential quantifier".

Shapiro's second objection to my use of possible worlds semantics

As I have emphasized several times already, the possible worlds semantics I gave, in explaining how the constructibility quantifier functions and how the deductive system of the Constructibility Theory operates, was not used to define the meaning of the logical and modal terms of the system: it was set forth primarily as a didactic device—a kind of model to facilitate transmitting to the reader an understanding of certain logical features of my system—or as a heuristic instrument for presenting or investigating modal situations in a perspicuous manner. But Shapiro thinks that the modal notions I use, without the aid of the sort of model theory (and hence set theory) in terms of which my theory is described, are simply not determinate and precise enough to develop mathematics in the purely modal way I advocate. Thus, he writes:

I submit that we understand how the constructibility locutions work *in Chihara's application to mathematics* only because we have a well-developed theory of logical possibility and satisfiability. Again, this well-developed explication is not *primitive or pretheoretic*. The articulated understanding is rooted in set theory, via model theory.

Set theory is the source of the precision we bring to the modal locutions. Thus, *this (partial) account of the modal locutions is not available to an antirealist*....In short, we need some reason to believe that, when applied to the reconstruction of mathematics, constructibility quantifiers work exactly as the model-theoretic semantics entails that they do. (Shapiro, 1997: 232, italics mine)

There are several things that should be said in response to this objection. Consider the second sentence, in the quotation above, 'Again, this well-developed explication is not primitive or pretheoretic'. Here, Shapiro seems to be taking the word 'primitive' to mean essentially what 'pretheoretic' means. Thus, since I characterize my constructibility quantifiers as 'primitive', Shapiro concludes that they are "pretheoretic".

Here is what Shapiro says that indicates how he came to attribute to me the view described above.

Clearly, constructibility quantifiers are established parts of ordinary language, and competent speakers do have some grasp of how they work. For example, we speak with ease about what someone could have had for breakfast and what a toddler can construct with Lego building blocks. Moreover, there is no acclaimed semantic analysis of these locutions, model-theoretic or otherwise, as they occur in *ordinary language*. These observations seem to underlie Chihara's proposal that the locutions are "primitive." We use them without a fancy model-theoretic analysis. (Shapiro, 1997: 232).

In this quotation, Shapiro is contrasting his own view of how the constructibility quantifier could obtain sufficient precision to be applicable to mathematics (by the use of model theory) with the view he attributes to me according to which I supposedly maintain that my constructibility quantifier is a pretheoretic, "primitive" notion of ordinary language—it supposedly is a notion that is unrefined by the techniques and results of model-theoretic semantics.

Shapiro's understanding of my position is partially correct: I do hold that there are constructibility quantifiers which are "established parts of ordinary language" and I do think "competent speakers do have some grasp of how they work." But I do not maintain that the constructibility quantifiers of ordinary language are identical to the constructibility quantifiers of Ct—they are, as I see it, very similar in grammatical structure and function, but not completely identical.

Consider the relational expressions 'is a member of', which is also an established part of ordinary language. We say such things as:

John is a member of the class of 2008.

Tracy is a member of the gang that has been terrorizing the neighborhood.

A member of the pack of wolves seen in the neighborhood has just killed a chicken.

I believe that such uses of the expression have led some mathematicians to believe that school classes, gangs, packs, bunches, and flocks are really sets.[39] Certainly, the expression 'is a member of' of ordinary language has much in common, both logically and grammatically, with the term used in set theory. But it would be a mistake to suppose that school classes, gangs, packs, and so on should be *identified* with the sets spoken of by mathematicians. A pack of wolves, for example, is not thought to go out of existence just because some member of the pack is killed: "That same pack has returned to ravage the countryside", it might be said, even though a member of the pack has been shot. No mathematician would believe that a set which has as members all but one member of a set A is identical to A. In other words, the mathematician may use, in his or her theorizing, an expression that is a part of ordinary language but give it a special sense that is subtly different from that of the expression in ordinary usage. And these differences may very well be the result of sophisticated theoretical analysis and mathematical reasoning. Thus, when Zermelo formulated his axiomatization of set theory, he was not merely setting down in axiomatic form ideas already implicit in our ordinary use of such expressions as 'is a member of'. It would be the height of implausibility to suppose that Zermelo's axiom of choice was implicit in the ordinary everyday use of the expression 'is a member of'.

Shapiro thinks that my constructibility quantifier is a "pretheoretic" notion because of such passages as the following:

It needs to be emphasized however that the semantics of possible worlds, which is to be used here, is brought in merely as a sort of heuristic device and not as the foundations upon which the mathematics of this work are to be based: *the constructibility quantifiers are primitives of this system*, and the Platonic machinery of Kripkean semantics is used to make the ideas comprehensible to those familiar with this heavily studied area of semantics. (Chihara, 1990: 25, italics added)

Shapiro seems to think that, in calling the constructibility quantifier a "primitive" of my system, I am implying that this logical constant of my system is primitive, in the sense of pretheoretic, crude, unrefined, rudimentary, untutored, or naive. However, as I used the term 'primitive', it was as a relative

---

[39] See, for example, Halmos, 1960: 1.

term—relative to the system in which it is being used: a notion that is a primitive of one system may occur as a defined notion in another system. The membership relation is a primitive of set theory; it is a defined term in *Principia Mathematica*. Frege emphasized that not every term in one's theory can be defined: some notions in the theory must be undefined. In Frege's system, *concept* and *extension of a concept* are primitive notions; whereas *zero* and *successor* are defined. In setting up a formal system, one chooses the terms that are to be primitives and the terms that are to be defined terms.[40] It can be seen that I was using the term 'primitive' in the passage quoted above in the sense of "primary", "assumed as a basis", or "undefined and original". There is no suggestion that the constructibility quantifier is a rudimentary, untutored, or naive pretheoretic notion. Thus, although ordinary, non-philosophical speakers of ordinary English do say such things as "It is possible to construct houses made entirely of ice", it should be kept in mind that the kind of possibility expressed in that double-quoted sentence should not be assumed to be the kind of "conceptual" or "broadly logical" possibility that is required in my constructibility quantifier. I would certainly not claim that such a heavily philosophically studied notion of possibility is a "pretheoretic" or preanalytic notion of ordinary language.

What about Shapiro's suggestion that, without the use of possible worlds semantics (and hence set theory), the logic of the constructibility quantifier would not be sufficiently determinate and precise to carry out the sort of reasoning required to develop mathematics within the Constructibility Theory? At least part of Shapiro's reason for maintaining such a position is his mistaken belief that, in calling the constructibility quantifier a "primitive" of my system, I was taking this logical constant to be a crude "pretheoretic" notion of ordinary language. As I noted earlier, a primitive of a system can be a highly refined, widely studied, and carefully analyzed theoretical term of a sophisticated system. For example, the terms 'set' and 'is a member of' are primitives of Zermelo's 1908 axiomatization of set theory.[41] The terms are not given model-theoretic analyses or definitions. Yet it would be wrong to classify them as crude "pretheoretic" terms of ordinary language. For Zermelo's axiomatization was made as the result of a careful and deep study of

---

[40] Some readers may find it enlightening to ponder the sorts of consideration that go into choosing what is and what is not to be a primitive of one's system. See in this regard, Chihara, 1998: 79–81.

[41] Zermelo, 1967: 201. There is no suggestion that these terms should be taken to be crude or unrefined concepts of the theory.

the works of practicing mathematicians—especially their use of set theory—at least partially in response to the various antinomies of mathematics that had been discovered and also partially in response to objections that had been made by eminent mathematicians to his proof of the well-ordering theorem.[42] Essentially the kind of theorizing, analysis, clarification, and theoretical concept formation that went into Zermelo's axiomatization can be carried out for the case of modal reasoning as well. Model theory is not essential to such a process. It is a process of theorizing that philosophers have been carrying out since the time of Aristotle, long before model theory was invented. I can find nothing in what Shapiro has argued that precludes developing the logic of the constructibility quantifier to a sufficient degree of precision to carry out the sort of development described in Part I of this chapter.

Let us now investigate the question as to whether one can have a sufficiently sophisticated understanding of the logic of the constructibility quantifier, without making any appeals to model-theoretic notions or other aids from possible worlds semantics, to make the sort of applications to mathematics that is the concern of Shapiro in the quotation above. Let us consider the development of finite cardinality theory given in Sections 2 and 3 of this chapter. Notice that the exposition and discussions of this theory (including all the proofs of theorems) are given without any appeals to any model-theoretic notions or to results from possible worlds semantics. Deductions and inferences are made using modal reasoning and without any mention of set theory or set-theoretical results. Yet, it can be seen that standard theorems of number theory can be obtained within this system. Those who think, with Shapiro, that such "applications [of the Constructibility Theory] to mathematics" can only be made with the aid of set theory should try to find some specific theorem I have proved that requires, in their opinion, a hidden appeal to set theory. This would provide both sides of the dispute with a definite example to investigate, to see if at some point in the reasoning, there is a surreptitious appeal to some model-theoretic result. In the absence of any such specification, I find unconvincing Shapiro's rather unspecific objection that somewhere in my development of mathematics, I need set theory to carry out the reasoning.

It might be objected that I do appeal to Frege's development of finite cardinality theory in developing my constructibility version of the theory, and since Frege's development is Platonic in nature (presupposing as it does

---

[42] See, for details of Zermelo's motivation for his axiomatization, Bach, 1998.

the existence of extensions of concepts), one can argue that my version itself presupposes the existence of abstract mathematical objects. In answer to such an objection, it should be noted that Frege's theory of cardinality is only used as a sort of model—there is no theorem of Frege's whose truth is presupposed anywhere in my development of number theory. There is clearly no obstacle to using various Platonic theories (such as set theory) in this way.

Can a nominalist make use of model theory?

Shapiro relates that Burgess once argued, at a convention, that it is far-fetched for someone who learned much about logical possibility from a high-powered course in mathematical logic, using a text like Shoenfield's *Mathematical Logic* (Shoenfield, 1967), to go on to claim a primitive pretheoretic status for this notion. Shapiro adds: "The same goes for constructibility assertions" (Shapiro, 1997: 238 n.). This last point lies behind the previously quoted claim of Shapiro's that set theory is the source of the precision we bring to the modal locutions and that "this (partial) account of the modal locutions is not available to an antirealist".

It can be seen that the crucial presupposition of this criticism is the following "illegitimacy thesis":

> It is illegitimate for Chihara to make use of set theory in explicating or clarifying the logic of the constructibility quantifier.

Why should one accept the illegitimacy thesis? In stating his objection, Shapiro does not supply any explicit argument supporting this thesis. But it is not difficult to see what he is thinking. He knows that:

> (1) I do not think that the axioms of set theory, as standardly (Platonically) understood, are literally true assertions.

He also believes that:

> (2) one cannot use an untrue theory in explaining or explicating anything.

So he concludes that I cannot make use of set theory in explicating or clarifying the logic of the Constructibility Theory.

But as I argued earlier, I reject (2) completely, so this line of reasoning can also be seen to be unconvincing. After all, what is wrong with, say, using set theory or model theory as an aid to the conceptual clarification of some of the finer points of the constructibility quantifiers or of the broadly logical notion of possibility? Certainly if the use I make of set theory presupposes that its theorems, as standardly (Platonically) understood, are literally true assertions, then a legitimate complaint can be lodged against me. But that I have used set theory in that way has by no means been shown.

Can model theory be legitimately used by the anti-realist as a tool in clarifying, explicating, and investigating the logical features of the constructibility quantifier, without presupposing the existence of mathematical objects? I shall respond to this question in Chapter 9.

6. RESNIK'S OBJECTIONS

Resnik raises four main objections to my theory. In what follows, I shall take them up in the order in which they are given.

Resnik's induction objection

The first is the objection that I presuppose mathematics in justifying the axioms of the Constructibility Theory—the suggestion being that I thereby presuppose the existence of mathematical entities in providing this justification. The basis for this objection is the fact that I use a form of mathematical induction in my justifications of the axioms of the Constructibility Theory (Resnik, 1997: 61–2).

Resnik considers a possible reply. I might argue, he suggests, that the principle of induction I use should itself be understood as a derived principle of the Constructibility Theory, so that the inductive principle I use can be understood as concerned merely with the constructibility of open-sentences. To this, Resnik replies:

But this would call for a further constructibility theory—a metaconstructibility theory—since he is trying to justify his initial constructibility theory. It would then be fair to ask for the justification of this metatheory. Presumably, Chihara would use induction to justify this theory, and we then would press him to eliminate it. ... Chihara should be obliged to stop the regress at some point and give a neutral, non-mathematical justification of his system. (Resnik, 1997: 61–2)

I do not wish to contest the claim that the principle of induction I use is, in some sense, a variation on the principle of induction used in number theory. Nor do I wish to deny that this principle is, in some sense, mathematical in nature. What counts as mathematical is somewhat vague and little is to be gained by arguing about such matters. The crucial question we need to ponder here, however, is whether or not the use of this principle ontologically commits one to the existence of mathematical entities. So let us consider in more detail the way this principle is used.

I start with a specification of a rule for constructing the arabic numerals, '1', '2', '3', ... Now suppose that the following two things are proved:

[1] Every open-sentence ϕ of the level given by the numeral '1' is such that ϕ satisfies condition F.

[2] All arabic numerals n and m that it is possible to construct and all open-sentences ϕ and θ that it is possible to construct are such that if m immediately follows n according to the specification mentioned above and if ϕ satisfies condition F, then θ satisfies F.

The principle of inference I employ, then, allows one to infer that it is not possible to construct an open-sentence ϕ of any level such that ϕ does not satisfy F. Now such a rule of inference is not a rule about mathematical objects. There is no mention of abstract mathematical objects or any quantification over mathematical objects. Call it a mathematical principle if one likes (because of its similarity to the familiar principle of number theory), but calling it 'mathematical' should not blind us to the fact that it is basically a modal principle—a principle not about abstract mathematical entities but about what it is possible to construct.[43]

There is no need to appeal to a metatheoretic version of the Constructibility Theory to justify the use of this principle. Nor it is necessary to appeal to the standard principle of mathematical induction to see that this modal principle is valid—its validity can be grasped directly.[44]

Resnik on the Axiom of Choice

Resnik's second objection concerns the fact that I do not include the Axiom of Choice among the axioms of the Constructibility Theory. He comments: "[The axiom] is now part of standard mathematics, and is required for some theorems that are employed throughout science. Thus it should be given a correlate in his system" (Resnik, 1997: 62).

There are two reasons why I did not regard the absence of the Axiom of Choice from the list of axioms of the Constructibility Theory as a serious problem. First of all, I was confident that no use of mathematics in the empirical sciences requires that the Axiom of Choice be true. As will become apparent in later chapters of this work, I believe that the axiom in question is not the kind of axiom that needs to be true, especially when it is literally

---

[43] For a more detailed discussion of my use of this nominalistic form of mathematical induction, see Chihara, 1973: ch. 5, sect. 2.   [44] Cf. Chihara, 1973: 178–81.

construed, to be useful in applications. In any case, if Resnik wishes to push this line of attack, he should point to a specific application of mathematics which necessitates the *truth* of the axiom. Secondly, I intended the mathematics of the Constructibility Theory to be that of *Principia Mathematica*, which is generally regarded as a formalization of the classical analysis adequate for all applications in the sciences. Now as I mentioned earlier, Russell and Whitehead proceeded in *Principia* without justifying the Axiom of Choice. As I pointed out earlier, whenever they needed to prove a theorem ϕ that depended upon the Axiom of Choice, they proved in *Principia* a theorem of the form:

$$\text{CHOICE} \rightarrow \phi$$

—something I could do easily in the Constructibility Theory. Notice that if, for some reason, scientists find it useful or convenient to formulate one of their theories in terms of a kind of set-theoretical structure in which Choice holds, it would be a simple matter to apply to such structures the theorems of the Constructibility Theory proved in the above way as dependent upon Choice.

Resnik's objection to my justification of the Abstraction Axiom

Resnik describes the constructibility version of the Abstraction Axiom in the following way:

In a simplified form it postulates that for any object y and any condition '... x ... y ...' formulated in the constructibility theory, an open-sentence is constructible that is satisfied by just the (constructible) things w that are such that ... w ... y .... Of course, '... w ... y ...' is an open sentence. But it does not verify Chihara's axiom because it does not *mention* the object y. The letter 'y' occurs in it as a free variable. (Resnik, 1997: 63)

Now suppose that the variable 'y' refers to an object k. Here is what I wrote:

Then it is reasonable to maintain that there is some possible world in which the language of this theory is extended to include a name of the object k. Then the formula expressing the condition in question can be converted into the required open-sentence by replacing all free occurrences of ['y'] by the name of k, and surely it is possible to do this. (Chihara, 1990: 66)

There are two parts to Resnik's objection to this reasoning, the first of which concerns my talk of a possible world. He realizes that my talk of possible worlds is not intended to be taken literally and that this use of words is only a heuristic device, but he objects that I am unjustifiably using intuitions about possible worlds in my justification of the axiom.

The second part of Resnik's objection is specifically directed at the passage in the quotation above expressing the idea that in some possible world there would be an extension of the language of the theory which includes a name of k. Now, Resnik remembers a passage in my book in which, in answer to the question 'What would it be like for a token of a type to exist in a possible world?', I reply: "Here, we can imagine a possible world in which some people . . . do something that can be described as the production of the token" (Chihara, 1990: 40). He infers from this that what it means to say that some token of a certain sort is constructible is that it is possible for people (that is, humans) to do the constructing. With such a restricted view of constructibility (constructibility by humans), Resnik then can go on to respond to my justification of the Abstraction Axiom by claiming:

[I]t seems to me that there may be physical objects that it is not humanly possible to name, simply because it is not humanly possible to identify them with sufficient precision. They might be too small, too fast, or too fleeting. (Resnik, 1997: 64)

But here, Resnik has simply drawn the wrong conclusion from the example quoted above in which some people do the constructing of a token. It is clear from the context from which the quote is taken that the example was brought in to explain *what it would be like for a token to exist at a world*—the example makes it clear that it need not be the case that there be some entity (the token) that exists at the world. It could be sufficient that some intelligent being performs some act (say, waves a flag in a certain way). The point being made was that "we need not concern ourselves with questions about the ontological status of tokens: in particular, we need not worry ourselves over whether a series of hand signals is or is not an entity, or whether it is a physical object of some sort" (Chihara, 1990: 40). Thus, it would be a mistake to infer from this example that what 'it is possible to construct' means in the Constructibility Theory is: it is possible *for humans* to construct. I never intended any such restricted interpretation, and the quotation Resnik cites in his book does not justify any such interpretation.

It should be clear, in any case, from the S5 modal system used in the formalization of the Constructibility Theory and from the justifications of the axioms of the theory given in the book, that the constructibility quantifier '(Cx)' ought not to be understood in the restricted way Resnik understand it.

I suspect that Resnik, deep down, realizes that I had no such understanding of the constructibility quantifier, because immediately following his objection to my proof of the Abstraction Axiom, he says: "[I]f Chihara simply means that for any object k it is logically possible that some being tokens a name for it, then his intuition seems more plausible" (Resnik, 1997: 64). He then continues: "In the end, then, Chihara's epistemology amounts to the epistemology of logical possibility." So perhaps even Resnik did not take the above objection to be a serious one. Of course, what I meant by the passage in question was that, for any object k, it is possible, in the 'conceptual' or 'broadly logical' sense of "possibility" discussed earlier in this chapter, that some being tokens a name for it.

Let us now return to the first part of this objection in which Resnik complains about my appeal to intuitions about possible worlds. I simply brought in talk of possible worlds as a heuristic device to aid the reader to call up the modal intuitions needed. Possible worlds semantics is frequently used as a sort of heuristic aid—much as we use Venn diagrams in assessing reasoning with categorical syllogisms. For example, one can see that a certain modal argument is invalid, by constructing a possible worlds diagram from which one can specify a structure in which the premises are true and the conclusion is false. Resnik seems to think that this would be an illegitimate use of possible worlds semantics, since the anti-realist does not believe in possible worlds and yet is using his or her intuitions about worlds to make these inferences. But it takes very little sophistication in modal reasoning to see that such a possible worlds diagram can be converted into a specification of the meanings of the predicates and a description of *how the world could have been* such that the premises would be true and the conclusion false.[45] In other words, such diagrams can be used to give the essential elements needed to specify how the premises could be true and the conclusion false—all this without making use of any intuitions about what possible worlds exist.

Similarly, we need no intuitions about possible worlds to follow the train of thought about the Abstraction Axiom that is the target of the first part of Resnik's objection (although, we do need intuitions about what is possible). The first sentence in the above quote can be understood to say: "Then it is reasonable to maintain that there could have been beings who extended the language of the Constructibility Theory to include a name of the object k." All my talk about possible worlds can easily be translated into talk about what could have been the case, as anyone familiar with the modality in question could have inferred.

---

[45] The reader can obtain a clear idea of how this can be done by studying how this is done for the modal sentential logic in Chihara, 1998: ch. 6.

Resnik's skepticism about modality

All of the above is only a preliminary to his most fundamental objection to my Constructibility Theory: Resnik has doubts about modality in general. First, he is skeptical that we have the epistemological means to know complicated modal facts of the sort required by the Constructibility Theory. Second, he is even skeptical that any of the theorems of the Constructibility Theory are true, since he doubts that there are any modal facts at all—he is a "non-cognitivist" when it comes to modal statements, believing that no modal statement has a truth value.

I shall take up first his doubts that there are any modal facts. Here's how he states his position: "I doubt that there are any modal facts to be known—even when the modality is that of logical possibility" (Resnik, 1997: 64). When it comes to logical possibility, he expressed his doubts this way:

The view I am proposing is a restrained *logical non-cognitivism:* sentences of ordinary language that seem *categorically* to attribute logical necessity or other logical properties and relations actually perform other functions, and are neither true nor false. (Resnik, 1997: 167)

What are these "other functions" that such categorical sentences perform? Here's the sort of analysis Resnik provides. Suppose that we assert that Frege's axioms are contradictory or inconsistent. According to Resnik's analysis, our utterance informs our audience that we expect Frege's axioms "to be treated in a certain way": we show that we are confident that our audience can see for themselves that Frege should "retract or qualify" his axioms (Resnik, 1997: 168). A second function of such statements is to express a certain commitment: in this case, to the falsity of the axioms.

I am skeptical of this analysis. When I tell a class of students that Frege's axioms are inconsistent, I certainly do not intend to show the class that I am confident that they can *see for themselves* that Frege should retract or qualify his axioms. It has been my experience that most students who are not mathematics majors are simply not able to carry out the sort of derivation needed to see that the axioms are inconsistent.

Resnik's non-cognitivist analysis of this case seems to be the result of the sort of strategy adopted by some phenomenologists regarding statements "about physical objects": find some implications of the analysandum—for the modal non-cognitivist, implications that are non-modal—and then claim that the function of an utterance of the sentence is to inform the audience of these implications. One problem with this strategy is that the implications one comes up with may hold only in certain contexts or with certain kinds of audiences, when no such context dependency is to be found in the analysandum.

Imagine J. B. Rosser announcing to some of his colleagues that Quine's system of axioms of the first edition of *Mathematical Logic* is inconsistent.[46] Is this an announcement that shows that he is confident that his audience can see for themselves that Quine should retract or qualify those axioms? Surely not. It is questionable that Rosser would have expected even someone who had worked extensively with that set theory to have been able to see for himself or herself that Quine should retract or qualify his axioms.

Resnik admits that some statements in which modal terms occur may have truth values: these are statements in which modal terms occur in non-categorical contexts. Here's an example of such a non-categorical statement:

[@] Any theory implying a falsehood is false.

Resnik suggests that such statements should be understood "as tacitly referring to the norms that govern (or ought to govern) our inferential practices" (Resnik, 1997: 169). Thus, [@] is rendered by Resnik as:

[@'] Any theory from which we may infer a falsehood is false.

In this context, implication is analyzed by Resnik in terms of what we may infer.

I find this analysis, too, to be highly questionable. [@] and [@'] clearly do not mean the same things, as can be seen from the fact that [@], as ordinarily understood by anyone with even a bit of logical training, does not presuppose or make tacit reference to a system of inference rules, whereas (as Resnik himself suggests) [@'] does. Furthermore, consider how one might justify one's belief in [@]. From the standard intuitive explication of implication, one can prove (trivially) that [@] holds, whereas a proof of [@'], depending upon the particular system of inference rules one is tacitly referring to, will very probably be relatively substantial. I do not see how [@'] can be a plausible rendering of [@]. In any case, Resnik admits that he has no systematic method for dealing with such non-categorical examples and that they pose a "serious, but not fatal, difficulty" for his position (Resnik, 1997: 169–70).

Returning to Resnik's non-cognitivist position on categorical modal statements, let us examine some specific examples of modal statements that mathematicians have made. Here is how Fermat stated what became known

---

[46] See Quine, 1963: 302 for references and more details on what Rosser proved.

as his "Last Theorem":

It is impossible to divide a cube into two cubes, a fourth power into two fourth powers, and in general any power except the square into two powers with the same exponent.[47]

Evidently, Fermat did not accept Resnik's doctrine that there are no modal facts.

Now Resnik could respond to this objection by claiming that Fermat's statement of his "theorem" just means "There is no solution to ..."—a straightforward existential statement which poses no difficulty for his views.

I see substantial problems with such a reply. First, this response conflicts with Resnik's non-cognitivism. Thus, notice that Fermat's statement seems to categorically attribute an impossibility and hence that, according to Resnik's non-cognitivism, it actually performs other functions, and is neither true nor false. But surely, under the suggested synonymy, Fermat's statement is true.

Second, the reply cannot be that the modal statement of Fermat's is *necessarily equivalent* (or even "logically equivalent") to the existential statement, since first, that very statement of equivalence is a modal statement, and second, the equivalence implies that Fermat's modal statement is true (which contradicts his thesis that no categorical modal statements are true). No, the reply must be that what appears to be a modal statement is actually an existential statement—that Fermat actually made not a modal statement but rather a straightforward negative existential statement.

Such a position seems to me to be wildly implausible. Suppose that Fermat had said: "It is impossible for even God to divide a cube into two cubes, a fourth power into two fourth powers, and in general any power except the square into two powers with the same exponent." What would that modal statement mean, according to the supposed Resnik position? Or: "Not even Descartes could divide a cube into two cubes, a fourth power into two fourth powers, and in general any power except the square into two powers with the same exponent." Fermat would have been willing to make both assertions. But are all these apparently distinct modal statements one and the same negative existential statement? That would be linguistically very implausible. What these examples highlight is the fact that the above-suggested response to my objection rests upon a substantial (and counterintuitive) linguistic thesis. If Resnik were to make the suggested response, the burden of proof would be on him to provide convincing linguistic evidence supporting the

---

[47] Dorrie, 1965: 96.

underlying linguistic thesis. I doubt very much that he could come up with what is required.

Continuing the theme of whether mathematicians make modal claims, most mathematicians would affirm that Gauss was able to prove a special case of Fermat's "theorem", namely that it is impossible to divide a cube into two cubes.[48] Of course, such an affirmation strongly suggests that Gauss proved that the statement 'It is impossible to divide a cube into two cubes' is true—something that conflicts with Resnik's anti-modalist position.

I noted in Chapter 1 that Euclid's geometry was modal in character. It is not surprising, then, that many of the solutions to ancient geometric problems that were given in the nineteenth and twentieth centuries are stated modally. For example, the theorem that answered the ancient problem of whether it is possible to square the circle is stated:

It is impossible to draw with a compass and straightedge a square that is equal in area to a given circle.[49]

Resnik's position on categorical modal statements implies that this statement has no truth value. But how can that be if it has been proved? How can this theorem have no truth value if it has survived hundreds of years of attempts to find a compass and straightedge method of squaring the circle? Do we not have some evidence of the truth of the statement?

Consider the following problem posed in 1891 by Edouard Lucas:

How many ways can n married couples be seated about a round table in such a manner that there is always one man between two women and none of the men is ever next to his own wife?[50]

This modal problem was solved by several mathematicians, who gave an effective procedure for obtaining the number. Notice that the solution can be tested for relatively small n by actually making arrangements of possible seating charts. If the tests for $n = 6, 7, 8, \ldots, 100$ are all found to correspond to the answers provided by the solution, won't this result provide some evidence that the solution is correct?

Resnik thinks that such statements as 'It is possible to construct an open-sentence token that is true of my gold pen, my left thumb, and the moon' lack truth values. But I have actually constructed an open-sentence token that is

---

[48] Ibid.

[49] The theorem followed from the proof given in the nineteenth century of the transcendence of π. See Dorrie, 1965: 136.

[50] Dorrie, 1965: 27. A detailed discussion of this problem is given there.

true of my gold pen, my left thumb, and the moon, and that shows, according to standard modal logic, that the modal statement in question is true. I am much more confident of the truth of the modal statements I have listed here than I am of the many paradoxical and bizarre consequences of Resnik's principle of the nonsensicality of trans-structural identity (thesis [2]).

A view opposing Resnik's skeptical doubts may be appropriate here. Putnam vividly expresses such an opposition:

From classical mechanics through quantum mechanics and general relativity theory, what the physicist does is to provide mathematical devices for representing all the *possible*—not just the physically possible, but the mathematically possible—config-urations of a system. Many of the physicist's methods (variational methods, Lagran-gian formulations of physics) depend on describing the actual path of a system as that path of all *possible* ones for which a certain quantity is a minimum or maximum. ... It seems to us that 'possible' has long been a theoretical notion of full legitimacy in the most successful branches of science. ... It seems to us that those philosophers who object to the notion of possibility may, in some cases at least, simply be ill-acquainted with physical theory, and not appreciate the extent to which an apparatus has been developed for *describing* 'possible worlds'. (Putnam, 1979: 71)

How do we know the axioms of the Constructibility Theory?

Let us now consider Resnik's contention that we lack the epistemological means to know complicated modal facts of the sort required by the Constructibility Theory. Resnik's contention is based upon his enumeration of the means we have for "determining logical possibilities": (1) modal logic; (2) inferences from what is actually the case to what is possibly the case; and (3) logical intuition (Resnik, 1997: 64). The first two, he tells us, are "probably too weak to provide all the knowledge" that is required (Resnik, 1997: 64). As for the third, he has little confidence in logical intuitions, "since often even the intuitions of professional logicians conflict" (Resnik, 1981: 64).

Before tackling this objection head on, a few preliminary points may be useful:

[1] Why should the fact that the intuitions of professional logicians sometimes conflict lead one to lose confidence in logical intuitions? True, there are some rather spectacular cases of disagreements among logicians about such basic logical principles as the law of excluded middle. But it is worth noting that these disagreements concern only the application of the "law" to infinite totalities—there is no disagreement when dealing with finite totalities.[51] Furthermore, the

---

[51] See, for example, Brouwer, 1967: 336, which is quite explicit about this.

disagreement is frequently based not upon a conflict of "logical intuitions", but rather on highly theoretical reasoning, perhaps involving theories of how we learn the logical connectives.[52] Besides, the cases of agreement among the logical intuitions of professional logicians far outweigh the cases of disagreement. After all, we don't lose confidence in our intuitions about the grammaticality of strings of English words just because fluent speakers of English sometimes disagree about whether a particular string of English words is or is not grammatical. It is striking that even the intuitionists do not advocate abandoning all our logical intuitions.

[2] I am uncomfortable applying the word 'know' to my beliefs about the axioms of the Constructibility Theory, since it is not clear to me just what is required in order to know such things. I see no compelling reason to claim to know that these axioms are true. It is sufficient that we have plausible grounds for believing the axioms. In this respect, my epistemological position vis-à-vis the axioms of the Constructibility Theory is surely no worse than that of the Platonic set theorists vis-à-vis the axioms of set theory: even those Platonists who rely upon indispensability arguments to support their ontological beliefs do not claim to know that the axioms of ZF are true.

As for Resnik's threefold classification of our means for determining possibility, he simply omits what is perhaps the most fruitful means: theoretical reasoning. Just as one can theorize about physical laws and principles, about logical laws and principles, and about grammatical laws and principles, one can also theorize about modal laws and principles. One can construct theories about possibility and test these theories, for example, to see if our logical, linguistic, and scientific theories come out as expressing possibilities or to see if the theory conflicts with any theories determined to be possible by the methods of (1), (2), and (3) described by Resnik. One can also theorize about what it would be possible for intelligent beings to do, given that these beings have such and such capacities and such and such a language, by extrapolating from what we humans are able to do. Much of the justifications of the axioms of the Constructibility Theory are of this theoretical sort. I see no reason why such reasoning should be considered to be inappropriate for the aims I have in mind.

---

[52] Recall the argument in favor of rejecting excluded middle that Dummett once gave (which was discussed in Chapter 3).

8

Constructible Structures

In this brief chapter, I develop an account of a kind of structure (or ersatz structure) in order to show how various mathematical theories are realizable (in some appropriate sense).

1. A PROBLEM FOR THE STRUCTURALIST

According to Parsons, the first problem facing the structuralist arises in attempting to state what structuralism is.[1] As I noted in Chapter 4, the problem arises from the fact that in order to state the view, one needs to say what a structure is. But it is not clear how one can say what structures are without either presupposing familiar mathematical entities (such as sets) or involving oneself in some sort of circularity (Parsons, 1996: 275).

This is a problem for the structuralist who is putting forward a far-reaching metaphysical overview of all of mathematics, but it does not arise for my view since I am not putting forward a general account of what mathematical theories, literally construed, are about, nor do I postulate some sort of metaphysical entity that forms the subject matter of all of mathematics.

2. REALIZATIONS WITHOUT COMMITMENT TO MATHEMATICAL OBJECTS

For several purposes—for example, in order to respond to what Shapiro calls the main stumbling block of the eliminative program (to be discussed in the next chapter)—it will be useful to be able to assert that, in some appropriate sense, a theory is realizable (or satisfiable). Thus, in what follows, I shall explain how, according to the view being developed here, mathematical theories (such as first-order Peano arithmetic) can in a sense be *realized*, even though there are no such metaphysical entities as structures.

---

[1] I discussed this problem briefly in connection with Shapiro's characterization of structure in Chapter 4.

An example of a constructible model of a first-order theory

Let us first consider a simple example of a kind of model of a first-order theory in which nothing actual exists in the domain of the model. The theory in question has a vocabulary consisting of only the following two predicates:

$$T^1, L^2$$

The assertions of the theory is the deductive closure of the following axioms:

A1: $(x)-L^2xx$

A2: $(x)(y)(z)((L^2xy \ \& \ L^2yz) \rightarrow L^2xz)$

A3: $(\exists x)T^1x$

A4: $(x)(T^1x \rightarrow (\exists y)L^2yx)$

Now a constructible model of this theory will be given by first specifying how the domain is to be conceived and then giving a natural language interpretation of the predicates.

Imagine that I have purchased some plastic block letters, each magnetized so that they can be used to construct words on, say, the door of a refrigerator. The domain of the NL interpretation to be specified will consist of (tokens of) *English words that it is possible to construct from these plastic letters.* (Strictly speaking, of course, nothing actual is in the domain at all, since no words have been constructed, but the above specification shows how the constructibility quantifiers are to be understood.)

Here is how the predicates are to be interpreted:

T^1: ① is a three-letter word

L^2: ① consists of more letters than ②

Under this interpretation, all of the quantifiers are taken to be constructibility quantifiers, so that axiom A3 asserts not that there exists a three-letter English word made from the plastic letters, but rather that it is possible to construct a three-letter English word using the plastic letters. Axiom A1 asserts that every English word that it is possible to construct from the plastic letters does not have more letters than it itself has. Axiom A4 asserts that any three-letter word *x* that it is possible to construct from these plastic letters is such that it is possible to construct an English word from these letters that is made up of more letters than *x* is. Axiom A2 asserts the transitivity law for the English words that it is possible to construct from these letters. It can be seen that, under this interpretation, all the axioms come out true. So we have a kind of model of the theory in which nothing is asserted to actually exist.[2]

Realizations without ontological structures

First-order structures that are models of a theory T are, according to standard definitions, some sort of set-theoretical entity. Can one understand how such theories can be realized (or satisfied) without presupposing the existence of some such mathematical entities as sets? Shapiro and Resnik think we can, of course, in terms of the notion of structure. But as I have already emphasized, I see no reason to believe that there are any such entities as structures of the sort characterized by Shapiro and Resnik.

My idea is to come up with some sort of "thing" which, if constructible, would do the mathematical "work" of structures, at least for such areas as model theory and foundations of mathematics. The sort of "thing" I have in mind would more or less fit what I had earlier called (in Chapter 3) "the mathematician's conception of structure". The goal is to find such "things" which are also nominalistically acceptable, so that they can be used as the "realizations" of mathematical theories without requiring the background theory to carry a commitment to the sorts of metaphysical entities that led to so much trouble for the structuralists discussed earlier.

Let us begin by consolidating the picture slightly by taking formal theories, such as first-order Peano arithmetic, to be special cases of what I call "structural descriptions".[3] The idea is that, given any first-order theory T, one can always generate the structural description 'a structure that is a model of T', so to treat such theories as structural descriptions will result in no significant loss of generality. I can then concentrate on the single task of finding appropriate things to serve as realizations of structural descriptions.

To aid our intuitions for the task at hand, consider the following example of the sort of common everyday kind of structural description that Shapiro talks about:

> On the very first play of the game, the starting offensive team of the Ohio State Buckeyes lined up in a single-wing formation.

---

[2] What I have given, in the above example, is an NL interpretation under which the assertions of the theory are all true. Recall that such interpretations were discussed in some detail earlier, in Chihara, 1998: ch. 2.

[3] Obviously, I intend 'structural descriptions' to apply to more than just first-order theories. For example, a structural description might consist of English sentences which describe a type of structure.

Here, we have an assertion that describes the Ohio State team as forming a kind of structure, a "single-wing formation".

Return to Resnik's characterization of structures as consisting of one or more objects that stand in various relationships, the objects being such that they have no identity or distinguishing features outside a structure. Resnik's characterization can serve to provide us with a rough idea of how what is happening on the football field can be regarded as *realizing* the above description. The members of the team are lined up on the field, forming a particular sort of geometric configuration that is called 'a single-wing formation'. Here, there is no suggestion that, in addition to the football players on the field, there is another entity—a structure—which is being described. No, common sense tells us that it is the offensive team itself that is being described when one specifies how the members are related to each other spatially. The relationships that obtain among the football players on the offensive team that are relevant here are the spatial ones viewed from a "bird's-eye vantage point". Notice, however, that I have omitted Resnik's requirement that the "objects" in question (the members of the team) have no identity or distinguishing features "outside the structure". However, a structural description of the single-wing formation need not mention any identity or distinguishing features of the individual players in the formation that are "outside the structure"—a feature of what I called (in Chapter 3, Section 1) "the mathematician's conception of structure".

3. CONSTRUCTIBLE REALIZATIONS

To be able to speak meaningfully about realizations of structural descriptions, we need some way of singling out both the objects in the domain of the realization and also the relevant relationships that these objects are to be in. Thus, to specify a *realization of a structural description* one need only:

(1) specify, by means of an appropriate open-sentence, the objects to be talked about or described by the structural description (thus giving "the domain of the realization");[4]

and

(2) specify, by means of appropriate open-sentences, those relationships among the objects specified in (1)—that is, those relationships among the

---

[4] The things in "the domain of the realization" will be just those things that satisfy the specified open-sentence.

things in the domain of the realization—that are to serve as the "realization relationships" singled out in the structural description;

all of this such that

(3) the objects in the domain are in the realization relationships exactly as described by the structural description.

Let us consider in detail an example to see how this might be done.

An example of a structural description

The domain of this structure consists of seven objects.

These objects are related by one binary relation R^2 that is a linear ordering of the domain and which satisfies the following two laws:

[1] $(\exists x)-(\exists y)R^2yx$

[2] $(\exists x)(y)(y \neq x \rightarrow (\exists z)R^2yz$

How might one show that the above structural description is realizable? Imagine that I have before me a stack of seven pennies which I call 'Pen'. I then construct the open-sentence:

x is a penny in Pen

to serve as a specification of the domain of the realization I have in mind. I then specify that the realization relationship of R^2 in the structural description is to be given by the open-sentence:

y is immediately on top of (and touching) x.

It is a simple matter to show that this relationship is a linear ordering and that the two laws given above are satisfied. So one can conclude that the above structural description is satisfiable.

It should be clear from the description I have given that, in order to show that a structural description is realizable, it is not necessary to make reference to mathematical objects or to the kind of metaphysical entities postulated by Shapiro and Resnik. All that the above requires (ontologically) is that there be the stack of pennies and that the pennies be in the realization relationships exactly as described by the structural description. There is no requirement that there be another entity—the mysterious structure—that is being described and whose properties are as paradoxical as the ones being attributed to them by the above-mentioned structuralists.

What is the realization?

Metaphysicians may wish to know just what thing, object, or entity is to serve as the above realization. It would be natural to think of Pen, the stack of pennies, as the "thing" that is the realization of the structural description. But since we would want the realization to determine not only what the domain is, but also what the relevant "relationships of the structure" are to be, I suggest that we take the realization to be, for the case under consideration, an ordered pair consisting of an open-sentence that is satisfied by the elements of the domain ('x is a penny in Pen') and an open-sentence that is satisfied by the structural relations of the realization ('F is a relation satisfied by an ordered pair $\langle x, y \rangle$ iff x is immediately on top of (and touching) y').

It might be objected by some that ordered pairs of open-sentences are mathematical objects and hence that my conception of a realization presupposes the existence of mathematical objects. Most readers of this work, however, would realize that all my talk of ordered pairs can be understood in terms of the constructions of my Constructibility Theory, as the following discussion will make obvious. It should also be kept in mind that, by this analysis, ordered pairs are constructible. Hence, the sort of realization described above will be constructible.

Constructible realizations (first approximation)

For purposes of explanatory simplicity, I shall not attempt to give a detailed explanation of how the sorts of realizations I have in mind are to be specified in full generality. Since operations can be regarded as relations, I shall take these realizations to "consist in" only a domain and some relations on this domain. Thus, I shall regard a realization to be an ordered pair whose first element is a property D that is satisfiable by just the objects in the domain of the realization and whose second element is an open-sentence R satisfiable by just the relations of the realization.

A small difficulty

Suppose, for now, that we wish to specify realizations in which the domain of the realization is to contain only "objects" (that is, entities of the lowest level in Ct). Now it might be thought that, within this framework, a realization can be taken to be an ordered pair whose first element is a property D of objects and whose second element is a *condition on relations* R—an open-sentence satisfiable by one or more relations in whose field are only objects that satisfy D. But this raises a problem. Reconsider the definition of ordered pair given in the previous chapter. Notice that an ordered pair is an open-sentence. Now every kind of open-sentence talked about in **Ct** is only satisfiable by things of exactly one level (this is a requirement of the simple type theory). The above characterization of realization seems to require that the elements of the ordered pair be open-sentences of different levels, since the first element is a level 1 open-sentence and the second element is a level 4 open-sentence satisfiable by relations on objects. So we need to adopt some sort of "trick" to get the sort of ordered pair we need: one whose first and second elements are open-sentences of the same level but that can still do the work of both D and R.

We cannot reduce the level of R to that of D, but it is easy to go in the other direction and get the effect of raising the level D to that of R. This will be done as follows.

Relational mirrors

Suppose D is a property and suppose that **R** is a relation that is such that, whenever an object x satisfies D, then **R** is satisfied by every ordered pair that is coextensive with an ordered pair $\langle x, x \rangle$, while at the same time, **R** is satisfiable by no ordered pairs other than those required by the previous condition. Then **R** will be said to be a *relational mirror* of D. Clearly **R** will be of level 3 and it will do exactly what we want D to do. Furthermore, it is clear that, given any property D, it is possible to construct a relational mirror of D.

Realizations

We can now take a realization to be an ordered pair whose first element is a level 4 open-sentence satisfiable by all and only relational mirrors of D and whose second element is a level 4 open-sentence R satisfiable by one or more relations in whose field are only objects that satisfy D such that: (1) any relation coextensive with a relation that satisfies R also satisfies R, and (2) nothing else satisfies R.

Then, an object x will be said to "belong to the domain of the realization" iff x satisfies the property D. Notice that a realization, as defined here, will provide all the information that we want structures of objects to provide: they, in effect, determine a domain of objects and a totality of relations on this domain.

4. CONSTRUCTIBLE REALIZATIONS OF MATHEMATICAL THEORIES

How, it might be asked, can it be possible to construct realizations of the sort of structural description given in typical mathematics texts—unless, that is, there are mathematical objects that can serve as the objects "in the domain of the realization"? One cannot, of course, use pennies or other physical objects as the objects of more complex mathematical structural descriptions, since it is questionable that there are enough such physical objects to do the trick. We evidently need to make use of the Constructibility Theory to get the sort of realizations needed in mathematics. Realizations will thus be taken to be constructible ordered pairs and the "things" in the domain of the realization will also be open-sentences that are constructible.

Extensional identity

Using the terminology of the previous chapter, if properties F and G are satisfied by exactly the same objects, then F will be said to be *coextensive with* or *extensionally identical to* G. Obviously, coextensiveness is an equivalence relation (i.e. is reflexive, symmetric, and transitive).

Similarly, if attributes $\mathcal{F}$ and $\mathcal{G}$ are satisfied by exactly the same properties, then $\mathcal{F}$ will be said to be coextensive with $\mathcal{G}$. Furthermore, coextensiveness can be defined for open-sentences of levels 3, 4, 5, ...

Relations will be understood to be ordered pairs, defined as it was done in Chapter 7.

Statements about structures

Consider statements of the form

> Every structure of such and such a kind
> is such that ... holds in it

and

> If there were a structure of such and such a kind,
> then ... would hold in it.

The structures about which such statements are made in this work are to include the constructible realizations discussed above.

5. THE NATURAL NUMBER REALIZATION

Let us first consider how we might specify constructible structures whose domains (to speak with the Platonist) consist of natural numbers. Of course, within the framework of the Constructibility Theory, the natural numbers are taken to be the finite cardinality attributes discussed in the previous chapter, so that the domain will be given by some "property" of these attributes. The "relations" needed will be open-sentences of the appropriate level and kind.

It should be noted that the "objects" that are to be regarded here as "in the domain of the realization" are not things like pennies that actually exist, but instead are open-sentences that it is possible to construct. It should also be noted that it is possible to construct an infinity of, say, distinct zero attributes, whereas the typical structural description of the natural number sequence postulates a unique element of the domain that is zero. Hence, the distinguished element of this realization that corresponds to, and that is supposedly designated by, the term '0' is not a unique element of the domain: any zero attribute can be taken to be denoted by '0', since in defining the concept of realization of a structural theory or structural description, the identity relation in standard accounts will be replaced here by coextensiveness.

That the structure constructible in the above way is, indeed, a realization of what I called "Peano's theory" was, in effect, shown in the previous chapter. However, it should be recalled that, in order to get a realization of the theory, the lowest-level quantifiers of the language must be taken to be constructibility quantifiers used in asserting the constructibility of such tokens as numerals or lists of strokes: this, in effect, allows the acceptance of a "hypothesis of infinity" (corresponding to the Axiom of Infinity of standard simple type theory).

6. HIGHER-LEVEL CONSTRUCTIBLE REALIZATIONS

We can similarly specify constructible realizations whose domain consists of "real numbers", "sets of real numbers", "sets of sets of real numbers", and so on, giving us the sort of realizations studied in classical analysis and utilized in the mathematics applied so often in the empirical sciences.

To specify such constructible realizations, I will again need a "hypothesis of infinity", which will allow a more or less standard Simple Type-Theoretical development of classical analysis, in which the natural numbers are the finite cardinality attributes discussed earlier, integers are taken to be equivalence classes of ordered pairs of natural numbers, rational numbers are taken to be equivalence classes of ordered pairs of integers, and real numbers are taken to be Dedekind cuts of rational numbers.[5] Within this framework, we can get a

---

[5] See Chihara, 1990: 68–73 for more details on this way of interpreting the level 0 constructibility quantifiers, and 1990: 95 for a brief discussion developing classical analysis within this framework.

kind of standard model of the sort required in the empirical sciences, where it is assumed that representations of physical space require a realization whose domain is isomorphic to the set of all ordered triples of real numbers.

In discussing realizations of this "constructible" sort, it should be kept in mind that the domains of such realizations may not have any members that actually exist at all. Remember, too, that at the lowest level the quantifiers are all constructibility quantifiers and hence do not range over actually existing objects. However, for realizations of this sort, it is possible to construct tokens that belong to the domains of such a realization, even if no such tokens have ever been constructed. The development of classical analysis, within the type-theoretical constructibility framework described above, clearly generates the kind of realization required by the standard mathematical representations of physical space.

7. REALIZATION OF FIRST-ORDER THEORIES

What is it for a structure of the sort being discussed above to be a realization of a theory formalized in the first-order predicate calculus?

(1) If S is a constructible structure and Φ is a sentence of the theory, then in order for S to be a realization of Φ, Φ must turn out to be true when:

    (i) occurrences of the quantifiers in Φ are relativized appropriately to the domain of S;

    (ii) the predicates and terms of the theory are taken to have the extensions given by the appropriate relations in S.[6]

and

    (iii) the identity predicate is taken to be coextensiveness.

Proposition (i) requires some explanation. Relativizing an existential quantifier to the domain of a realization is straightforward in the case in which the domain is given by an open-sentence that picks out actual objects (the variables of quantification are then just restricted to the objects that satisfy the open-sentence). This is standard procedure. But as I noted earlier, the open-sentence determining the domain of the realization may not pick out things

---

[6] Strictly speaking, there would have to be some way given for matching the predicates and terms of the theory with the relations in S. This can be done in a variety of ways, and I omit such considerations for the sake of simplicity of exposition. Perhaps the simplest way would be to define not 'is a model of' but rather 'is a model of (relative to a correlation between the predicates and terms of the theory and the relations in S)'.

that actually exist—it may only characterize things that can be constructed (recall the example of the realization whose "domain open-sentence" picks out finite cardinality attributes). In that case, occurrences of the existential quantifier in Φ should be taken to be (or "replaced by") occurrences of the constructibility quantifier. It follows that, for a realization to be a model of a sentence of the theory, there is no need to assume the actual existence of any mathematical object. In many cases, then, it is only the constructibility of tokens of various kinds that is needed.

I shall not give an example of a constructive realization of any first-order theory here, since in the next chapter, I shall discuss in detail a version of Peano arithmetic and specify a realization of this first-order theory.

9

Applications

One test of a philosophical view of mathematics is to see whether it helps us to understand, at least in a general way, how mathematics is used or "applied" in the empirical sciences, logic, and everyday life. An adequate view of mathematics should facilitate obtaining insights into the role mathematics plays in our theorizing about the physical world and about valid scientific reasoning—*it should not make scientific and everyday uses of mathematics a complete mystery*. This chapter, then, is concerned with the sort of Big Picture first described in the Introduction: we want to see how mathematics fits in with science, logic, and everyday reasoning. We wish to see if the account of mathematics provided thus far allows us to see how mathematics is suited to play the sort of role it has been given in science, logic, and everyday life.

In this chapter, then, I shall be conducting a metatheoretic investigation, in which some part of classical ("real") mathematics, some part of science, and the structural account will be regarded as the object theories. This investigation will be carried out from the perspective of an anti-realist, utilizing the Constructibility Theory as a sort of investigatory tool to facilitate seeing how the mathematical theory can be applied in science, if the structural account is correct, without presupposing either the truth of the theorems of the mathematical theory or even a specification of what the theorems (literally) mean.[1]

Maddy's mystery

The structural view I have championed in this work does face a mystery of sorts. I have been suggesting that mathematical theorems, literally construed, need not be taken to be true. Yet, as Penelope Maddy has argued: "if mathematics isn't true, we need an explanation of why it is all right to treat it as true when we use it in physical science" (Maddy, 1990: 24).

---

[1] It has been objected that, in this chapter, I do not take up all the problems about the applicability of mathematics that philosophers have tackled. I believe, however, that discussing all such problems would detract from the goals of this work and would make the book much too long. After all, some of these problems are not especially relevant to the structural account I am putting forth here.

What does it mean to say that, when mathematics is used in science, it is treated as true? Maddy is not perfectly clear about this point, but I believe that what she has in mind comes to something like the following. In the course of scientific reasoning, we frequently use mathematics to justify an inference to some conclusion about the physical world. The mathematics used in such cases may amount to some mathematical theorem or axiom that functions in the reasoning very much like a premise of an argument. For example, to take a very simple case in which all the premises used are made explicit, we may infer

There are twelve coins on table A at time t

from the premises:

[1] There are five dimes on table A at time t.
[2] There are seven quarters on table A at time t.
[3] A coin is on table A at time t iff it is either a dime or a quarter.
[4] Nothing is both a dime on table A at time t and a quarter on table A at time t.
[5] $5 + 7 = 12$.

Here, we seem to be using '$5 + 7 = 12$' as one of the premises of the reasoning—we seem to be taking the arithmetical theorem to express a fact in the way that 'There are five dimes on table A at time t' does. Furthermore, we can cast doubt on the conclusion of the inference by disputing [5]. Of course, it is difficult to see how one might dispute '$5 + 7 = 12$', but suppose the example had been such that [5] was the sentence '$1,487 + 9,326 = 10,813$'. One can imagine a dispute arising about such a case. Of course, when the mathematical premise is an abstruse theorem of analysis, it is even easier to see how a dispute might arise about the "truth" or "correctness" of the premise.

So I have a problem—something I shall call 'Maddy's mystery'. This is the mystery of why, given my view that the theorems of mathematics (literally construed) need not be true, it is all right to treat them as if they were true in science and everyday life. Although I have already hinted what my response will be to such a potential objection to my view (e.g. in Chapter 2), much more needs to be said. In this chapter, I shall attempt to dissolve this mystery and other related "problems of applications" of mathematics, sketching along the way an account of how mathematics is used in science from the perspective of my structural-constructibility framework. In addition, I shall reexamine various aspects of the indispensability argument, and show the dubiousness of certain premises of the argument—premises that have not been seriously questioned either by extreme Platonists or by defenders of nominalism.

I shall begin my investigations with a study of various applications of number theory, and then take up applications of analysis, before finishing up with a discussion of the use of set theory for the study of the semantics of logic and for the clarification of the logic of the Constructibility Theory (a topic I deferred in Chapter 7).

1. Applications of Constructibility Arithmetic

The constructibility account

I shall begin by analyzing applications of arithmetic using the Constructibility Theory, without any appeal, in this section, to the structural account. The Constructibiity Theory can be used to show how all our common everyday inferences involving the operations of addition and multiplication are indeed valid. As an example of an intuitively correct arithmetical inference that might be made in an everyday situation, consider again the case in which one concludes:

There are twelve coins on table A at time t

from the five premises given earlier. The Constructibility Theory can be used to establish the soundness of this inference in the following way: for each of the premises [1]–[4], as well as the conclusion, there is a corresponding sentence of the Constructibility Theory (which I shall call "the c-version of the sentence") such that the sentence of the natural language is true iff its c-version is true. It can be shown that if the c-versions of the premises [1]–[4] are all true, then the c-version of the above conclusion must also be true. We can thus see that, first, if the premises are all true, the conclusion must also be true; and second, nowhere in this justification is it assumed that [5] must be literally true.[2]

In the above account, I am not claiming that the c-versions of the premises give the meaning of the premises. My Constructibility Theory is not meant to justify a form of *hermeneutic nominalism*. However, as a native speaker of English, I am in a position to know that each of the above premises is true iff its c-version is true, even if I cannot confidently give a semantic analysis of the premise. Taking a page from G. E. Moore's philosophy, my position is that

---

[2] See Chihara, 1990: 89–92 for additional details of this reasoning.

I can know, for example, that *this* (pointing at my right hand) is a hand iff *that* (pointing at my left hand) is a hand, even when I am not in a position to give any precise semantic analysis of the statements I make using the sentence 'This is a hand'.

The semantical problem of applicability

Philosophers have uncovered two problems of applicability regarding such inferences as the above. First, there is what Mark Steiner calls the "semantical problem of applicability": In [5], the numerals '5', '7', and '12' seem to refer to mathematical objects (the numbers five, seven, and twelve, respectively). On the other hand, the occurrences of 'five' and 'seven' in [1] and [2] seem to be functioning as predicates of coins. It has been thought that this equivocation destroys the validity (or at least the correctness) of the inference in question. The philosophical problem, then, is to come up with a semantical account of these sentences in terms of which the above inference can be seen to be a correct one.[3]

Let us see if the "equivocation" described above undermines the constructibility justification that has just been given. Let ϕ be the property

$$x \text{ is a dime on table } A \text{ at } t$$

and let θ be the property

$$x \text{ is a quarter on table } A \text{ at } t.$$

Then the c-version of [1] can be taken to be:

$$\phi \text{ satisfies a 5 attribute}$$

and the c-version of [2] can be taken to be:

$$\theta \text{ satisfies a 7 attribute.}$$

One can then infer, using the definition of addition in the Constructibility Theory and the c-version of [4], that

$$(\phi \vee \theta) \text{ satisfies a } 5 + 7 \text{ attribute.}$$

From this, the c-version of premise [3], and the theorem of the Constructibility Theory corresponding to [5], it is a simple matter to draw the desired conclusion (as I will explain in more detail below). Thus, there is no problem

---

[3] I am here giving Steiner's account of his "semantical problem" of applicability, which is discussed in Steiner, 1995: 132. Steiner attributes the formulation of this problem to Carl Posy.

of equivocation in the constructibility justification given, and this justification validates the inference in question without presupposing the truth of [5] (literally and Platonically construed).

The metaphysical problem of applicability

The second problem was raised by Michael Dummett with the words: "[H]ow can facts about [immaterial objects] have any relevance to the physical universe we inhabit—how, in other words, could a mathematical theory, so understood, be *applied?*"[4]

Steiner understands Dummett's underlying reasons for these "complaints" to be captured by the following argument:

(1) On the Platonist's view, physical laws and theories must express relations between mathematical and non-mathematical objects.
(2) Every relation in physics is a causal (or spatio-temporal) relation.
(3) Mathematical objects do not participate in causal (or spatio-temporal) relations.

Therefore,

(4) On the Platonist view, all physical laws and theories are false. (Steiner, 1998: 21)

Steiner's response to this argument is to point to Frege's analysis of mathematics, according to which "mathematical entities relate, not directly to the physical world, but to concepts; and (some) concepts, obviously, apply to physical objects" (Steiner, 1998: 22). In other words, according to Frege's analysis, premise (2) is simply false: some relations expressed in physics relate physical objects to concepts, and concepts are related to mathematical objects. For Steiner, then, the mystery "vanishes without a trace". Steiner attempts, in this way, to show that the above argument does not constitute a reductio ad absurdum of the Platonist's position after all.

I doubt, however, that Steiner's argument accurately captures the underlying reasons for Dummett's problem or that any such argument is implied by what Dummett says. I don't see why Dummett's metaphysical problem should be taken to arise from an argument with the premise that every relation in physics is causal (or spatio-temporal). If Dummett had intended any such argument, one would expect him to have given reasons for accepting a premise as controversial as (2), but I can find nothing in the section of

---

[4] Dummett, 1991: 301. Cf. Steiner, 1995: 135–7.

Dummett's book from which the above quotation was taken expressing any such reasons.

Let me suggest an alternative way of understanding Dummett's reasoning. If we understand Dummett to be talking about the above number-theoretic inference, for example, we can regard him as making the following point: the use of such theorems of number theory as [5] to draw conclusions about coins is mystifying because [5] is taken to express a theorem of pure mathematics about how the abstract "other-worldly" Platonic objects 5, 7, and 12 are related to one another—a fact that seems to be utterly irrelevant to "this-worldly" facts about little round metal discs on table A.[5] Thus, Dummett criticizes Platonism on the grounds that "it leaves unintelligible how the denizens of this atemporal, supra-sensible realm could have any connection with or bearing upon conditions in the temporal, sensible realm that we inhabit" (Dummett, 1993: 430–1). Michael Liston puts it "in a nutshell" as follows: "[W]hy should appeals to mathematical objects whose very nature is non-physical make any contribution to sound inferences whose conclusions apply to physical objects?" (Liston, 2000: 191).

Of course, this "metaphysical problem" is not a problem for my con-structibility justification, since I do not insist that the numerals '5', '7', and '12' refer to mathematical objects. However, I do hold that one ought to be able to infer the conclusion in question from what we are given above, and I allow that the theorem of the Constructibility Theory that corresponds to [5] provides the reasoner with relevant information that facilitates the reasoning. Thus, as analyzed in Ct, there is little mystery about how the constructibility version of [5] can be relevant to the cardinality facts in question. Having concluded, from the c-versions of premises [1]–[4], and the definition of addition, that the property 'x is a coin on table A at time t' satisfies a $5+7$ attribute, we are allowed to draw the desired conclusion in Ct because the constructibility version of [5] tells us that any property that satisfies a $5+7$ attribute also satisfies a 12 attribute (and conversely). Thus, we can infer that 'x is a coin on table A at time t' satisfies a 12 attribute and hence that there are twelve coins on table A at time t. So there is no mystery of how the theorem corresponding to [5] can facilitate the drawing of the conclusion in question.

The constructibility analysis also makes clear that Steiner's suggested Fregean solution to his version of the problem can also be applied to the

---

[5] Shapiro has raised just such a problem for Platonists, writing: "What does the realm of numbers and sets have to do with the physical worlds studied in science? How can such items shed light on electrons, bridge stability, and market stability?" (Shapiro, 1997: 245). Steiner classifies this problem among the "metaphysical problems" of applicability (Steiner, 1995: 134–5).

problem as I understand it: the reason such appeals to mathematical objects make a contribution to sound inference is because, within the Fregean framework, mathematical objects are related to concepts (cardinal numbers being extensions under which concepts fall), and concepts are related to physical objects. The essential Fregean relationships that are relevant in this example are all mirrored in the relationships given by the constructibility justification. Of course, no references to mathematical objects are needed when the Constructibility Theory is used in place of the Fregean system. One might say that the constructibility justification brings out the essentials of the Fregean solution without the unwanted ontological baggage.

Quine's Platonism reconsidered

Dummett's mystery may lead one to reconsider the Quinean doctrines we discussed at length in Chapter 5. Recall that Quine held that we should believe in the existence of mathematical objects because the postulation of these entities is indispensable for science. Quine thought that we can legitimately postulate the existence of mathematical objects for basically the same reason we postulate the existence of microscopic physical objects (such as atoms and molecules): in both cases, we postulate the unobservable entities in question in order to produce the best (simplest, most fruitful, etc.) account of our sensory experiences. But reconsider our grounds for believing in molecules and atoms. As was noted in Chapter 5, scientists postulated such entities in order to account for such specific phenomena as Brownian motion. Now what if it were learned that, contrary to what Jean Perrin and many other scientists had believed, the molecules of water that they had postulated in order to explain Brownian motion turned out not in fact to exist. Then it would be agreed by all that something else must account for Brownian motion: it could not be the motion of molecules that was causing it. So if it were discovered that molecules of water do not exist after all, then the scientific community would again be faced with the mystery of what is causing the pollen particles to be in constant Brownian motion. In short, the *existence* of the molecules is required in order for many of the scientist's explanations of specific phenomena to work.

Now suppose that there are no mathematical objects. Unlike the molecules postulated by physicists to explain Brownian motion, the mathematical objects postulated to exist by Quineans are not thought to causally interact at all with physical entities. The Quinean never provides us with any scientific explanations that depend upon any hypothesized causal powers of mathematical entities, and there is no specific phenomenon (such as Brownian motion) that mathematical objects are postulated to explain. Evidently, no scientific explanations of specific phenomena would collapse as a result of any hypothetical discovery that no mathematical objects exist. Then why is there any pressing need to postulate the *existence* of mathematical entities? What explanations would collapse as a result of a discovery that there exist no mathematical entities after all? Does not this line of thought strongly suggest that mathematics is not, as Quine had claimed, "on a par with the physics, economics, etc., in which mathematics is said to receive its applications", but functions in science in a way that differs considerably from how these other scientific theories function?

It might be replied that what would collapse are various Platonic *philosophical* explanations. In particular, the Fregean explanation of the validity of the above number theoretic inference would collapse. To such a hypothetical response, I would note that it has just been shown that such inferences can be explained in an alternative way, without presupposing the existence of mathematical objects. In particular, it was shown above how, using the Constructibility Theory, the correctness of the number-theoretic inference in question can be explained. So losing the Fregean Platonic explanation (which requires the existence of such metaphysical entities as concepts and extensions of concepts) is, from my point of view, no tragedy.

Is '5 + 2 = 7' an empirical generalization?

I have heard it claimed that '5 + 2 = 7' is not "completely true", on the grounds that adding 5 gallons of alcohol to 2 gallons of water does not yield 7 gallons of liquid.[6] Apparently, such a view roughly parallels the sort of doctrine advanced by John Stuart Mill, who, according to Donald Gillies, held that theorems of arithmetic are "based on a number of facts which may be verified by experiment and observation such as the following. If I count out 2 apples and put them in an empty box, and count out 2 more apples and put them into the box, then, if I count the apples in the box, I will arrive at the figure four" (Gillies, 1982: 25).

The kind of empirical "fact" described above, however, can hardly be used to justify most theorems of arithmetic. If we chose different things to count, say ants instead of apples, and if we used much larger numbers, the results would frequently not match the corresponding arithmetical theorem.

---

[6] This was once argued by a mathematician during a meeting of the Cambridge University Moral Science Club which I attended during the 1972–3 academic year. Cf. what Michael Detlefsen has said about "the arithmetical behavior of actual physical objects" in Detlefsen, 1986: 33.

(Imagine replacing the two occurrences of the numeral '2' in the above sentence about counting with occurrences of the numerals '13,578' and '34,897', while also replacing the occurrence of 'the figure four' with '48,475'—I dare say, the resulting sentence will frequently not be verified, depending upon who is doing the counting and under what circumstances.)

Interestingly, Frege criticized Mill's view of arithmetic with the words: "That if we pour 2 unit volumes of liquid into 5 unit volumes of liquid we shall have 7 unit volumes of liquid, is not the meaning of the proposition $5 + 2 = 7$, but an application of it".[7] Suppose the following premises are true:

[1] I have 5 gallon cans each filled with alcohol and 2 gallon cans each filled with water.

[2] I pour each of the gallon cans of alcohol into an empty vat; and then, not removing any of the alcohol, I pour each of the gallon cans of water into this vat.

Can I conclude from arithmetic that I will have 7 gallons of liquid in the vat? If so, I should expect all seven of the gallon cans to be filled with liquid when I pour the contents of the vat back into the cans. But, in fact, it will be found that the seventh of the gallon cans, into which the mixture is poured, is only partially filled.

Would such a little "experiment" refute some theorem of arithmetic or logic? Does it show that arithmetical theorems are empirical generalizations after all? Using the analysis of cardinality of the Constructibility Theory, what we can infer from premises [1] and [2] (using logic alone) is that 7 gallons of liquid had been poured into the vat—something that is obviously true and not contradicted by experience.[8] What we cannot infer is that there will remain 7 gallons of liquid in the vat or that each of the 7 gallon cans will be completely filled if the contents of the vat is poured back into the gallon cans. The latter can be only inferred if we are given the added premise that mixing these liquids results in no loss of volume—a premise that is now known from physics to be false. In other words, the fact that adding 5 gallons of alcohol to 2 gallons of water does not yield 7 gallons of liquid does not refute any law of logic or arithmetic but only a mistaken physical assumption about the conservation of volumes of liquids when mixed. Thus, the constructibility

---

[7] Frege, 1959: 13e. Steiner thinks that Frege's objection is "too tame" on the grounds that the given example is not even an application of arithmetic. See Steiner, 1998: 27.

[8] See Chihara, 1990: 93–4 for details.

analysis of cardinality provides us with the correct view that '5 + 2 = 7' is in no way contradicted by the empirical facts cited above, and this should give us one more reason for placing our confidence in the analysis.

Kitcher's objection to classical nominalism

The above points are relevant to an objection that Philip Kitcher has raised to the "classical nominalist", who is accused by Kitcher of facing "the puzzle of why studying the properties of physical inscriptions should be of interest and of service to us in coping with nature" (Kitcher, 1984: 115). I can understand how there could be a puzzle of how studying the *physical* properties of inscriptions can be of general and wide-ranging service to us in coping with nature. But I know of no contemporary nominalist who holds that any mathematical or logical theory is the study of the physical properties of inscriptions.

Consider the nominalistic view that some formula ϕ of a first-order theory T is a theorem of T iff it is possible to construct a numbered sequence of tokens of sentences of T such that each token sentence is either an axiom of T or is inferable from earlier numbered token sentences in the sequence by the rules of inferences of the system. It would be grotesque to classify this view as a theory about the *physical* properties of inscriptions (such as what the inscription is made of, what its chemical composition is, what its mass is, etc.), since nowhere in the justification of such a biconditional will the physical properties of the tokens even be mentioned: what physical properties the tokens in the proof may have are not of any interest and will not enter into any of the considerations. The view is only "about inscriptions" in so far as it is concerned with the constructibility of inscriptions: the emphasis should be on "constructibility". The fact that such a sequence is constructible gives us important information about some token of ϕ, much in the way that the fact that certain operations with scientific instruments are performable can give us important information about some substance or solution. It should be clear to all who have studied the finite cardinality theory of Chapter 7 that the constructibility of an "inscription" that is a one-one relation correlating some property F to some property G can be of service to us in coping with nature.

The Burgess–Rosen question

Let us reconsider the Burgess–Rosen question: what is accomplished by such nominalistic reconstructions of mathematics as is to be found in the

Constructibility Theory? What uses do such reconstructions have? Are they only of use for imagining what the science of alien intelligences might be like, as Burgess and Rosen have suggested?

The above use of the Constructibility Theory to validate our intuitive reasoning about cardinality does not fit the science fiction mold. In the justification given above, it is not being claimed that ordinary people are using constructibility quantifiers when they draw arithmetical inferences or that the Constructibility Theory should actually replace the arithmetical theories we use in making everyday inferences involving cardinality. The Constructibility Theory was only used to show us that the standard arithmetical inferences we draw are indeed correct. The role of the Constructibility Theory was to validate our accepted arithmetical reasoning—not to improve on it. The Constructibility Theory was being used as a sort of tool of the metatheory to facilitate the uncovering of the logical correctness and rationality of our intuitive theorizing about cardinality. This is not science fiction.

Burgess's refutation of the causal theory

At this point, it may be enlightening to see how the Constructibility Theory can also usefully be employed in the assessment of a philosophical argument. Consider Burgess's attempted refutation of a version of the causal theory of knowledge. In his article "Epistemology and Nominalism", Burgess notes that the statement:

Avogadro's number is greater than 6×10^{23}

has been judged by scientists to be not only true, but even known to be true (Burgess, 1990: 6). He also points out that the statement seems to imply that there are numbers. Evidently, we have scientific grounds for asserting that it is known that there are numbers. Consequently, Burgess suggests, science provides us with grounds for rejecting the version of the "causal theory of knowledge" according to which any statement or theory implying that there are objects of a certain sort cannot be known to be true unless some objects of that sort causally interact directly or indirectly with us.

I do not find such objections to the causal theory at all convincing. Certainly, scientist do claim to know such things as:

The number of planets whose orbit is smaller than that of the Earth is 2.

The question is: are these scientists claiming to know something that implies the actual existence of abstract mathematical objects? Suppose we put to these scientists the question: "What justification do you have for making this claim?" I am confident that whatever scientific grounds these scientist would provide in response to our question would amount to no more than grounds for asserting that the open-sentence 'x is a planet whose orbit is smaller than that of the earth' satisfies a 2 attribute. In other words, the grounds that the scientist would supply would be no more than grounds for asserting the constructibility version of the cardinality statement.

Would any of the grounds that a scientist might supply in answer to our question justify, over and above the constructibility cardinality statement, the proposition that there exist entities with which we are in no direct or even indirect causal relations? Not likely. I am confident that scientists would provide a reasonably cautious person with no good reason for asserting: "Now I know that there exist in the actual world abstract entities from which we are completely and totally cut off causally."

This example again illustrates how the Constructibility Theory can be a significant aid to the philosopher—in ways that cannot be accurately characterized as science fiction.

2. PEANO ARITHMETIC

In this section, I shall consider arithmetical theories whose *theorems* are used to draw conclusions about empirical matters. The central question to be investigated will be: must these theorems be regarded as true if they are to be so applied? *The structural account will now be brought in to answer this question.*

Let us consider an arithmetic theory of standard mathematics. For specificity, take as the object of our study first-order Peano arithmetic. It was shown earlier (in Section 3 of Chapter 7) that constructibility versions of Peano's axioms can be proven in **Ct**. It is a simple matter to construct an effective translation function f that takes any sentence ϕ of first-order Peano arithmetic into a sentence $f(\phi)$ of **Ct** such that ϕ is a theorem of the former iff $f(\phi)$ is a theorem of the latter. Such a result enables us to see why it is legitimate to apply the theorems of Peano arithmetic to everyday and scientific reasoning in order to draw conclusions about ordinary objects like tables and coins.

But there is another way we can regard the situation in which first-order Peano arithmetic is applied—one that is more in keeping with the structural account of mathematics of this work. It will be useful at this point to specify a particular version of first-order Peano arithmetic for consideration.

The formal theory PA

Theory **PA** is a deductive theory formalized in the first-order language $\mathcal{L}'$ of Mates's *Elementary Logic* (Mates, 1972).[9] The vocabulary of **PA** consists of the non-logical constants: e, s^1, f^2, and g^2. According to what is called "the intended interpretation" of the theory, e stands for zero, s^1 for the successor operation, f^2 for addition, and g^2 for multiplication. I introduce the following abbreviations in order to make the sentences easier to read:

τ' for $s^1\tau$

$(\tau + \upsilon)$ for $f^2\tau\upsilon$

$(\tau \cdot \upsilon)$ for $g^2\tau\upsilon$

I will now give the axioms of **PA**, using the above abbreviations and omitting parentheses in accordance with standard mathematical practice, again to increase readability.

A1. $(x)(y)(x' = y' \rightarrow x = y)$

A2. $(x)(x' \neq e)$

A3. $(x)(x + e = x)$

A4. $(x)(y)(x + y' = (x + y)')$

A5. $(x)(x \cdot e = e)$

A6 $(x)(y)(x \cdot y' = x \cdot (y + x))$

A7. For every formula Φ, an axiom is any closure of

$$(\Phi\alpha/e \ \& \ (\alpha)(\Phi \rightarrow \Phi\alpha/\alpha') \rightarrow (\alpha)\Phi$$

where α is a variable.[10]

Theory **PA** can be shown to assert all the standard theorems of intuitive arithmetic (when appropriate definitions are added). Furthermore, it can be shown, by standard metamathematical techniques, to be incomplete and undecidable.[11]

I shall now specify a "constructive" model (realization) of the theory, using "entities" of the Constructibility Theory as the things that are to make up the domain of the structure. It will facilitate describing this model if a recursive

[9] See Chapter 9, Section 3, for details.

[10] The notation '$\Phi\alpha/e$' is used to refer to the formula that results from replacing all free occurrences of the variable α in Φ with occurrences of the constant 'e'. I am using the notational convention of Mates, 1972: 50–1.

[11] For details, see, for example, Mendelson, 1987: ch. 3.

rule is given for constructing the natural number attributes that are to be in the domain. I shall assume the standard recursive rule for producing the sequence of arabic natural number numerals starting with '1'. Then the rule for constructing natural number attributes in a standard form will be given by:

First construct the open-sentence:

$$F \text{ is equinumerous with } 'x \neq x'.$$

Then construct the open-sentence

$$F \text{ is equinumerous with } 'x \neq x \vee x = 1'.$$

If one has constructed an open-sentence ϕ according to this rule, then to construct the next one, simply construct an open-sentence that consists of first erasing the right quotation mark in ϕ and then appending to what results the sequence of symbols

$$\vee \, x = \theta'$$

where θ is the arabic numeral that, according to any effective rule for constructing in standard order the finite cardinal arabic numerals, is to be constructed right after constructing the rightmost arabic numeral occurring in ϕ.

Call this rule for constructing natural number attributes, rule R^*.

Rule R^* can be regarded as generating the domain of the structure I am specifying.[12] Of course, we will not speak of the elements of the domain as existing—we shall say, instead, that they can be constructed. The non-logical constants of **PA** can then be interpreted using the definitions and terms of the cardinality theory of the Constructibility Theory as follows:

e: the open-sentence 'F is equinumerous with '$x \neq x$''

s^1: the successor of ①

f^2: ① plus ②

g^2: ① times ②

We know from the cardinality theory of Chapter 7 that, given any natural number attribute $\mathcal{G}$, it is possible to construct a natural number attribute $\mathcal{F}$ according to rule R^* that is a successor of $\mathcal{G}$ and that will be coextensive with any successor of $\mathcal{G}$ that it is possible to construct. Thus, any successor of ① will

---

[12] For many applications of the theory, one would want the domain to include more than what is given by rule R*.

be "unique" in the sense that any two successors of ① would have to be coextensive. We also know that, given any two natural number attributes α and β, it is possible to construct a natural number attribute $\mathcal{F}$ according to rule R^* which is coextensive with an $(\alpha + \beta)$ natural number attribute and that any open-sentence that is an $(\alpha + \beta)$ natural number attribute will be coextensive with $\mathcal{F}$. Thus, it is possible to construct a "unique" $(\alpha + \beta)$ natural number attribute. The obvious analogous thing can be shown to be the case for the product of natural number attributes.

Given what was proven in Chapter 7, it is a relatively straightforward (but tedious) matter to show that each of the axioms of **PA** would come out true under this interpretation (using constructibility quantifiers). We could then conclude that all the theorems of **PA** come out true under this interpretation. These truths would give us information about how open-sentences that are constructible according to rule R^* are mathematically related to other open-sentences that are so constructible. And this information can be applied directly to the cardinality theory of Chapter 7 to enable us to draw conclusions about what must hold in everyday situations and in science. This would allow us to draw, within the framework of the cardinality theory of **Ct**, all the standard conclusions about finite cardinality that Peano arithmetic is thought to establish.

Finally, a few additional definitions would be required to obtain the sort of constructible realizations that were discussed in Chapter 8. We would need to specify the open-sentences D and R that would define the domain and the relations of the realization. Given the above, I trust that the industrious reader could produce the required specification without much trouble.

Applying **PA**

Let us now suppose that the logical language in which **PA** is expressed is expanded to a first-order many-sorted language so that the quantifiers for the natural numbers are distinct from the quantifiers for empirical objects. (The mathematics of this new language will be restricted to the mathematics of **PA**.) This will allow empirical statements, including finite cardinality statements, to be more easily expressed. Assume also that this language is given an NL interpretation $\mathcal{I}$ under which all the members of Γ are true (and not merely true in a structure).

There is one question about $\mathcal{I}$ that should be cleared up. What specific "entities" must be assigned by $\mathcal{I}$ to the natural number terms? Consider, for example, the empirical sentence 'There are five apples on the table'. What must be assigned to 'five' if the sentence is to be true? What has emerged from the detailed work of the structuralist is that the truth value of the above sentence does not depend upon what specific thing is assigned to 'five': anything will do so long as the appropriate structural relations among the things in the domain of the number quantifiers are all preserved—any "standard model" of PA will do as the natural numbers. Thus, having chosen a model in which the domain of the natural numbers is an ω-ordered set, a number term (such as 'five') can be regarded as standing for that object in the domain that satisfies the appropriate structural conditions (is the successor of the successor of the...first member of the ordered set).[13] One can thus describe the above-chosen NL interpretation as one in which the truth values of the sentences of the language will remain the same under all changes in what is assigned to the cardinal number terms and the arithmetical operations, so long as what the cardinal number quantifiers range over is the domain of a standard model of PA and the natural number terms and the arithmetical operation symbols are understood structurally, so that, for example, the terms refer to appropriate "places in the standard model".

Applying PA *without assuming its truth*

Given the situation described in the previous section, let ϕ be a theorem of PA, and suppose that ϕ is used to infer θ from the empirical premises Γ. Can the validity of such an inference be justified without assuming that ϕ is a true statement?

Since ϕ is a theorem of PA, ϕ must be true in all models of PA. We can thus infer that ϕ is true in all models of PA that are consistent with $\mathcal{I}$ and that are "standard" in the sense described in the previous section. It was shown earlier that it is possible to construct a realization of PA that is standard in the above sense. Since all the members of Γ are true statements (under the given interpretation of the language), we can infer that it is possible to construct such standard models of PA that are consistent with the interpretation $\mathcal{I}$. Since θ was inferred from ϕ and Γ, we can infer that θ would be true under any interpretation of the language that yielded a model of both ϕ and Γ. It follows that θ must be true, no matter how the given interpretation of the language might be filled out to make specific assignments to the arithmetical terms, so long as the assignment was consistent with $\mathcal{I}$ and yielded a standard model of PA. But this implies that θ must be a true sentence. Now in all this, it is never

---

[13] This was argued very convincingly by Benacerraf in Benacerraf, 1965.

assumed either that there are natural numbers or that ϕ is a true statement; it is only assumed that ϕ is true in all models of **PA**.

It might be objected, however, that the above reasoning is not admissible to an anti-realist, since some model-theoretic results of first-order logic are assumed. Do not such results presuppose the existence of mathematical objects—just what is rejected by the anti-realist? The anti-realist, however, does not have to appeal to set theory to supply the model theory (and hence the semantic results) needed in the above reasoning: the Constructibility Theory can be used instead. Thus, the anti-realist accepting the Constructiblity Theory can avoid the need to refer to mathematical objects.

It should be noted that, in the above anti-realist strategy, it is not being assumed that the original reasoner (say, a scientist) who infers θ makes use of the Constructibility Theory. (There is no reason to suppose that ordinary non-philosophical reasoners are concerned with the ontological commitments of all their assertions or beliefs.) The Constructibility Theory is required by the anti-realist only when the question of the ontological presuppositions of all the steps in the reasoning come into question. The anti-realist only appeals to the Constructibility Theory to show that it is not necessary to assume the existence of mathematical objects at any step in the reasoning in order to draw the desired conclusion θ.

Notice also that the Constructibility Theory is not being used by the anti-realist in the metatheoretic investigation described above to aid us in imagining how aliens might reason scientifically. We are concerned here with the arithmetical inferences normal people (engineers, scientists, ordinary laborers, etc.) actually make. These investigations thus participate in Frege's project of casting light upon "what mankind has done by instinct [in reasoning arithmetically]" (Frege, 1959: 2e). This is not science fiction.

3. Putnam's If-Thenism

Those readers familiar with standard views in the philosophy of mathematics will probably find what I have just been suggesting to be similar, in important respects, to Hilary Putnam's "if-thenist" account of mathematics.[14] Putnam

---

[14] Putnam proposed this account in Putnam, 1967, a paper that was written as a sort of tribute to Bertrand Russell. It has been very influential in the philosophy of mathematics, especially among nominalistically inclined philosophers of mathematics. I would list Field, 1980 and Hellman, 1989 as showing the influence of Putnam's paper.

once proposed, under the influence of Russell's *Principles of Mathematics*, that "pure mathematics consists of assertions to the effect that *if* anything is a model for a certain system of axioms, *then* it has certain properties" (Putnam, 1967: 294). Putnam did not provide anything like a justification for this view—he did not, for example, provide linguistic or mathematical evidence that pure mathematicians *always* make such assertions. His strategy was to present this Russellian if-thenist view and then to rebut some possible objections and criticisms of it.

One objection that he considers is the following: suppose that we are working in a system S that does not, in fact, have any models. Then the if-then sentence that the pure mathematician asserts about any theorem of S will be vacuously true. Now suppose that S is, say, Zermelo-Fraenkel set theory (that is, ZF) and that ZF does not have any models. Then for every sentence φ of ZF, the if-then sentence about φ would seem to be (trivially) true and hence assertable. This would then trivialize all the assertions of ZF.

Putnam's refutation of this objection rests upon his contention that "the essential business of the pure mathematician may be viewed as deriving logical consequences from sets of axioms" (Putnam, 1967: 302). This is a view that mirrors the doctrine he attributes to Russell that "mathematicians are in the business of showing that *if* there is any structure which satisfies such-and-such axioms (e.g. the axioms of group theory), *then* that structure satisfies such-and-such further statements (some theorems of group theory or other)" (Putnam, 1967: 281). Consequently, the above objection does not raise a genuine problem for his account of pure mathematics since "the business of the mathematician is not in discovering *materially* true propositions of the form 'If *M* is a model of *T* then so-and-so', but in discovering *logically true* propositions of that form" (Putnam, 1967: 291). In other words, Putnam was maintaining that the assertions of pure mathematics consist of logically true propositions of the form 'If *M* is a model of *T* then so-and-so', so even if ZF has no models, that would not imply that any sentence of the form 'If *M* is a model of *T* then so-and-so' would be an assertion of ZF.

But how is pure mathematics applied if, as Putnam claimed, it consists only of such if-then statements? I shall not go into Putnam's attempts to answer this question. Suffice it to say that he responded to this question only insofar as it concerned applications of elementary arithmetic. Such a reply can hardly be considered adequate, since arithmetic is such a small part of mathematics. Perhaps it was when he came to realize that his suggested solution to the problem of application of arithmetic could not be expanded, in any straightforward way, to show how analysis and set theory are applied that he abandoned his deductivist position entirely.[15]

How my account differs from Putnam's if-thenism

My own account of mathematics can be better understood by seeing how it *differs* from the if-thenist position of Putnam's. It should be emphasized right from the start that I have not been proposing any sort of account of what the theorems of pure mathematics, literally construed, assert, and to that extent I have not been advocating the if-thenist analysis of Putnam. Remember that I have disavowed giving any sort of analysis of the meaning of the theorems or assertions of mathematics (literally construed). What I have been suggesting is that, regardless of what a theorem of a mathematical theory may actually assert, a proof of the theorem can be seen to give us roughly the above sort of if-then information. (Recall the notion of "structural content" that was sketched in Chapter 4.) Thus, if a set theorist proves a theorem of **ZF**, then we are in a position to make an assertion that is similar in certain respects to the sort of assertion that Putnam claimed all of pure mathematics consists of. But *I am not making any general claims about the form of all the assertions of pure mathematics, as did Putnam.*

Structural content

It is time to take up again the notion of "structural content" and explicitly specify just what, according to my view, can be inferred from a proof of a theorem of mathematics. Let us start with the precise case in which a theorem ϕ is proved in a theory of first-order logic, say **PA**. Then I claim that we can infer from the proof:

Any realization of **PA** would have to be a realization of ϕ.

The displayed sentence expresses (or gives) what I called earlier the *structural content* of ϕ. Notice that the displayed sentence is a modal sentence; it is not a material conditional of the sort that led to the sort of difficulty described above that Putnam addressed.

There are a few other points of clarification that should be made. I do not wish to restrict information of the modal kind I have in mind to only cases in

---

[15] Field writes (in Field, 1980: 113): "Putnam takes back the view put forth in the earlier paper (Putnam, 1967), claiming in effect that the account given of the application of number theory couldn't possibly be extended to an account of how the theory of functions of real variables is applied to physical magnitudes" (here, Field refers the reader to Putnam, 1979: 74–5).

which a theorem is proved in an axiomatized formal theory such as **PA**. I want to allow that Newton and Leibniz, for example, proved theorems even though they were reasoning in an unformalized and unaxiomatized system of mathematics.[16] They were, as I see it, theorizing about a kind of structure, and hence their proofs were providing mathematicians with structural information. Thus, a proof they gave of some theorem ϕ would provide the following sort of information:

Any structure of the kind about which the mathematician(s) producing the proof is (are) theorizing would have to be a realization of ϕ.

Thus, given a structure **S** of the kind about which the mathematician producing the proof is theorizing, each quantifier in ϕ would have as its range the domain of **S** and each individual constant in ϕ would denote an appropriate element of the domain of **S** and each n-ary predicate would denote an appropriate n-ary relation among the elements of the domain of **S**—all of this such that ϕ is true in the structure. For cases like the Newton–Leibniz one, there is admittedly a great deal of vagueness and unclarity, but for such cases, one should not expect the precision and definiteness of first-order logic (more on this in the next chapter). In general, the structural content of any theorem ϕ of mathematics can be given in the above way, but obviously the earlier one should be used wherever possible. The advantage of dealing with formalized axiomatic theories is that it allows much more clarity and precision.

Finally, I should emphasize that by "a proof of a theorem", I do not intend to restrict what are to count as proofs to only formal proofs or to proofs that only establish what logically follows from a set of axioms. In this respect, too, my position on proofs differs from what Putnam advocated at this time, since he specifically asserted:

[I]n pure mathematics, the business of the mathematician is not in discovering *materially* true propositions of the form 'If M is a model for T then so-and-so', but in discovering *logically true* propositions of that form. Even if a proposition of the form in question is true, if it is only 'true by accident' ..., then it will not be provable by purely formal means, and hence will not be asserted by the mathematician.[17]

---

[16] Just what specific kind of structure these pioneers had in mind is not easy to specify with confidence. I shall discuss this topic in the next chapter.

[17] Putnam, 1967: 291. It should be emphasized, however, that Putnam has since changed his position on what a genuine proof in pure mathematics should be (as will be indicated in the next chapter).

I want to allow that Newton, Leibniz, and Euler gave genuine proofs of mathematical theorems, even when their proofs could not be translated into a formal proof or could not be said to yield the sort of logical knowledge that, say, a proof in first-order logic establishes. (I shall say more about this point in the next chapter.)

4. FREGE AND DUMMETT ON WHY MATHEMATICAL THEOREMS MUST EXPRESS THOUGHTS

Maddy's mystery is closely related to another potential objection to my view that is due to Frege and which has been discussed by Dummett. As I have been emphasizing, according to my structural account of mathematics, a mathematical theory (such as PA) does not have to be true to be used or applied in science and everyday life. Indeed, I even suggested that the theorems of such a mathematical theory need not even express anything that can be said to be true or false (since they can be uninterpreted sentences of a formal language). Such an account has been attacked by Frege on the grounds that a mathematical theorem that is applicable must express thoughts—indeed, true thoughts.

Why did Frege think that, in order to apply a theorem ϕ, ϕ must express a thought? Here's how Dummett explains Frege's thinking:

[Frege] takes the application of a mathematical theorem to be an instance of deductive inference. It is possible to make an inference only from a thought (only from a true thought, that is, from a fact, according to Frege): it would be senseless to speak of inferring to the truth of some conclusion from something that neither was a thought nor expressed one. (Dummett, 1991: 256)

Before I could take this Fregean objection to formalism as a serious objection to my views, I would have to be convinced that the premises of this argument are all reasonably supported. Now one premise of this argument is not supported at all. Why should we accept the Fregean view that any application of a mathematical theorem must be an instance of deductive inference? Dummett tells us that Frege takes applications in this way, but that is no strong reason for us to accept the view. In philosophy, we do not accept the premises of some argument simply because some well-known philosopher assumes that it so.

Also, I can see no plausibility at all in Frege's doctrine that one can make a deductive inference only from a *true* thought.[18] Imagine that Sherlock

---

[18] There are similarities between this Fregean doctrine about deductive inference and Bernard Bolzano's views about the ground–consequence relation. Bolzano held, for example, that any

Holmes is told certain things about the physical condition of a client by Dr Watson and that, from what he is told, Holmes deduces that this client could not be the murderer. But suppose that Watson had made an honest mistake (the client had some medical tests faked) and that the "information" given to Holmes turned out to be wildly inaccurate. Are we to say that Holmes did not make any deductions from what he was told after all? That it is impossible to make an inference from a false thought? I find such a view incredible.[19]

Still, there is much to Frege's objection that needs consideration. So let us investigate further.

An example in which a theorem of PA is applied

I shall describe a case in which PA is used to draw some conclusions about how the diamonds in a particular jewel case are to be apportioned to the members of a gang of jewel thieves. Suppose that the leader of this gang is a nominalist who has studied the arithmetic of the Constructibility Theory (of Chapter 7) and imagine that he knows:

(1) The open-sentence 'x is a diamond in the case' satisfies a k attribute (a natural number attribute) and the open-sentence 'x is a member of the gang' satisfies a j attribute.

and

(2) Any k attribute is a P-descendant of any j attribute.

Suppose further that this leader proves a theorem ϕ of PA which is such that one can infer (5) below. He then reasons as follows:

(3) Since ϕ is a theorem of PA, every realization of PA must satisfy ϕ. (This is, basically, the "structural content of ϕ".)

(4) The constructive realization described in Section 2 above must therefore satisfy ϕ.

---

proposition that was the ground of another proposition had to be true. For more on Bolzano's views about the ground–consequence relation and its similarities to Frege's views about proof, see Chihara, 1999. It should be noted that the above Fregean doctrine can be interpreted in a way that results in a more plausible view: interpret the doctrine to be about "deductive proof" instead of "deductive inference". I can understand someone arguing that there can be no deductive proof that proceeds from false premises, since if the premises of an argument are false, then one doesn't have a proof.

[19] It was suggested to me by William Goodwin that reductio ad absurdum proofs pose another problem for this Fregean doctrine.

(5) Given the theorem of **PA** that was proven, we can conclude from (4) that every k attribute must be a "prime-number" attribute.[20]

(6) From (2) and (5), it follows that there is no way of apportioning the diamonds in such a way that each member gets the same number of diamonds as every other member.

One feature of the above reasoning should be emphasized: φ is never assumed to be a true sentence. Since **PA** is not given an interpretation, φ is an uninterpreted formal sentence and as such cannot be true. Still, φ has a structural content telling us that every realization of **PA** is a realization of φ. Thus, we have a case in which a theorem of **PA** is applied to draw a conclusion about a practical matter even though the theorem itself is not assumed to be true or even meaningful.

Looking beyond the described example

It might be objected that the above reasoning about apportioning the diamonds is such an unreal example that one cannot infer anything about the uses of mathematics in science and engineering. After all, how many scientists work with formalized systems of mathematics such as **PA** or know anything about the Constructibility Theory?

Yes, the reasoning described above is, in some respects, not at all typical. But the point of the example is to show that *it is possible* for a mathematical theory or system to be used in an everyday situation or in a scientific investigation to draw practical inferences, even though the theorems of the system are not assumed to be true or even meaningful. Such examples show that *the way a mathematical theory system is used* to draw practical or scientific inferences may be crucially relevant both to the question of whether an application of mathematics presupposes the truth of the mathematics used and also to the question of whether a confirmation of a scientific theory involving the use of mathematics thereby confirms the mathematics used.

More importantly, the above example is not as atypical as it may seem: the mathematical theorems applied by empirical scientists are not usually uninterpreted formal sentences, but as was shown earlier for the case of **PA**, these theorems do not have to be assumed to be true when they are applied. Furthermore, it will also be shown later that theorems of analysis do not have

---

[20] In calling certain attributes "prime-number" attributes, I do not wish to suggest that I am making reference to numbers (that is, mathematical objects). I only wish to make clear, by the terminology, that such attributes have the sort of divisor properties that one attributes to prime numbers.

to be taken to be true sentences in order to be applied in science and engineering. Thus, in general, theorems of mathematics do not have to be taken to be true to be applied.

A possible Fregean response

A Fregean might reply to my example of the jewel thieves by claiming that we do not have here a case in which a mathematical theorem that does not express a genuine thought is being applied; it is a case in which the Constructibility Theory is being applied and **PA** is only being used "instrumentally". In response to this possible objection, I would note: (a) if this defense of Frege is accepted, then it would seem that the same might be said about Frege's own account of the applications of arithmetic. By parity of reasoning, one could say that, in concrete cases of the sort he analyses, genuine theorems of arithmetic are not being applied; rather, theorems of Frege's own system are being applied. (b) In one ordinary sense of the word 'apply', the theorem of Peano arithmetic in question *is* being applied in the above example—it is not being 'applied' in perhaps a restricted technical sense. (c) Nothing of crucial importance hangs on whether or not, in some sense of the term 'applied', the theorem in question is being applied in the case considered; for my position is not that theorems of mathematics need not be true to be applied, but rather it is that theorems of mathematics need not be true to be *used* in scientific theorizing. (d) Steiner claims that, "for Frege, applying meant 'deducing by means of'" (Steiner, 1998: 25). If that is what 'applying' meant, then one could say that the theorem of Peano arithmetic in question was applied (in that sense) to arrive at the conclusion about apportioning the diamonds—it could be said that the leader of the gang *deduced* that the diamonds could not be apportioned equally *by means of* the theorem.

A more general response to the Fregean objection

My general position has been that theorems of mathematics, whatever their literal meaning may be (assuming that they have a "literal meaning"), do not have to be true to be justifiably used by scientists to draw the inferences they do in their scientific work. This is because each such sentence ϕ has a structural content that allows scientists to make these inferences. Thus, according to my account, when a theorem ϕ of mathematics is applied, ϕ itself may or may not "express a genuine thought" (in the Fregean sense of that expression), but another sentence which expresses the structural content of ϕ does express a genuine thought. Hence, it can be allowed that, even in the Fregean restricted sense of 'apply', there are genuine deductive inferences being made when a mathematical theorem is applied.

5. A Reexamination of Various Indispensability Arguments

I shall now give grounds for doubting the soundness of premise (3) of Resnik's H-N version of the indispensability argument and premise (iii) of Resnik's alternative argument (discussed in Chapter 5). This is the premise that says:

> [*] We are justified in drawing conclusions from and within science only if we are justified in taking the mathematics used in science to be true.[21]

Resnik gives no justification for [*], perhaps because he thinks it is just obviously true. However, I have heard philosophers argue for [*] along the following lines:

> If the mathematics used in science and engineering were not true, how could we, in good conscience, use it in the way we do to build bridges and to design rockets? If the mathematical theorems upon which we base our scientific inferences and theories were false, would not our bridges collapse and our rockets go off target more often than not? So we have good grounds for believing that the mathematics used in science and engineering is true.

The results of this chapter undermine this reasoning: contrary to what Resnik assumes, mathematical theorems do not have to be taken to be true sentences in order to apply them in science and engineering. Thus, one can now give concrete reasons for questioning the fundamental assumption of the holistic version of the indispensability argument (discussed in Chapter 5) that *any* confirmation of a scientific hypothesis thereby confirms *mathematics*, that is, all the mathematics that is used in science. For example, since the version of "Peano arithmetic" mentioned in the above paragraphs need not even be given an NL interpretation, the use of such a version of arithmetic to draw inferences in science can hardly result in a confirmation of the truth of the formal version of arithmetic being used. Such uninterpreted systems can be said to be "realized", "true in a structure", or "true under an

---

[21] See Resnik, 1998: 234, where the quoted sentence expresses a premise of an argument that is formed by coupling his indispensability argument with a pragmatic argument. It also expresses a premise of his "indispensability argument for mathematical realism that separates questions of indispensability from questions of confirmation" given on 233.

interpretation", but they cannot be correctly said to be true. Hence, they can hardly correctly be said to be confirmed.

These points reinforce my earlier doubts about the fundamental assumptions of the holistic version of the indispensability argument. As I noted in Chapter 5, a premise I found especially questionable asserts that it does not matter how mathematics may be used in carrying out a scientific confirmation or even what particular branch of mathematics may be involved in the scientific theorizing, since any confirmation of the scientific theory thereby confirms *mathematics*. In the example (about the apportioning of the diamonds), a theorem of Peano arithmetic plays a role in carrying out the deductions, even though the theorem itself is not confirmed. I find it hard to place any credence in an account of confirmation implying that, however a bit of mathematics may be used in the confirmation of a scientific theory, the experiment or experience confirming the theory also confirms not only that bit of mathematics, but all the mathematics used in science. What an incredible thesis about confirmation these Quineans accept!

I shall now turn to the applications of more complex mathematical theories than number theory: *analysis*.

6. SHAPIRO'S ACCOUNT OF APPLICATIONS

My account of how the mathematics of the analyst is applied in science can better be appreciated by comparing it with an account of applications put forward by Shapiro in the last chapter of his *Philosophy of Mathematics* (Shapiro, 1997). Here is how Shapiro attempts to describe (from his *ante rem* perspective) a class of cases in which the mathematician's structural information is directly applied in the empirical sciences. The empirical scientist is pictured as finding a physical object exemplification of a structure studied by mathematicians—an exemplification that is a system of observable physical objects. The mathematician's theorems can then provide the scientist with information about this exemplified structure, since the "properties of a structure apply to any system that exemplifies it" (Shapiro, 1997: 249). In other words, applications of the sort Shapiro has in mind are made when the scientist discovers in nature a system of physical objects that are related in the way the points in a mathematical structure are related; in that case, the mathematician's theorems about this structure can be, and in many instances are, directly applied by the scientist to this system. As Shapiro sees it: "Mathematics is to reality as universal is to instantiated particular ... the "universal" here is a pattern or structure, and the "particular" is a system of related objects."[22]

Although Shapiro is able to describe a few specific cases of applications of mathematics of this sort, he is forced to admit that mathematics is not, in general, applied in this simple way. For in order to have a system of physical objects that exemplifies some mathematical structure, every place in the structure must be filled with a physical object from the system (Shapiro, 1997: 249). Since most non-trivial applications of mathematics in physics involve structures with uncountably many places (for example, the continuum of space-time points), there seems to be little chance of accounting for many non-trivial applications in the above way—one cannot expect to find systems of observable physical objects with uncountable domains. Thus, much of Shapiro's discussion of how mathematics is applied consists of various attempts to extend the above simple model of application to cover more realistic examples of applications. However, the simple model of applications remains, in Shapiro's work, as the underlying paradigm of how mathematics is applied.

Shapiro's principal idea for expanding his paradigm account of applications is to add modal notions to the mathematical ideas in his theory of structures. This allows him to substitute systems of "possible physical objects" in place of the systems of physical objects, which his simpler account had required. Unfortunately, his ideas on this kind of expansion are only presented in a very sketchy and incomplete fashion. I have only the vaguest of ideas as to how this expansion is supposed to be carried out. For example, here is one way in which Shapiro claims that mathematics is applied:

[T]he theorist describes a class of mathematical objects or structures and claims that this class represents the structure of all possible systems of a certain sort. Relations among the objects or structures represent relations among the possible objects. If the claim is (more or less) correct, the theorems about the class of structures will correspond to facts about the possible systems—about what is and what is not possible. Such claims can be tested in practice. (Shapiro, 1997: 250)

The suggestion is that such claims of the theorist can be more or less true. But if so, then it would seem that relations among the mathematical objects (the "positions") in some structure will have to correspond to relations among

---

[22] Shapiro, 1997: 248. The view of application Shapiro describes above seems to be what Resnik had in mind when he wrote of deductivism: "[I]t appears to account nicely for the applicability of mathematics, both potential and actual; for when one finds a physical structure satisfying the axioms of a mathematical theory, the application of that theory is immediate" (Resnik, 1980: 118).

"possible objects". But how can this be unless there are such things as relations and "possible objects"? One wants to know just what sort of ontology is being presupposed in the metatheory that is being used to explain these applications of mathematics. Up to this point, Shapiro seems willing to include only structures and positions in structures in his "background ontology", writing:

[O]n any structuralist program, *some* background theory is needed. The present options are set theory, modal model theory, and *ante rem* structure theory. ... *Ante rem* structuralism is more perspicuous in that the background is, in a sense, minimal. On this option, we need not assume any more about the background ontology of mathematics than is required by structuralism itself. (Shapiro, 1997: 96)

This explanation seems to require that his "background ontology" include two other sorts of entities: relations and "possible objects". If so, given his acceptance of the Quinean doctrine "no entity without identity" (recall Chapter 4), he needs to have a criterion of identity for both relations and "possible objects". Needless to say, Shapiro never supplies such criteria (and it is doubtful that he could).

Consider, in this regard, the following quotation from Shapiro's book:

Classical mechanics entails that there is at least a continuum of possible configurations of physical objects. There is even a continuum of possible pairs of point masses. We do not have to reify the "possibilities"; we speak of their structures instead. (Shapiro, 1997: 250).

Here, Shapiro seems to be suggesting that, in carrying out theoretical reasoning in classical mechanics, we do not have to refer to (or quantify over) possible configurations of physical objects or possible pairs of point masses: we simply speak instead of "their structures". Fair enough, but how is one to do that? Remember, Shapiro is trying to explain to his fellow philosophers how mathematics is applied in classical mechanics. And his actual written explanation makes heavy reference to "possible configurations of physical objects", "possible systems", "possible pairs of point masses", "all possible physical objects", and "possible collections of physical objects". How are such explanations to be understood if no references to such things are to be made? It is not clear to me.

In summation, I do not contest Shapiro's thesis that there are some applications of mathematics that fit his paradigm model of applications of mathematics. However, the vast majority of cases, and even the typical cases, in which mathematical analysis is applied in the physical sciences do not fit

Shapiro's model. And it is not clear to me just how Shapiro is proposing to explain the more typical cases of applications.

To see how my account of applications of analysis differs from Shapiro's, let us turn to an exciting event in the very recent history of mathematics.

7. FERMAT'S LAST THEOREM

In 1993, the world of mathematics was stunned by the announcement that Fermat's Last Theorem had been proven by the Princeton mathematician Andrew Wiles. News of this result flashed around the world by email and telephone, followed by reports in leading newspapers heralding the event.[23] Why the fuss? Fermat had written, in one of his notebooks, what was in effect the following (natural) number-theoretic assertion:

For all natural numbers, $x, y, z, p > 2$, $x^p + y^p \neq z^p$,

commenting in the margin: "I have a truly marvelous demonstration of this proposition which this margin is too narrow to contain" (Singh, 1997: 62). This happened some time around 1637. Since then, many of the greatest mathematical minds and countless amateur mathematicians have attempted to prove the proposition, but it remained unconquered for these many hundreds of years until Wiles came upon the scene.[24] It should be mentioned, however, that even before Wiles's proof had appeared, mathematicians had been able to prove a number of impressive theorems bearing upon the plausibility of Fermat's conjecture. For example, it had been already been proved that for any counterexample $< x, y, z >$ to the conjecture, x would have to be at least seventeen million digits long (Brown, 1999: 166).

What was so remarkable about Wiles's proof of Fermat's statements was its use of high-powered twentieth-century mathematics to solve this purely number-theoretic riddle. Wiles had proved the theorem via a proof of the Taniyama-Shimura conjecture—a proof involving theorems about modular

---

[23] The date was 23 June 1993. At a conference held in Cambridge, England, Wiles gave a lecture in which he sketched his proof of the theorem. It turned out, however, that the original proof submitted by him had a serious flaw, which could only be repaired after many months of hard work. For more details, see Singh, 1997: ch. 7. For a nice discussion of the history of this problem before Wiles solved it, see Laubenbacher and Pengelley, 1999: ch. 4.

[24] Some of the great mathematicians who attempted to prove the "theorem" were: Euler, Legendre, Gauss, Dirichlet, and Kummer. (Dorrie, 1965: 96).

forms and elliptic curves, which involves theorizing about the complex plane and a four-dimensional space called hyperbolic space.[25]

There are many aspects of the Wiles proof that may puzzle philosophers, but given the nature of this present work, in which mathematical structures have emerged as its central topic, it seems natural to take up the following question: why were investigations into the nature of such conceptually complicated mathematical entities as modular forms in hyperbolic space needed to solve a problem about positions in the much simpler natural number structure? One might think that a question about the natural numbers would be best answered by reasoning directly about the natural number structure. *Why was it necessary for mathematicians to theorize about positions in the much more complicated structure of hyperbolic space in order to prove the theorem?*

To see how this can come about, consider the case in which the theory of functions of complex variables is used to prove the Fundamental Theorem of Algebra:

> Every polynomial of positive degree, with complex coefficients, has a complex zero.[26]

It turns out that the Fundamental Theorem can be used to prove in a relatively simple way the following corollary:

> Every polynomial of degree n, with only real coefficients, can be factored into a product of real linear and real irreducible quadratic polynomials.[27]

The latter is a theorem about the continuum—the set of real numbers. Yet, the proof of the corollary takes us into the realm of complex numbers.[28] This can be understood to be a case in which the mathematician obtains important information about the type of structure that forms the subject matter of the algebraic study of the field of real numbers, and she does this by investigating a more complicated structure in which is embedded the type of structure in question. One can see how reasoning about the larger, encom-

---

[25] The Berkeley mathematician Ken Ribet had earlier proved that the Taniyama-Shimura conjecture implied Fermat's last theorem.

[26] See Birkoff and MacLane, 1953: 107 for a discussion of this theorem. For an elegant proof of the Fundamental Theorem, using the theory of functions of complex variables, see Knopp, 1945: 113–14.

[27] See Birkoff and MacLane, 1953: 110. It should be noted that the theorem can be made more informative by adding to the above statement: "with negative discriminant". I omitted mention of this fact in the above statement because it is not essential for the point I wished to make.

[28] For details, see Birkoff and MacLane, 1953: 110–13.

passing structure can yield information about the embedded structure, since the embedded structure will be related mathematically to the larger structure in countless ways.

Still, it may be wondered why it was useful to theorize about the more complex structure in order to discover various truths about the "entities" in the embedded structure. I am sure there are many explanations that can be given in answer to such wonderment, but one thing stands out as evident: theorizing about features of the more complex structure may make apparent to the "mind's eye" a number of mathematically significant relationships, features, and regularities involving "entities" in the embedded structure that are difficult, if not practically impossible, to "visualize" if one is restricted to reasoning only in terms of the simpler structure. The reason may be that certain sorts of relationships, features, and regularities concerning the "entities" of the simpler structure can only be comprehended when these "entities" are seen in their relationships with the "entities" of the more complex structure. This is a bit like the case in which certain traits of character of a certain member of a family can only be easily noticed in contexts in which this member interacts with a more extended and more diverse group of people than those only in the family.

A mathematical example that illustrates the above idea may be helpful here. Consider the following "diagrammatic proof" of the Pythagorean theorem (see Figure 1). The reasoning would proceed as follows:

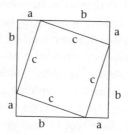

FIGURE 1

[1] The area of the outer square $= (a + b)^2$.
[2] The area of the outer square is also $= c^2 + 4 \times$ (area of triangle abc)—as is obvious from the figure.
[3] But from algebra, we know that:

$$(a + b)^2 = a^2 + 2ab + b^2.$$

[4] Hence, from [1], [2], and [3], $a^2 + 2ab + b^2 = c^2 + 4 \times$ (area of triangle abc).

[5] The area of triangle abc $= 1/2$ ab.

[6] So $4 \times$ (area of triangle abc) $= 2ab$

[7] From [4] and [6], and simplification, we get:

$$a^2 + b^2 = c^2.$$

The role of the diagram is clear. One can regard this diagram as representing the "structure" (pattern) of the triangle abc as part of a larger, more complex encompassing "structure" (pattern), which then makes apparent a number of geometric and algebraic relationships, involving the sides of the triangle abc—relationships that are essential to the above reasoning but that would not be apparent in the absence of the more complex "structure".[29] (This example also indicates one way in which diagrams and pictures can play a significant role in establishing mathematical theorems and in generating mathematical proofs—something to be discussed in the next chapter.)

Returning to the discussion of Wiles's use of very high-powered theorems of analysis to prove Fermat's theorem, this can be seen to be a case in which reasoning about very complicated and sophisticated structures is used to draw conclusions about the much simpler embedded structure of the natural numbers—something that was found in the cases involving the Fundamental Theorem of Algebra and the Pythagorean theorem.

Why such inferred structural knowledge can be applied to actual physical operations and situations is easily understood by the anti-realist making use of the Constructibility Theory. For this kind of structural information can be transferred to the constructibility structures studied in the Constructibility Theory and then applied to situations involving physical processes or events to draw verifiable conclusions. Thus, if certain sorts of real-life situations can be usefully represented in or by a type of mathematical structure, and if this

---

[29] Taking diagrams to be structures is something that Resnik has advocated with his pattern theory of structures. See in this regard Resnik, 1997: 206. In opposition to the widely held view that diagrams and pictures are only heuristic devices that have no place in a mathematical proof, Jon Barwise and John Etchemendy write (in connection with the above proof of the Pythagorean theorem): "Once you have been given the relevant diagram, the rest of the proof is not difficult to figure out. It seems odd to forswear nonlinguistic representation and so be forced to mutilate this elegant proof by constructing an analogous linguistic proof, one no one would ever discover or remember without the use of diagrams" (Barwise and Etchemendy, 1991: 12).

type of structure can be seen to be embedded in a kind of structure studied in analysis, then the mathematical knowledge obtained by analysts can be more or less directly applied to the real-life situations represented. Of course, all of this reasoning could be carried out using the "structures" of the Constructibility Theory.

8. APPLICATIONS OF ANALYSIS: SOME GENERAL CONSIDERATIONS

In the previous section, we have seen in a general way how the structural information that is obtained from the branch of mathematics known as 'analysis' can be of use in science and in everyday life. One reason explanations of the applications of analysis tend to be messier and more complex than the explanations of applications of arithmetic is that applications of analysis frequently require significant idealizations and simplifications of actual empirical situations for purposes of mathematical modeling—something that is not frequently present in applying finite cardinality theory. (I shall discuss such idealizations in more detail later in this chapter.) However, the general outlines of how applications of analysis can be understood should now be coming into focus.

An alternate model of application of analysis

I believe that Shapiro's paradigmatic model of applications should not guide us in our quest for an understanding of how mathematics is applied. My account of the applications of analysis in physics differs from Shapiro's paradigm case in several respects. First of all, I see no reason why the real-life situations being modeled or represented by the mathematician should form a physical-object exemplification of the mathematician's structure. Furthermore, I would prefer to describe the sort of real-life situations being modeled in terms of mathematical structures as those in which *the scientist finds it useful and fruitful to represent certain sorts of occurrences, phenomena, or relationships in terms of the mathematician's structure (or type of structure)*. I shall give reasons for describing such situations in the way I do shortly.

For a model of applications of analysis in physics that sharply contrasts with Shapiro's paradigm case, consider how the physicist Robert Geroch views the use of mathematics in his field:

What one often tries to do in mathematics is to isolate some given structure for concentrated, individual study: what constructions, what results, what definitions, *what relationships are available* in the presence of a certain mathematical structure ...? But this is exactly the sort of thing that can be useful in physics. .... Thus *mathematics can serve to provide a framework within which one deals only with quantities of physical significance, ignoring other, irrelevant things.* (Geroch, 1985: 1, italics mine)

Geroch then adds:

The idea is to isolate mathematical structures, one at a time, to learn what they are and what they can do. Such a body of knowledge, once established, can then be called upon whenever it makes contact with the physics.

Notice that there is no talk here of looking in nature for a system of physical objects (or even "possible objects") that are related in the way the points in some mathematical structures are related, that is, there is no suggestion that physicists look for a "physical-object" or even "possible physical-object" exemplification of mathematical structures. That is just not the sort of thing that the physicist tries to do. Rather, he seems to be looking for abstract frameworks for expressing physically significant features and relationships—frameworks that can be studied independently of the physics and about which much detailed information can be obtained from the mathematician.[30]

The differences between my account of applications and Shapiro's will be made more evident in the following section in which mathematical modeling will be discussed in some detail. Let us keep in mind Geroch's model of applications of mathematics in physics in pondering the discussion to follow.

9. Mathematical Modeling

In their textbook *Mathematical Models and Applications*, Daniel Maki and Maynard Thompson write: "An interest in mathematics by scientists is often generated by a desire to form and study abstract models for real situations" (Maki and Thompson, 1973: p. xi). Their text was designed to be an introduction to the theory and practice of building such models.

They picture the construction of such a model typically taking place following an observation or series of observations, in a laboratory or some sort of relatively controlled situation, which gives rise to a theoretical problem. (For example, a psychologist may observe an unexpected type of behavior in rats

---

[30] Recall, in this connection, the quotation in Chapter 2 from Jaffe and Quinn: "It is now mathematicians who provide [physicists] with reliable new information about the structures they study" (Jaffe and Quinn, 1993: 3).

running a maze.)[31] The scientist then attempts to make the problem as precise as possible, so that a clear and definite statement of the problem can be expressed linguistically. We are told that this stage "typically involves making certain idealizations and approximations" (for example, it may be decided that the color of the rats or the number of compartments in the maze can be omitted in the statement of the problem, but not the fact that all the rats are siblings).[32] The resulting description of the situation is called a "real model".

The real model stage is preliminary to the obtaining of a description of the situation in relatively precise mathematical terms, whereby "the real model becomes a *mathematical model* in which the real quantities and processes are replaced by symbols and mathematical operations" (Maki and Thompson, 1973: 2). What Maki and Thompson envisage, however, is not just a straightforward mathematical description of the situation, but rather a kind of generalized implicit description of the situation, via a full-blown axiomatic system with undefined terms, obtained by "abstraction and quantifying the essential ideas of the real model" (Maki and Thompson, 1973: 2).

The final stage is the construction of a "logical model", which is "an association of real objects with the undefined terms" of the axiom system, resulting in verifiable statements about the real world (Maki and Thompson, 1973: 2). Essentially, what they are calling a "logical model" is what I have called, for the case of first-order logic, an NL interpretation. It is via the logical model that the mathematical model can be seen to make contact with the real model.

Must a mathematical model be "perfect" in order to be useful? That is, must a mathematical model have a logical model that yields only true statements? The authors respond:

It often happens that a mathematical model is very useful even though it does not have an appropriate logical model in the real world. Instead, it *almost* has an appropriate logical model. Here *almost* can mean ... that the mathematical model is true in most real cases, but not in all, or it may mean that even though the mathematical model is never precisely correct, it is almost correct all the time. (Maki and Thompson, 1973: 26).

---

[31] This example is given explicitly in Maki and Thompson, 1973: 2.

[32] Maki and Thompson, 1973: 2. In a more advanced text on mathematical modeling, A. C. Fowler writes: "It is important to realize that all models are idealizations, and are limited in their applicability. In fact, one usually *aims* to over-simplify; the idea is that if a model is basically right, then it can subsequently be made more complicated, but the analysis of it is facilitated by having treated a simpler version" (Fowler, 1997: 3).

Notice that, if a mathematical model has a logical model "in the real world", then we have what is, essentially, a *model* in the familiar sense used by philosophical logicians, that is, an interpretation under which the axioms of the mathematical model are all true.

Now, model building frequently involves the use of differential equations and concepts of continuous mathematics, that is, analysis. In that case, known mathematical results from analysis can be applied to gain information about the mathematical model and a fortiori about the real model:

The results of the mathematical study are theorems, from a mathematical point of view, and predictions, from the empirical point of view. The motivation for the mathematical study is not to produce new mathematics, ... [but] to produce new information about the situation being studied. (Maki and Thompson, 1973: 3)

I shall not present any specific examples of model building, since the text is filled with such examples. One point to notice is that what Maki and Thompson describe does not fit Shapiro's simple paradigm at all. That the model builder should look for physical object exemplifications of mathematical structures conflicts with the model builder's strategy of simplifying, idealizing, and approximating.

Applications of analysis

In what follows, I shall consider applications of theorems of analysis in mathematical modelling. The reasoning will be similar to the deductions given earlier in investigating the applications of **PA** to infer empirical cardinality statements from other empirical statements. (The reader may wish to review the earlier example from Section 2.) Suppose that, after a real model has been devised, a mathematical model M is constructed so that, with the specification of a logical model L, all the axioms of M can be seen to be true (and not merely true in a model)—this, despite the fact that the logical model does not assign specific entities to the mathematical terms of the language, it being understood that, so long as the relevant structural relationships required by the mathematical system are all preserved by the assignment, any specific assignment will do. Let us suppose that, in this situation, a theorem ϕ of analysis is used to infer, from M and L, that some empirical statement θ is true. Can this inference be justified without assuming that ϕ is a true statement?

Since ϕ is a theorem of analysis, we can infer that ϕ must be "true in all models" of the axioms of analysis. Now the logical model L makes assignments to all the "empirical" non-logical constants of the language in such a way that, under this assignment, all the members of M become true statements.

What about the mathematical terms occurring in the sentences of M? We have seen from our structural analysis that the truth value of the members of M will remain the same so long as the referents of the mathematical terms are fixed to yield a model of the axioms of analysis that is standard and consistent with the assignments of L. Now it was shown earlier that it is possible to construct realizations of the axioms of analysis that are standard. Since all the axioms of M, under interpretation L, are true statements, it must be possible to construct standard models of analysis that are compatible with L. Obviously, ϕ is true in all standard models of analysis that fit L. Since θ was inferred from ϕ and M, θ too must be true in all standard models of analysis that fit L, and hence, like the sentences of M, θ must also be a true statement. Now in all the above, it is never assumed either that there are real numbers or that ϕ is a true statement; it is only assumed that ϕ is true in all models of analysis.

The above discussion shows how applications of mathematical analysis in mathematical modeling can be analyzed by means of the structural account. What about applications of analysis in full-blown scientific theories? To this, I shall now turn.

10. ALBERT'S VERSION OF THE MATHEMATICS OF QUANTUM MECHANICS

Since quantum mechanics was singled out by many critics of Field's instrumentalist account of mathematics as posing a serious problem for his program of nominalizing science, I shall examine here the mathematics of quantum mechanics.[33] One advantage of discussing the applications of such a structurally complicated example is that, if the anti-realist can make good sense of such applications, then it will appear much more plausible that the anti-realist's account will also cover the more typical (and simpler) examples of everyday applications of mathematics in science. It will forestall the objection that my account works only for simple cases of application of mathematics.

---

[33] Such an objection was raised by D. Malament in Malament, 1982. More recently, Geoffrey Hellman has written: "[T]here seem to be serious obstacles in the way of Field's program as a strategy for nominalizing modern physical theories, especially quantum mechanics, in which the domains of the models are already highly abstract, and which do not lend themselves readily to a space-time reformulation" (Hellman, 1989: 116 n.).

I shall take up the mathematics of quantum physics in two stages, beginning with David Albert's simplified version. For purposes of getting to the essentials of what he considers the central problem of quantum mechanics, "the problem of measurement", Albert describes a very elementary version of the "mathematical formalism" used in quantum mechanics. This "formalism" is said to be "an algorithm ... for predicting the behaviors of physical systems" (Albert, 1992: 17). The basic mathematical model he has in mind consists of a vector space (called the "state space") with a collection of operators on the vector space.[34]

What is a vector? Vectors are frequently said to be "directed line segments" in order to emphasize two features of vectors: they have both direction and length. Mathematically speaking, in a (real) Euclidean vector space, three-dimensional vectors can be taken to be ordered triples of real numbers. The direction of a vector will then be given by specifying that the ordered triple gives one the position in space of the "head" of the vector (using the standard Cartesian coordinate system), the "tail" being specified to be at the origin; the length of the line segment from the origin to the position of the "head" (using the standard metric) will give one the length of the vector.[35]

Addition of vectors can be defined in terms of addition of the components. Thus, for the three-dimensional Euclidean case,

$$< a, b, c > + < x, y, z > = < a + x, b + y, c + z > .$$

It can be seen that addition of vectors is commutative and associative.

Another important vector operation is scalar multiplication (the multiplication of a vector by a number), which can also be defined in terms of the multiplication of the components. Thus, for the three-dimensional case, scalar multiplication by a real number r is defined:

$$r < a, b, c > = < ra, rb, rc > .$$

It can be seen that the scalar multiplication of a vector by real number r always yields a vector with the same direction as the vector operated on but with a length r times the length of that vector.

---

[34] The spaces actually used in quantum mechanics are Hilbert spaces (the definition of which will be given in the next section). Albert works with the simpler vector spaces in explaining the measurement problem in order to avoid getting bogged down in mathematical technicalities that are not needed to see how the problem arises.

[35] An alternative approach to vectors and vector spaces takes vectors to be certain sorts of mappings or "translations" of sets onto itself. For an elementary treatment of vectors along these lines, see Vaughan and Szabo, 1971.

In the following, the notation '$|A>$' will be used to denote vectors; scalar multiplication will be denoted by '$r|A>$'. It is easy to prove that scalar multiplication is distributive over vector addition, that is:

$$r(|A> + |B>) = r|A> + r|B>$$

A vector (or linear) space is a totality of vectors closed under addition and scalar multiplication of vectors. Every vector in the state space is called a "state vector". Each operator is a mapping of the vector space into itself. Every physical system is associated in the algorithm with some particular vector space. The various physical states of any system of particles correspond to the state vectors. Thus, the vector space is taken to represent all the possible physical states of the particles in question.

One kind of operator, the *linear operator*, plays a central role in this "algorithm". A linear operator O is an operator that satisfies the following conditions:

For any vectors $|A>$ and $|B>$, and any real number c,

$$O(|A> + |B>) = O|A> + O|B>$$

.

and

$$O(c|A>) = c(O|A),$$

where $O|\phi>$ is the vector that results from applying O to $|\phi>$.

The measurable properties of the physical states of the particles of a system are to be represented by linear operators (of a certain sort) on the vector space.

Suppose, for example, that the system to be discussed concerns the states of some particular electron and that the measurable property of that electron to be represented by some linear operator O is the angular momentum with which the electron is spinning about an axis through its center and which runs along the x-direction.[36] Suppose that state vector $|s>$ represents some possible physical state of the electron in question. A central idea of the

---

[36] The angular momentum is called "spin", even though the electron is not literally rotating; however, in some ways, it acts as if it were. See Steiner, 1998: 84–6 for a nice discussion of "spin" (of an electron) and its mysterious property of always being in only one of two possible states called "spin-up" and "spin-down". (Note: spin is only up or down only for the electron and other so-called "spin-1/2 systems".) In these pages, Steiner also describes another mysterious feature of electron spin.

formalism is very roughly the following: If $O|s>$ happens to be $|\phi>$, this would tell us that a measurement of that angular momentum carried out when the electron is in that state will yield result $|\phi>$

As a step in making this rough idea more accurate, the notions of an *eigenvector* of an operator O and also of *eigenvalue* are explained as follows: If $O|B> = k|B>$, where k is a real number, then $|B>$ is an eigenvector of O with eigenvalue k. In other words, if O operates on vector $|B>$ and does not change its direction, then the resulting vector is an eigenvector of O with eigenvalue equal to its length (relative to the length $|B>$).[37] Then, the rough idea above can be given the more accurate statement: if $O|s>$ happens to be an eigenvector with eigenvalue ρ, this would tell us that a measurement of that angular momentum carried out when the electron is in the state represented by $|s>$ will yield result ρ.

What if $O|s>$ happens to be a vector that is not an eigenvector of O? It is here that probability theory enters into the mathematics of the system: quantum mechanics tells us that the outcome of such a measurement is a matter of probability. Suppose that we are dealing with a system in a state represented by state vector $|A>$, and suppose that we wish to know the value of a measurement of property B on that system in that state. Finally, suppose that the eigenvectors of the property operator O for B are:

$$|B = b_i > \text{ with eigenvalue } i.$$

The theory says that the probability that a measurement of B will yield the outcome that this property is b_i is the square of the absolute value of the *inner product* of $|A>$ and $|B=b_i>$, where the inner product (vector product or scalar product) of any vectors $|C>$ and $|D>$ is the length of $|C>$ times the length of $|D>$ times the cosine of the angle between $|C>$ and $|D>$. If the vectors are in a real Euclidean space, represented in terms of coordinates (as ordered n-tuples of real numbers), the inner product of $|C>$ and $|D>$ can be calculated as the sum of the products of the corresponding coordinates. For example, the inner product of $<1, 2, 3>$ and $<4, 5, 6>$ is $(1 \times 4) + (2 \times 5) + (3 \times 6) = 32$.[38] Clearly, the length of any vector $|C>$ is equal to the square root of the inner product of $|C>$ and $|C>$ (since the cosine of 0 is 1).

---

[37] See Albert, 1992: 29 for a fuller explanation of the notions of eigenvectors and eigenvalues.
[38] Borowski and Borwein, 1991: 521. See Albert, 1992: 35–6 for more details on how probability enters into these calculations.

Every state vector, subject to various forces and physical constraints, changes over time according to a system of dynamical laws known as the 'Schrödinger equations', and the changes in the state vectors determined by these laws are always changes in direction and not of length.

Up to now, I have been discussing only *real* vector spaces, vector spaces in which the vectors are regarded as ordered *n*-tuples of real numbers and in which the scalars are all real numbers. It turns out that the vector spaces of quantum mechanics are actually *complex* vector spaces, that is, vector spaces in which the state vectors can be regarded as sequences of complex numbers and in which complex numbers are used in scalar multiplication. Thus, if $|B>$ is a state vector, and c is a complex number, then the vector $c|B>$ is also a state vector.

Since the vector spaces used in quantum mechanics are complex vector spaces, it is easy to see why the measurable properties can be represented only by certain kinds of linear operators on the vector space. It turns out that "the values of physically measurable quantities are always real numbers" (Albert, 1992: 40). But not all linear operators on a complex vector space yield only real number eigenvalues when operating on state vectors. Thus, the measurable properties can be represented only by *Hermitian* linear operators, since, crudely put, Hermitian operators are linear operators all of whose eigenvectors have only real number eigenvalues.[39]

Instead of discussing more of the technical details of Albert's version of the mathematics of quantum mechanics—these details are not needed for the points I wish to make—let us consider the task of specifying, within the framework of the Constructibility Theory, the mathematical structures needed by the scientific theory.

Albert's system within the Constructibility Theory

Enough has been sketched of Albert's version of this mathematics to see that there would be no great difficulty in developing such a mathematical system within the Constructibility Theory. Taking the natural numbers to be the natural number attributes described in Chapter 7, one can define "integers", "rational numbers", and "real numbers" in the way described earlier (in Chapter 8, Section 6). Then the "complex numbers" (complex number open-sentences) can be defined in the Constructibility Theory to be, in

---

[39] The reader can find a more mathematically detailed explanation of what Hermitian operators are and why they are needed to represent measurable properties in Albert, 1992: 40–1.

effect, equivalence classes of ordered pairs of "real numbers" that satisfy certain conditions. The vectors of a complex vectors space can then be regarded as sequences of the "complex numbers" for which addition and scalar multiplication of these sequences are so defined as to satisfy certain other conditions—all of which can be straightforwardly specified within the Constructibility Theory. Since the vector spaces of Albert's mathematics is a totality of these vectors satisfying certain closure conditions, such totalities are easily constructible within the framework of the Constructibility Theory.

11. THE MATHEMATICS OF QUANTUM PHYSICS

The vector spaces needed in quantum mechanics are complex vector spaces with an *inner product*, which (as it was described earlier) is a mapping of the Cartesian product of the space with itself into the set of complex numbers, satisfying certain other conditions.[40] To specify such mappings is, of course, not difficult within the Constructibility Theory, and such spaces are easily constructible in the theory. A complex vector space with an inner product is called a "pre-Hilbert space" (Blank, Exner, and Havlicek 1994: 4). A Hilbert space is a pre-Hilbert space that is "complete", that is, every Cauchy sequence of elements of the pre-Hilbert space converges to an element of the space.[41] There is no serious problem of showing, within the framework of the Constructibility Theory, the constructibility of pre-Hilbert spaces that are complete in the relevant respect. The introduction of probability theory would not pose any additional serious problems for the present task of specifying a constructibility model of the mathematics needed in quantum mechanics, since the mathematics of the needed probability theory can be easily reproduced within **Ct**.

It is easy to see, from the above, that the mathematics of quantum mechanics can be developed within the Constructibility Theory—a result that

---

[40] See Blank, Exner, and Havlicek, 1994: 4 for more details.

[41] Barrett, 2001: 251. Von Neumann required a Hilbert space to satisfy an additional requirement: it had to be "separable", which according to Jeffrey Barrett, "was meant to prevent the space from being so large that one loses some of the structure that allows one to understand its elements as representing quantum-mechanical states" (Barrett, 2001: 251). Mathematicians will be familiar with the *HarperCollins Dictionary of Mathematics* definition of a Hilbert space as "a real or complex LINEAR SPACE on which an INNER PRODUCT is defined, and in which all CAUCHY SEQUENCES converge to a limit" (Borowski and Borwein, 1991: 267). A "linear space" is said to be a term for vector space, "especially one consisting of ordered *n*-tuples of real or complex numbers" (Borowski and Borwein, 1991: 346). For a more mathematically advanced discussion of the Hilbert spaces used in quantum mechanics, see Blank, Exner, and Havlicek, 1994: ch. 2.

is significant for our investigations for two reasons: first, we can be confident that the mathematical reasoning needed in quantum mechanics can be validly carried out without requiring an appeal to (or quantification over) mathematical objects; and second, we can be assured that the sort of constructibility realization described earlier for the case of arithmetic can be specified for the case of the mathematics of quantum mechanics—a realization that can be described as a standard model, in which natural number attributes play the role of the natural numbers so that the standard applications of finite cardinality can be logically carried out in this framework.

The models of quantum mechanics

In his book *The Scientific Image*, Bas van Fraassen notes that in many textbooks and treatises on quantum mechanics we find what are described as 'axioms of quantum theory'. For example, we find such "axioms" as:

(1) To every pure state corresponds a vector, and to all the pure states of a system, a Hilbert space of vectors.
(2) To every observable (physical magnitude) there corresponds a Hermitian operator of this Hilbert space.
(3) The possible values of an observable are the eigenvalues of its corresponding operator.
(4) The expectation value of observable A in state W equals the trace $Tr(AW)$.[42]

According to van Fraassen, we should view these axioms as a "description of the models of the theory plus a specification of what the empirical substructures are".[43]

The idea that scientific theories should be understood and analyzed in terms of the models they characterize is a central theme of an important group of philosophers of science, all of whom adopt a "semantical approach" to theories (see Suppe, 1974: sect. V.C). The following analysis does not depend upon the correctness of the model-theoretic approach to quantum theory, but my analysis is consistent with van Fraassen's account.

---

[42] Van Fraassen, 1980: 65. It should be noted, however, that the notion of "observable" used by the physicist does not correspond to the special sense van Fraassen gives to the term in his book.
[43] Van Fraassen, 1980: 65. I disagree with van Fraassen, however, that such a description gives a specification of "what the empirical substructures are" (see the previous note).

Applications of Hilbert space theory

Let us consider how a scientist might apply a mathematical theorem φ of Hilbert space theory to infer an empirical conclusion θ from Γ, which consists of true empirical statements of quantum theory. As before, the question is: must φ be assumed to be true to justify such an inference?

As for the case of mathematical modeling described earlier, we can assume that all the members of Γ will be true no matter what things are assigned as the referents of the mathematical terms of the language, so long as the assignments yield a model of the Hilbert space theory that is standard and consistent with the characterization of the models of quantum theory given by the axioms. Since φ is a theorem of Hilbert space theory, we can infer that φ is true in all the standard models of Hilbert space theory that are consistent with the structural characterization given by the axioms of quantum theory. Earlier, it was indicated how one can construct a standard realization of Hilbert space theory. We can infer that it is possible to construct standard realizations of Hilbert space theory that are consistent with the structural characterization given by the axioms of quantum theory. Since θ can be inferred from φ and Γ, we can conclude that, as was the case for the members of Γ, θ too must be true no matter what things are assigned as the referents of the mathematical terms of the language, so long as the assignments yield a standard model of the Hilbert space theory that is consistent with the structural characterization of the quantum theory. Thus, we can infer that θ must be true, without making the assumption that φ is a true sentence or that there are mathematical objects.

Mixed statements

Let us take stock at this point and compare the kind of justification just given of the inferences scientists draw in quantum mechanics using the mathematics of Hilbert space theory with my earlier justification of the inferences we draw about cardinality using Peano arithmetic. In the number-theoretic case, theorems are able to yield valuable information about real world situations through the intermediary of the kind of connecting empirical statements that Frege called "statements of number"[44]—statements of the form:

There are five apples on the table.

The number of enrolled students in the class is 75.

There are seven fewer candy bars than there are children.

---

[44] See Frege, 1959: Part IV.

Notice that such statements include terms that are neither logical nor number-theoretical. Such "mixed statements" connect (or "bridge") the structural mathematical theory with things in the physical world. Obviously, there will be such "mixed statements" connecting the structural mathematical theory of quantum mechanics with things in the physical world. But we should not expect the latter kind of "mixed statement" to have all the nice features of the former kind. For the case of statements of number, one can state, within the framework of the Constructibility Theory, relatively simple and precise methods of determining the truth value of such statements. Obviously, we cannot expect to find such simple and precise methods for determining the truth values of the mixed statements needed in quantum theory.

What do the points in the constructive realization represent?

Let us now survey, with Geroch's comments in mind (see Section 8, above), the general features of the mathematical formalism described above. This formalism, in effect, describes certain mathematical structures, whose points are vectors and operators on vectors. The physicist uses this formalism to represent such things as the relationships among the measurable properties of particles and the development of states of particles over space and time, all this within a framework in which one deals only with "quantities of physical significance, ignoring other, irrelevant things". There is a great deal that is known about these structures, since mathematicians have studied them in depth, and the physicist can make use of this knowledge in her analyses of physical systems.

Obviously, it would be a mistake to regard the points or positions of such structures as representing physical objects. When the mathematical formalism is applied by the physicist to determine, say, the results of a particular measurement of some particular state of some particular electron, the physicist does not expect to find physical objects corresponding to the points in the mathematical structures described by the formalism—physical objects which are supposed to be related to one another in just the way the points are related to one another as was required by Shapiro's paradigm model. It would be better, surely, to think of the points in the structures as representing states, properties, relationships, and operations.[45] Thus, among the measurable

---

[45] Furthermore, the particular system of mathematics used by the quantum physicist is not so much *discovered* as *chosen*. We do not have here a case in which scientists discover in nature a system

properties represented by the linear Hermitian operators of quantum mechanics are the "coordinate-space properties", that is, such properties as position, momentum, velocity, and energy (Albert, 1992: 49 n.).

Notice that the metatheoretic discussion of the "coordinate-space properties", which are supposed to be "represented by the linear operators", does not require a special ontology of abstract entities, since the discussion can be carried out by means of appropriate open-sentences of the form:

x is the position of y at time t,

x is the momentum of y at time t,

x is the velocity of y at time t,

x is the kinetic energy of y at time t,

etc.,

where the variables occurring in the open-sentences are taken to be appropriate variables of the Constructibility Theory. Thus, consider the open-sentence

x is the position of y at time t.

The variable x can be specified to range over ordered triples of real numbers, y can denote some specific physical particle α, and z can be specified to range over real numbers—all of this within the framework of the Constructibility Theory. Using constructibility quantifiers, we can make specific assertions about α, such as

$<\pi, 2.76, 12>$ is the position of α at time 3.98,

---

of physical objects that exemplifies the Hilbert space structure, since not every place in the structure is filled with a physical object from the system. Some features of the mathematical structure used in this representation were chosen for methodological or pragmatic reasons. For example, it has been argued that the rationale for taking this Hilbert space to be separable is partially based on convenience ("it allows us to use countable orthonormal bases" (Blank, Exner, and Havlicek, 1994: 255) and partially because of "a heuristic argument" (see Blank, Exner, and Havlicek, 1994: 380), remark 11.1.1 (b)). The rationale for taking it to be complete has been thought by some physicists to be a matter of "mathematical simplicity". Thus, Blank writes: "The requirement of mathematical simplicity leads us to the assumption that the state space is complete, i.e. a Hilbert space." (Blank, Exner, and Harlicek, 1994: 254). Peressini has noted these less than decisive grounds for the physicist's choice of the particular mathematical structure used in quantum theory in questioning whether all the features of the "mathematized physical theory" are as indispensable as one might think (Peressini, 1997: 225 n. 19). Even the choice of a *complex* Hilbert space has been said to be "to certain extent arbitrary" (Blank, Exner, and Harlicek, 1994: 255).

by writing:

(Cφ)(Cφ)(φ satisfies condition – – – and θ satisfies condition ∗ ∗ ∗ such that
φ is the position of α at time θ

where condition – – – (written out in the notation of Ct) guarantees that φ
is the ordered triple < π, 2.76, 12 > and condition ∗ ∗ ∗ (written out in the
notation of Ct) guarantees that θ is the real number 3.98.[46]

In other words, the detailed metatheoretic discussion of the application of
analysis does not require an ontology of properties and relations in order to
make sense of the above talk of operators on vectors representing properties
and relations.[47]

What still needs to be done is to explain how the truth values of such
"mixed" mathematical statements as

500 miles per minute is the present velocity of the rocket

can be determined or at least estimated in certain situations. This will be done
in what follows.

Mixed mathematical statements and lengths

A mixed mathematical statement is one that contains both mathematical
terms and also words or expressions from common everyday language or
scientific discourse that are not found in the sentences of pure mathematics.
I have already discussed one sort of mixed mathematical statement, namely
what might be called "mixed cardinality statements". For example, the
statement

There are five apples on the table

is a mixed mathematical statement the c-version of which was analyzed
earlier in Chapter 7.

---

[46] The terms 'π', '2.76', '12', and '3.98' in the above explanation should not be taken to refer to
Platonic mathematical entities but instead should be understood within the framework of the
Constructibility Theory.

[47] There are, of course, other properties represented by the linear Hermitian operators of
quantum mechanics that are not coordinate-space properties, as are the properties expressed by
such open-sentences as 'x is the mass of y at time t', 'x is the charge of y at time t', and 'x is the spin
of y at time t'. Clearly, these do not raise any major new difficulties for the nominalistic view being
developed here.

I shall concentrate here on mixed mathematical statements that have the form:

A is in position y.

Such statements can be analyzed in terms of statements of the form

A is r meters from B,

where A and B are taken to be points on a straight line segment and r is a variable ranging over real number open-sentences, as defined in the Constructibility Theory.

For this analysis, I shall make use of what was done in an earlier work of mine. In chapter 6 of *Constructibility and Mathematical Existence* (Chihara, 1990), I set out a system within which "length" was defined for certain idealized entities that I called "straight line segments". These entities can be taken to be the straight lines segments constructible in Euclid's geometry. They are idealizations of the "straight lines" we draw on paper with straight-edges. The main idea was to assume, as primitive notions, *equal in length* and *greater in length*, so that, on the basis of a few basic axioms about these notions—it was assumed, for example, that *equal in length* was an equivalence relation and that *greater in length* was transitive and irreflexive—one could develop a substantial body of theorems about length if one made use of the vocabulary and logic of mereology. Only a few intuitively plausible principles about parts and wholes are needed from mereology, since the Constructibility Theory (especially the cardinality portion of the theory) is used to develop much of the mathematics of the theory of length. It turns out that if one believes, as Hartry Field does, that the space-time points of physics exist, as do arbitrary regions of these space-time points, then the mereological postulates of this system can be taken to be obviously true.

Within this system, by taking some line segment to be of unit length, truth conditions can be given of open-sentences of the form

x is r units long

for r a rational number open-sentence of the Constructibility Theory.

A line segment that satisfies such an open-sentence is called a "rational line segment". It was shown that if x and y both satisfied the same open-sentence of the above form, then x would have to be equal in length to y. Consequently, the totality of rational line segments can be distributed into mutually exclusive "equivalence classes"[48] that completely exhaust this totality—each rational line segment satisfying one and only one of these classes, all the "members" of which are equal in length. Thus, in basically the way cardinality attributes were defined above in Chapter 6, *rational lengths* are defined in this system to be "equivalence classes" of rational line segments. Addition and multiplication of rational lengths are so defined that these operations on rational lengths can be seen to structurally mirror the standard arithmetical operations on non-negative rational numbers. From the Platonist's realistic perspective, the structure sketched here, whose domain consists of the rational lengths and whose operations mirror addition and multiplication of rational number attributes, can be seen to form a model of the classical theory of non-negative rational numbers.

The system of rational lengths can then be expanded to include *real lengths*, where a real length is defined to be, essentially, a Dedekind cut of rational lengths. Again, the operations of addition and multiplication of these real lengths can be defined pretty much as they are defined in standard analysis texts, so that, from the Platonic position, we can be seen to get a model of the classical theory of non-negative real numbers.

Strictly speaking, of course, these developments truly yield such models only if the postulates and assumptions of the theory are genuinely true—something that many Platonists would tend to accept, but which I do not. Instead, I treat the theory of lengths described here much as physicists treat scientific theories that appear to postulate the existence of certain ideal entities, such as extensionless particles, completely rigid bodies, and absolutely incompressible liquids—fictional entities which real existing things only approximate.[49] The differences between the observable behaviors of the postulated ideal entities and those of the actual real entities theorized about may be negligible, for many purposes for which the scientist's theories were devised, but careful theoreticians would expect to have to adjust their calculations to take account of these relatively small differences, depending upon the situation and goals.[50] Of course, applied mathematicians

---

[48] Strictly speaking, what I am calling an "equivalence class" is to be an open-sentence discussed in the Constructibility Theory. I use the Platonic language for didactic purposes, since most readers can be expected to follow the train of thought above much more easily when I use the terminology of the Platonist.   [49] See the discussion in Chihara, 1990: 101–6 for more details.

[50] The task of taking account of such differences would be similar, in some respects, to the task of estimating errors that Suppes discusses in Suppes, 1967, and hence would undoubtedly be a very complex problem in many cases.

and scientists are well aware of the fact that "real world situations" that have been mathematically modeled through idealizations and approximations cannot be expected to fit exactly the mathematical model constructed— a point to be emphasized below in the section on the Fundamental Theorem of Calculus.[51] Since the fit between such mathematical models and the "real world situations" modeled is frequently less than perfect, one cannot expect to find the sort of "exemplification" relation that is required by the simplified Shapiro paradigm of applications.

Let us now turn to the lengths of actual "physical straight lines". Instead of ideal straight line segments, I shall concern myself here with such things as *lines drawn on paper, edges of boards, streaks made by beams of light*, and so on, which can be considered to only approximate the ideal straight line segments discussed in Euclid's geometry. We shall need physical criteria or genuine methods of measurement for determining the (approximate) truth of "atomic statements" of the form

$$x \text{ is equal in length to } y,$$
$$x \text{ is greater in length to } y,$$

where x and y are taken to be such physical line segments.

We obviously do have a variety of methods for determining the approximate truth of such statements, using, for example, such instruments as meter sticks, tape measures, and optical devices (such as range finders). These criteria can also be used to determine the approximate truth of statements of the form

$$x \text{ is a part of } y$$

and

$$x \text{ is a line segment that is disjoint from } y$$

allowing us to estimate the truth of statements of the form

$$A \text{ is } r \text{ meters from } B,$$

where A and B are physical points on a physical straight line segment and r is a variable whose values are rational or real number open-sentences of the Constructibility Theory.

---

[51] See for example Maki and Thompson, 1973: a whole textbook dedicated to the mathematical modeling of such "real world situations".

Our understanding of such statements would rely upon methods of approximation and estimation derived from the ideal truth conditions of statements of the form

The length of *x* is *r* units,

which were given in the ideal theory of rational and real lengths (described above).

From these beginnings, it is possible to extend the above treatment of ideal lengths to obtain a theory of ideal positions in space. This can be done by reference to a trio of ideal straight lines, each of these lines being perpendicular to a plane determined by the other two lines. These lines, the *x*-axis, *y*-axis, and *z*-axis, allow us to take an ideal point in space to be determined by an ordered triple of real numbers, the first number giving the distance in units from the origin along the *x*-axis, the second giving such a distance along the *y*-axis, and the third giving such a distance along the *z*-axis. In this way, one can obtain what is essentially a "foundation" for the standard treatment of motion in space using Cartesian coordinates and the classical mathematics of motion as pioneered by Newton and Leibniz.

We saw earlier (in Chapter 5), however, that Penelope Maddy had assembled a case for questioning the actual continuity of physical space. So it needs to be emphasized that I am not here claiming to justify the assumption that physical space is actually continuous. The above developments were conspicuously based upon assumptions and axioms that were said to be idealizations. So what is being justified is the use of the classical treatment of motion in space as an idealization of actual motions. On the other hand, I am assuming that we have various empirical criteria and methods for determining whether the scientist's analyses of motion, based upon such idealized assumptions, are workable and accurate for practical purposes.[52]

Other mixed mathematical statements

What about the other types of mixed statements listed above? I shall not go into the details of explaining how scientists have developed methods of empirically determining or estimating the truth values of such statements, within the framework being developed here. The foundations for such a development are to be found in standard texts of measurement

---

[52] Again, it should be emphasized that the analysis given above is not meant to explain how ordinary people and scientists actually reason. Recall my comments at the end of Section 2.

theory, coupled with the theory of functions of real and complex variables, developable within the Constructibility Theory.

I turn now to a more detailed treatment of the way scientists have developed practical and technically feasible procedures for doing such things as drawing lines and curves, constructing perpendiculars, tangents to curves, bisections of angles, and so on, of which the constructions talked about in geometry can be regarded as idealizations.

12. THE FUNDAMENTAL THEOREM OF THE INTEGRAL CALCULUS

A theorem of classical analysis that is known to every engineering student is the Fundamental Theorem of the integral calculus.[53] It states:

> For any real-valued function of a real variable, $f(x)$, which is continuous in the interval (a, b), if $F(x)$ is an indefinite integral of $f(x)$, then
>
> $$\int_a^b f(x)dx = F(b) - F(a)$$

I can imagine a mathematical realist arguing against my position as follows:

> We all know what this theorem says and we know that it is true: not only have we have proved its truth, we have even verified its truth in countless ways. Take, for the sake of simplicity, the case in which the function $f(x)$ is positive throughout the interval $[a, b]$. We know that the definite integral $\int_a^b f(x)dx$ gives us the area under the graph of $f(x)$ in the interval $[a, b]$. We have graphed a large number of such functions on engineering paper and empirically estimated what $\int_a^b f(x)dx$ is by counting squares and, sometimes, making estimates of the area under the curve in a particular square. In each case, by calculating $F(b) - F(a)$, we have determined that we get a value that is close, if not identical, to our calculation of $\int_a^b f(x)dx$. We have also found that, by making our estimates more carefully and more accurately, we can get an even closer fit. This kind of checking of

---

[53] This theorem was discovered by one of the originators of the calculus, Gottfried Leibniz. See Boyer, 1985: 442. Most students of the calculus would probably not realize from Leibniz's statement of the theorem that the Fundamental Theorem of the calculus was being expressed. See Laubenbacher and Pengelley, 1999: sect. 3.5 for a clear description of how Leibniz was led to his discovery of the theorem and how he stated the theorem. In sect. 3.6 of that work, there is a more easily recognizable version of the theorem stated in the terminology of Cauchy's version of the calculus.

the theorem has been carried out so many times and by such a wide variety of students and scholars that it is hard to see how the theorem could not be true.

I agree with the above objection that the Fundamental Theorem does provide us with very useful information and that understood or interpreted in a certain way, what it "says" is true. What I deny is that, understood *literally and Platonically*, it is true. Thus, in what follows, I shall indicate how an anti-realist can account for all of the incontestable facts mentioned above by the realist, without assuming that the Fundamental Theorem, literally and Platonically construed, is true.

The objection is obviously closely related to the question of how analysis is applied, in general, in science and engineering, and this question in turn is connected to the question of how geometry is applied in these disciplines. So reconsider how I explained (in Chapter 2) how Hilbert's geometry can be applied in science and engineering even though the (interpreted) theorems are not, when Platonically construed, strictly speaking, true statements.

My explanation began with the fact that the axioms of Hilbert's geometry characterize a type of structure and that, at some point, physical space itself was represented as having a mathematical structure of the sort that is characterized by the axioms—a structure isomorphic to the set of all ordered triples of real numbers, ordered in the familiar way.[54] Under this representation, a position in this space corresponds to an ordered triple of real numbers, a straight line in space would be a set of such ordered triples that satisfies equations of a certain kind, and the space itself would correspond to the totality of all such ordered triples of real numbers.

Now the fact that physical space is represented in the above structural way by both empirical scientists and mathematicians is no accident. The use of a coordinate system of numbers to represent positions in space is conceptually related to a whole system of practical and technically feasible procedures for doing such things as drawing lines and curves on flat surfaces (such as flat sheets of paper), constructing on such surfaces perpendiculars, tangents to curves, and bisections of angles, laying out borders and boundaries on tracts of land, and measuring distances and lengths of physical things. These procedures were developed, over a huge stretch of time, by a great many very talented people who were motivated by a variety of practical goals and who had to deal with a multiplicity of practical problems. Euclid's geometry clearly

---

[54] Of course, all this structural talk can be understood within the framework of the Constructibility Theory.

resulted from an abstraction from, and idealization of, these practically performable procedures.[55] In turn, the use of a coordinate system of real numbers to represent lines, curves, and points in space can also be seen to be an abstraction from, and an idealization of, these practically performable procedures involving both geometry and measurement. Thus, one can see that the analytic structural representation of space is linked conceptually to a fund of practical methods of measuring and comparing lengths of physical lines, as well as to procedures for determining and estimating lengths, areas, and volumes.

I have been stressing, above, the development of practical and empirical methods of measuring and estimating such things as lengths, areas, and volumes. Clearly, there had to be an equally important development of mathematical theory before the kind of mathematical representations of physical space being discussed could be rigorously made. How such a development took place will be indicated in what follows. Several coordinate systems were known to such seventeenth-century mathematicians as

[55] David Sherry attributes to A. Seidenberg the view that Euclid's constructions were developed from solutions to practical problems: "Thus, restrictions to ruler and compass construction is no mere coincidence, but a holdover from peg and cord methods used in religious rites as well as surveying and engineering" (Sherry, 1999: 33–4). Cf. Felix Klein's comment that the "fundamental concepts and axioms [of geometry] are not immediately facts of perception, but are appropriately selected idealizations of these facts. The precise notion of a point, for example, does not exist in our immediate sense perception, but is only a fictitious limit which, with our mental picture of a small bit of shrinking space, we can approach without ever reaching" (Klein, 1939: 186–7). In Proclus's *Commentary on Euclid*, the following passage is relevant to the above observation: "[W]e say, as have most writers of history, that geometry was first discovered among the Egyptians and originated because the Nile overflows and obliterates the boundary lines between their properties. It is not surprising that the discovery of this and other sciences had its origin in necessity, since everything in the world of generation proceeds from imperfection to perfection. Thus, they would naturally pass from sense-perception to calculation and from calculation to reason" (Artmann, 1999: 11–12). Cf. also: "When the floods receded, many landowners' boundary marks had inevitably been washed away, so it was important that surveying should be carried out immediately. ... The priests inaugurated this rapid re-survey of the land, which had to be ready for winter cultivation. ... But by means of exact area measurements and verbal descriptions, the status quo was re-established. Graeco-Roman writers from Herodotus (c. 484–c. 420 BC) right down to Cassiodorus (c. AD 490–c. 583) attribute the origins of geometry, literally 'measuring of the earth', to this practice" (Dilke, 1987: 7–8). Ernest Adams, commenting on "the relation of geometrical concepts and propositions to practical procedures", notes that "while geometrical theory can be regarded as being about *idealized* drawings on *idealized* surfaces, its propositions inform us about real drawings on real surfaces ..." (Adams, 2001: pp. xvi–xvii). How methods of locating positions in space, and in particular, positions on a line in physical space, as well as methods of measuring and comparing lengths of things, are related conceptually to the ideal structure described above is indicated in the discussion in Chihara, 1990: ch. 6.

Descartes and Fermat. Indeed, Fermat used a coordinate system to specify or refer to a point or set of points in space (Boyer, 1985: 380–2). In (about) 1671, Newton wrote a work in which eight new types of coordinate systems were suggested. There, he gave a formula for radius of curvature in both rectangular and polar coordinates; he also gave equations for the transformation of rectangular to polar coordinates (Boyer, 1985: 448–9). By 1728, Euler had made the use of coordinates in two and three dimensions "the basis of a systematic study of curves and surfaces" (Boyer, 1985: 503). Boyer also notes that, ever since Albert Girard (1590–1633), "it had been generally known that the real numbers—positive, negative, and zero—can be pictured as corresponding to points on a straight line" (Boyer, 1985: 548). Unfortunately, there was no definition of 'real number'. It was not until the nineteenth-century "arithmetization of analysis" culminated in the specifications of several mathematically adequate definitions of 'real number' (for example, by Weierstrass and also by Dedekind) that this lacuna was filled (Boyer, 1985: ch. 25).

As for the development of the calculus, it needs to be noted that from the time of classical Greece, mathematicians had sought to develop methods for calculating the areas of regions with curved boundaries. The above developments in analytic geometry, as well as important research carried out in the seventeenth century on finding tangents to curves and calculating "maxima and minima", set the stage for the spectacular work of Newton and Leibniz, resulting in their independent discoveries of the calculus as a method for calculating areas and volumes, as well as Leibniz's discovery of the Fundamental Theorem.[56]

Thus, the fact that the mathematical methods of the calculus yield results that are approximated by the results we get using a variety of empirical methods for estimating such things as distances, lengths, areas, volumes, densities, and velocities (utilizing what our scientific theories tell us about such things as the nature of light and optical instruments) is not a surprising and inexplicable coincidence that just happens to have been empirically verified in countless ways. Given the way mathematical analysis has been developed, this is just what one would expect.

Let us now reconsider in a general way the problem raised by the Fundamental Theorem. What does the proof of the theorem establish? The proof proceeds from certain assumptions, and utilizes rules of inference that preserve "truth in a structure". In the case at hand, the proof establishes facts

---

[56] For an overview of the historical development of the calculus, see Laubenbacher and Pengelley, 1999: ch. 3.

about every structure of the sort that forms the subject matter of the theory of functions of a real variable (structures which I will call, for purposes of reference, 'structures of kind $f[RxR]$'). In particular, it establishes facts of the form: each structure of kind $f[RxR]$ is one in which the statement of the Fundamental Theorem (when reinterpreted so that the sentence is taken to be "about" that structure) holds. In short, it in effect tells us that the structural content of the theorem holds.

The printed lines of the engineering paper approximate the ideal straight line segments of geometry.[57] We have only an approximation since the printed lines, no matter how fine, have a certain thickness and they are not absolutely straight (as can be seen when the lines are put under a microscope).[58] These printed lines form a kind of grid consisting of "atomic squares" (that is, squares containing no interior lines), each square being, to the naked eye, equal in size to every other atomic square. Taking the length of a side of an atomic square to be of unit length, and choosing one of the horizontal lines to be the x-axis, one can regard each position x on the x-axis as corresponding to a real number r—where, according to the constructibility analysis discussed above, the length of the line segment from the origin to x is r units. Similarly, picking a vertical line to be the y-axis, one can regard each position y on the y-axis as corresponding in the above way to a real number s, so that one can regard any position on the page as corresponding to an ordered pair of real numbers $<r, s>$. This correspondence rests upon a number of idealizations, but this is as it should be, since, when we attempt to confirm the Fundamental Theorem by means of actual graphs drawn on engineering paper, we are dealing only with approximate positions and distances on the paper. Thus, the various positions on the engineering paper used to graph continuous functions of real variables form the domain of a sort of approximation model of a submodel of $f[RxR]$. Since the structural content of the Fundamental Theorem holds of the idealized structure, to the extent to

[57] Such marks are classified by Adams as "surface features": included in this classification are "bumps, dents, scratches, sticky spots, and, preeminently, visible marks and figures like the letter 'S'" (Adams, 2001: 5). Adams does not regard surface features as genuine three-dimensional objects and hence as physical objects (Adams, 2001: 9).

[58] Adams develops a theory about such "surface features", writing: "[T]he theory developed in the present essay ... explains why small dots are not geometrical points, but the smaller they are the closer they come to instantiating points because they come closer to satisfying the requirement that any two things that touch them must touch each other" (Adams, 2001: 14). Regarding the lines we draw on flat surfaces, he writes: "[N]o observable thing, even a two-dimensional feature on a surface, is thin enough to be a geometrical line, but it follows from the theory developed in Part II ... that the *edges* of observable things ... are one-dimensional" (ibid.).

which the positions on the engineering paper and the practically determinable spacial relations on this domain approximate the mathematical features of the idealized structure—specifically, to the extent to which the spatial relations between the positions in the graph we draw on the engineering paper approximate the idealized structural relationships holding between the values of the function $f(x)$ for each argument in the interval $[a, b]$—the theorem will provide us with accurate information about mathematical relationships that obtain on the graphing paper.[59] It is not surprising, then, that we will find our estimations of the definite integral to be close to our calculation of $F(b) - F(a)$, since the methods that we use to construct the lines on the engineering paper and the curve that represent the "graph" of $f(x)$ are part of a whole system of practical methods of measuring and estimating distances and areas from which our mathematical coordinate system, our graphical method of representing functions of real variables, and our theory of the calculus itself were obtained by abstraction and idealization. Of course, none of this account requires that the statement of the Fundamental Theorem, literally and Platonically construed, be true.

Some simple examples

Consider the following example. Let f be the constant function $f(x) = 5$, for every real number x. $F(x) = 5x$ is an indefinite integral of f. So, for $a = 0$ and $b = 15$, we can conclude from the Fundamental Theorem,

$$\int_a^b f(x)dx = F(b) - F(a)$$

that the definite integral above is 75. Now consider a representation of the graph of f drawn on engineering paper. The definite integral is given by an area bounded by the printed horizontal lines $y = 0$ and $y = 5$, and the printed vertical lines $x = 0$ and $x = 15$.[60] Now, the parts of the surface of the paper that are the square areas formed by the grid made by the printed lines can be

---

[59] Cf. Suppes's explanation of how the notion of model enters "in a natural and explicit way" in discussions of the nature of measurement:

> Given an axiomatized theory of measurement of some empirical quantity such as mass, distance, or force, the mathematical task is to prove a representation theorem for *models* of the theory which establishes, roughly speaking, that any empirical model is isomorphic to some numerical model of the theory. ... What we can do is to show that the structure of a set of phenomena under certain empirical operations is the same as the structure of some set of numbers under arithmetical operations and relations. (Suppes, 1967: 58–9)

[60] The definite integral described above gives the area.

regarded as the elements of the domain of an empirical structure—a structure that is isomorphic to a substructure of $f[RxR]$. So it is easy to see how the structural content of the Fundamental Theorem can be applied to the empirical structure. Indeed, it is easy to see, from the isomorphism mentioned above, that much of what holds of the region of RxR bounded by the horizontal lines $y = 0$ and $y = 5$ and the vertical lines $x = 0$ and $x = 15$ will hold of the empirical structure described above. In particular, one can conclude that the cardinality of the totality of the square parts of the surface that constitute the domain of the empirical structure will be identical to the cardinality of the totality of corresponding unit square regions of RxR that are in the bounded mathematical region described above. Thus, counting the relevant square parts of the sheet and coming up with 75 provides one with confirmation that one has applied the Fundamental Theorem correctly and has made the correct calculations—this despite the fact that the empirical structure has a domain with "objects" that result from idealizing assumptions and large simplifications. Here, the idealizations and simplifications clearly do not invalidate the inference we draw about the relevant features of the mathematical substructures.

An indication of how closely the graphs and figures we actually construct approximate the ideal graphs and figures of our mathematical theories can be seen by carrying out some basic Euclidian constructions. For example, by accurately constructing a triangle on a flat surface and then by accurately constructing the three medians of the triangle by drawing straight line segments connecting each vertex of the triangle with the midpoint of its opposite side, one can see how closely the features of the drawn figure approximate the features geometry attributes to ideal triangles: since it is a theorem of Euclidean geometry that the three medians will intersect in a single point,[61] the accuracy of one's construction can be determined by seeing how closely the constructed medians intersect in a single point.

13. THE BURGESS "TONSORIAL QUESTION"

For many years, John Burgess has been investigating the role, in philosophy and science, of the principle known as "Occam's Razor" ("Entities are not to be multiplied beyond necessity"). He is concerned with whether

---

[61] Adams points out that this theorem is implicit in a theorem of Archimedes' work *On the Equilibrium of Planes: Of the Center of Gravity of Planes*. See Adams, 2001: 242 n. 2. On that page, Adams also discusses this theorem in some detail.

Occam's Razor, restricted to abstract entities, is a rule of scientific methodology. This is what he calls the *"tonsorial* question" (Burgess, 1998: 197). Burgess thinks the tonsorial question is highly relevant to the Platonism–nominalism controversy in philosophy of mathematics because he is convinced that one of the principal supports of contemporary reconstructive nominalism (as discussed in Chapter 5) is the conviction that a system of mathematics that avoids a commitment to mathematical objects is scientifically preferable to one that makes such a commitment. Burgess believes that, if he can provide convincing evidence that the tonsorial question deserves a negative answer, he will have undercut one of the principal arguments for contemporary nominalism in the philosophy of mathematics.

Whether the avoidance of a commitment to mathematical objects is a scientific benefit is, according to Burgess, something that should be judged by scientists. Thus, the tonsorial question becomes, for Burgess, primarily an historical question: he suggests that "the claim that economy of mathematical ontology has weight with the scientific community ... must be tested directly against the evidence of past decisions of physicists" (Burgess, 1990: 11). As a result of his investigations of such past decisions, Burgess arrives at a negative answer to the tonsorial question,[62] and he advances the explanatory thesis that "Ontological concern [in general] is foreign to the scientific culture" (Burgess, 1998: 213). Part of his reason for putting forward such a controversial explanation is to be found in some work done by his colleague Gideon Rosen. Burgess quotes with strong approval the following passage from Rosen's doctoral dissertation:

My conjecture is that from the scientist's standpoint—and here I mean the speculative theorist and not just the experimenter or the engineer—*the fact that* [a] *new theory manages to eschew commitment to abstract objects and so achieve a striking ontological economy constitutes no reason at all to prefer it.* I would change my mind if the nominalist ... could point to a single substantial instance in the recent history of science to suggest that *this sort* of economy has ever mattered to scientists. After all, if it were truly important we should expect to find professional physicists engaged in the nominalizing project. But of course we don't. And this suggests (at least to me) that ... purging physics of a commitment to numbers serves no goal recognized either implicitly or explicitly in the practice of science. (Burgess, 1998: 204)

---

[62] In Burgess, 1990, Burgess gave a more cautious answer, writing: "[T]he burden of proof seems to be more on those who would insist that ontological economy of mathematical apparatus is also a weighty scientific standard. This burden of proof has not yet been fully met" (12).

These views about scientific methodology can be seen to be supported by conclusions about scientific practice that Penelope Maddy drew from her studies of Feynman's lectures, especially her conclusion that "physicists seem happy to use any mathematics that is convenient and effective, without concern for the mathematical existence assumptions involved" (Maddy, 1997: 155). Recall, from my earlier discussion of Maddy's views about physicists, that she, too, was led to doubt that natural science is concerned with assessing mathematical ontology, writing: "If it were in that business, it would treat mathematical entities on an epistemic par with the rest, but our observations clearly suggest that it does not" (Maddy, 1997: 157).

My doubts, which arose in pondering these arguments, are not about the accuracy of the historical studies upon which the arguments are based, but rather are about the correctness of certain conclusions that were drawn. Let us grant, for the sake of argument, that empirical scientists rarely, if ever, express any concern at all about the ontological commitments of the mathematical theories they employ in their scientific work. Can we confidently infer that this lack of concern is due to what Burgess believes lies behind it, namely: "Ontological concern is foreign to the scientific culture"?

Let us compare positions. Burgess believes that the mathematical theories scientists use, when literally construed, are ontologically committed to mathematical objects. He also believes that, to apply these theories in their scientific work, scientists must accept the literal truth of the theorems of these theories. I, on the other hand, have not taken a stand on what the theorems of mathematics mean, and hence have not committed myself to the doctrine that the mathematical theories scientists use are ontologically committed to mathematical objects. I have also maintained that, to apply these theories in their scientific work, scientists do not have to accept the literal truth of the theorems of these theories—it is enough to accept the structural content of these theorems. So if I am right, the scientists are justified in not taking seriously the ontological commitments of the mathematics they use.

Of course, that does not mean that Burgess is wrong in thinking that ontological concern is foreign to the scientific culture. But it can be seen that further empirical studies are required to justify such a belief. The scientist's lack of concern about the ontological commitments of the mathematical theories they employ may be due to a number of possibilities. For example, it may be due to the scientist's "gut feeling" that the use they make of mathematics does not require of them a commitment to the literal truth of the theorems of mathematics they use—which, according to the above analyses, would be correct. Or it may be that scientists do not believe that the theorems of mathematics have the metaphysical implications that Burgess attributes to them. I have been struck by the fact that, in discussing the nominalism–Platonism dispute with non-philosophers, the scientists among the discussants are frequently surprised, if not downright amazed, to learn that there are many eminent contemporary mathematicians and logicians who maintain not only that the axioms of mathematical theories such as set theory and number theory are genuine assertions about the real world, but also that there exist in the real world such things as sets, functions, numbers, spaces, and the like. The impression I have gotten, as a result of such discussions, is that it had never occurred to these scientists that mathematical axioms—especially the existential ones—can be reasonably understood to have such metaphysical implications. It may be the case (although I would not attempt to establish it) that many scientists understand the existential assertions of mathematics as something akin to the assertions of what I call *"existence in a kind of structure"*. What I mean by that can easily be understood by way of examples. Consider the existential axiom of group theory that asserts the existence of an identity element. Hardly anyone would understand this axiom to be the assertion that in the actual universe there is such an identity element. Rather, all that is actually being asserted, one would think, is that, in any group, there is an identity element. So it seems to me possible that some scientists think that the axioms of set theory asserting the existence of sets of various sorts are actually only asserting the existence of sets of this sort in any set structure.

In any case, we certainly cannot infer, without a great deal of additional research, that the casualness exhibited by many scientists about the ontological commitments of the mathematics they use is due to their general lack of concern for ontological matters, as Burgess seems to have concluded.

14. APPLICATIONS OF SET THEORY IN LOGIC

It is generally agreed that enormous advances have been made in logic in this century, and that these advances owe much to the development of model theory—a branch of set theory. For example, the very definition of the consequence relation for first-order logic—one of the most fundamental notions of all of mathematical logic—is given in terms of sets. Such facts led Hilary Putnam to the following sort of argument for the indispensability of sets or classes for logic:

[T]he natural understanding of logic is such that all logic, even quantification theory, involves reference to classes, that is, to just the sort of entity that the nominalist wishes to banish. (Putnam, 1971: 32)

Even if one agrees with Putnam that the "natural understanding of logic" is such that logic seems to involve reference to classes, one is not thereby committed to the position of allowing that the use of logic requires the acceptance of the existence of classes. After all, the nominalist has been given no strong reason for maintaining that logic must involve reference to classes. The nominalist can respond that the standard set-theoretical model theory of logic can be done within the framework of simple type theory. Since the mathematics of simple type theory can be carried out within the Construct- ibility Theory, such a model theory (and hence much of standard mathe- matical logic) can be done within the framework of the Constructibility Theory. In this way, the apparent reference to classes can be avoided.

There is another, closely related, response to Putnam's argument that brings to mind a question raised by Shapiro's claim that set theory (in particular model theory) is the source of the precision we bring to the modal locutions. Since I reject set theory as false when it is literally and Platonically construed, Shapiro in effect argued that I cannot legitimately make use of the model- theoretic account I gave of the modal locutions and, in particular, of the constructibility quantifiers. The question I shall now take up is the one I deferred in Chapter 7 for later consideration: can set theory or, more speci- fically, model theory be legitimately used by the anti-realist as a tool in clar- ifying, explicating, and investigating the logical features of the constructibility quantifier, without presupposing the existence of mathematical objects?

Let us imagine making use of set theory in order to investigate the basic logical features of the Constructibility Theory. Suppose that set theory is to be used to set up, for heuristic purposes, the model-theoretic semantics of the constructibility quantifiers. We saw earlier that set theory can be regarded as a theory of a kind of structure, so for the case at hand, we can imagine that a theory of such a structure is to be used to investigate the logical features to be attributed to the constructibility quantifiers by imagining that the con- structibility quantifiers are governed by logical laws that are linked to struc- tures of the sort that are substructures of the kind studied in set theory. In carrying out this investigation, theorems of set theory will play a role. So the question is: must the theorems of set theory be regarded as truths (true of the universe of abstract entities that, supposedly, set theory describes) for set theory to play such a role? Not at all. No good reason at all has been given by Shapiro (or anyone else that I know of) for accepting the affirmative answer to this question. The study being envisaged is a heuristic exploration with the aim of clarifying, explicating, and investigating the logical features of the constructibility quantifier. We may need to take note of theorems of set theory to carry out such an exploration, say, in order to see how the attribution of such and such a structural feature to a logical constant would affect the logical laws governing the constant, but for such a purpose, the theorems in question, literally and Platonically construed, would not have to be true—it would be sufficient that the structural content of the theorems be true. After all, what we need to know is how attributing such a structural feature to a logical constant would affect the logical laws governing the constant, and to know this, we may need to know what would have to hold in all structures of the relevant kind, but we would not also have to know anything about some mysterious hypothesized entities that do not exist in the physical world. As was the case for the application of mathematics to the empirical sciences, the truth of the structural content of the theorems is all that is needed for the heuristic investigation being envisaged to be successfully carried out—and for that, there is no need for a belief in classes.

Given the above reply to Shapiro, I trust that the reader can now reproduce the second kind of reply to Putnam's argument to which I alluded earlier in this section. Much of this is relevant to Maddy's mystery.

15. MADDY'S MYSTERY

Recall that this "mystery" was raised by Maddy with the words: "if mathematics isn't true, we need an explanation of why it is all right to treat it as true when we use it in physical science." Reconsider the example in which we infer

There are twelve coins on table A at time t

from the premises:

[1] There are five dimes on table A at time t.
[2] There are seven quarters on table A at time t.
[3] A coin is on table A at time t iff it is either a dime or a quarter.
[4] Nothing is both a dime on table A at time t and a quarter on table A at time t.
[5] $5 + 7 = 12$.

Why is it all right to treat [5] as one of the "true premises" from which the conclusion is drawn? The analysis of this reasoning given earlier shows that the conclusion can be validly inferred from premises [1]–[4] and that there is no need to assume that [5] is true. However, according to this analysis, the Constructibility version of [5] is true, so treating [5] as true in such contexts as the above will not lead us astray.

By a similar process of reasoning, we can explain the legitimacy of treating the theorems of number theory (expressed in a natural language) as true, without thereby committing oneself to the thesis that these theorems, literally construed, are true. A Constructibility version of the theorem is true and treating sentences as true when we apply the theorem will not lead one to falsehoods.

What about the theorems of analysis? We also saw, earlier, that what we need to rely upon, in applying such theorems as the Fundamental Theorem of the Calculus, is the information that they provide about the models of the theory (its structural content). Read in this structural way, the theorem is true. Thus, taking a theorem of analysis as providing us with this structural information, it is true—so we are justified in treating the theorems of analysis as true *in this way*. In a variety of ways, then, it can be explained why it is reasonable to treat the theorems of mathematics as true.

10

If-Thenism

I have been asked, sometimes by readers of my structural account and sometimes by attendees at one of my lectures on this material, how my view differs from this or that well-known position. The two accounts that are most often mentioned in this regard are *if-thenism* and *mathematical fictionalism*. The last two chapters of this work will take up these two rival views, with the aim of producing a deeper understanding of my structural account by way of contrast and comparison.

As I have indicated in several places already, my account of mathematics is similar to the view known as "if-thenism" or "deductivism".[1] This is a view that also attempts to account for the sort of puzzling features of mathematical practice that form the basis for the second and third of my five puzzles—two puzzles I shall attempt to resolve in this chapter.

Here is how P. H. Nidditch characterizes deductivism:

[I]t has become more and more widely accepted during the past hundred years, with the result that it is now the orthodox doctrine, that to say of a mathematical proposition p that it is true is merely to say that p is true in some mathematical system S, and that this in turn is merely to say that p is a theorem in S. ... This view of the nature of mathematical truth ... was first put forward with full explicitness and clarity by the Scottish philosopher Dugald Stewart. "Whereas, in all other sciences," he says, "the propositions which we attempt to establish express fact, real or supposed—in mathematics, the propositions which we demonstrate only assert a connection between certain suppositions and certain consequences. Our reasonings, therefore, in mathematics, are directed to an object essentially different from what we have in view, in any other employment of our intellectual faculties—not to ascertain *truths* with respect to actual existence, but to trace the logical filiation of consequences which follow from our assumed hypotheses." (Nidditch, 1960: 287)

Nidditch also quotes Charles Sanders Peirce as claiming: "Mathematics does not undertake to ascertain any matter of fact whatever, but merely posits

---

[1] If-thenism is characterized by Penelope Maddy as the view that mathematics "is the logical study of what conclusions follow from which premises" (Maddy, 1990: 25). It is criticized in Maddy's work and also by Michael Resnik in Resnik, 1980: ch. 3.

hypotheses, and traces out their consequences" (Nidditch, 1960: 288). Nidditch goes on to claim that "there is general agreement that a mathematical theory is, in the now popular phrase, a *hypothetico-deductive system*. It is a system in which there are certain *primitive propositions* (postulates) which are adopted without proof and from which alone all other categorically asserted propositions of the system have to be logically deducible" (Nidditch, 1960: 289).

The structural account I have been developing is especially similar to the *model-theoretic version* of if-thenism of Putnam's that was discussed in the previous chapter. Recall that this was the view in which Putnam maintained that "pure mathematics consists of assertions to the affect that *if* anything is a model for a certain system of axioms, *then* it has certain properties" (Putnam, 1967: 294). Now a number of objections have been raised to if-thenism—objections which could easily be thought to apply equally well to my structural view of mathematics because of its similarities to if-thenism.[2] However, before taking up these criticisms, I shall give my analyses and explanations of the remaining two of the five puzzles I presented in Chapter 1.

1. THE THIRD PUZZLE

The inertness feature attributed to mathematical objects gives rise to the following questions. (1) How are we able to refer to these inert mathematical objects? (2) How are we able to gain knowledge of these inert objects? (3) Why is it crucial for scientists to refer to and to discover relationships among these inert mathematical objects in order to discover facts about the physical universe? These are truly puzzling questions if one assumes, as many (if not most) contemporary philosophers of mathematics do, that (a) our mathematical theorems, literally construed, express true propositions; and (b) our mathematical theorems are about (make reference to) mathematical objects from which we are forever utterly cut off (causally).

The structural view put forward in this work provides us with a nice way out of these conundrums, since from the perspective of this view, unlike that of the mathematical structuralists discussed in Chapter 4, we are not committed to maintaining that most mathematical theorems, literally construed, are true. Hence, we are not committed to maintaining that mathematical theorems actually refer to causally inert objects. According to the view of the

---

[2] This idea has been expressed several times at talks at which my structural view of mathematics was presented.

present work, one need not adopt the position that mathematics, so analyzed, is required to be true and, hence, truly refers to mathematical entities. A fortiori, there is no problem of explaining why it is crucial for scientists to refer to and to discover relationships among these inert mathematical objects in order to discover facts about the physical universe. It can be seen that, no matter how one may analyze the literal meaning of mathematical sentences, one can make good sense of mathematical practice and the applications of mathematics in science without requiring mathematical theorems, literally construed, to be true. It is enough that sentences expressing the structural content of the theorems be true. Thus, the inertness problem never gets off the ground within the framework of this work.

Since my view is that one of the principal kinds of knowledge mathematics delivers is the knowledge of the form:

> any realization of the axioms (or, alternatively, any structure of the kind being investigated in the system) would have to satisfy——,

there is no special problem of how we gain such knowledge. In general, this knowledge is obtained by means of mathematical proofs of theorems. Thus, we obtain an obvious explanation of "the primacy of proof as the avenue to mathematical knowledge *par excellence*"—which Geoffrey Hellman feels "must receive a natural explanation, even if other avenues are left open" (Hellman, 1989: 4).

Other avenues to mathematical knowledge

What are these "other avenues" to mathematical knowledge to which Hellman alludes? I shall take up some specific examples of these other routes to mathematical knowledge in the next section. However, before beginning this topic, I should make a few comments about what I mean by 'mathematical proofs'. A proof, for me, is a chain of reasoning by means of which one can arrive at the rational conviction or knowledge that something is the case. But I have already maintained that what are generally called theorems need not be true. So how can I allow, with Hellman, that proofs are avenues to mathematical knowledge?

A proof of a theorem ϕ is ordinarily regarded as a proof that ϕ is true. However, for me, a proof of ϕ is a proof that the structural content of ϕ is true. Hence, a mathematical proof can indeed be a means of arriving at the knowledge of a truth, but the truth arrived at, in this case, is a modal truth.

I should also emphasize that, despite the frequent references I have made to proofs in first-order logic or to proofs in **Ct**, I do not mean to restrict mathematical proofs to just the sequences of formal syntactic objects studied by mathematical logicians (especially proof theorists). Such derivations do not, frequently, terminate in a formal sentence that is seen to be true: one frequently concludes that such and such is a theorem of some theory. So the knowledge that results from such a derivation is frequently not knowledge that the sentence derived is true, but rather the knowledge that the sentence derived is a theorem of some formal theory. In any case, mathematical proofs, for me, include what *practicing* mathematicians produce, publish, and cite in justification of the mathematical claims they make. They are the sort of thing the mathematician William Thurston had in mind in writing:

[W]e should recognize that the humanly understandable and humanly checkable proofs that we actually do [produce] are what is most important to us, and that they are quite different from formal proofs. For the present, formal proofs are out of reach and mostly irrelevant: we have good human processes for checking mathematical validity. (Thurston, 1994: 171)

These non-formal proofs can be, according to my view, perfectly rigorous and convincing.[3] So I agree with the characterization Arthur Jaffe and Frank Quinn give with the words:

[W]e claim that the role of rigorous proof in mathematics is functionally analogous to the role of experiment in the natural sciences. ... First, proofs provide a way to ensure the reliability of mathematical claims, just as laboratory verification provides a check in other sciences. Second, the act of finding a proof often yields, as a byproduct, new insights and unexpected new data, just as does work in the laboratory. (Jaffe and Quinn, 1993: 2)

---

[3] One feature of such mathematical proofs is the existence of gaps or "skipped steps". There is a story told of Norbert Wiener (probably apocryphal) that illustrates this feature. It seems that at one advanced seminar at MIT, Wiener was at the blackboard giving an especially difficult proof. At one point, one of the advanced graduate students attending the class interrupted the noted mathematician: "I'm afraid I don't see how you got from step (18) to step (19)," he said rather timidly. As this sort of interruption only rarely occurred in Wiener's classes, the father of cybernetics was taken aback (as no doubt were the other students). He hastily erased what was written after the '19' and seemed to be in deep thought as he mumbled something unintelligible. After an appreciable length of time, he wrote exactly what he had written before on line (19) and started to go on to step (20). Again, the brave student interrupted the proof production: "I'm sorry Professor Wiener, but I didn't see how you got step (19)". Wiener frowned, scowled, and stared fiercely at the student. Then he again erased what he had written after the '(19)' and stared at the board for an even longer period of time. Finally, he very carefully wrote exactly what he had written before on line (19) and started to go on to step (20). For the third time, there was the interruption: "But Professor Wiener, I didn't see how you got step (19)." At this, the esteemed mathematician blew up. "I don't know what to do with you," he bellowed, "I've worked it out three different ways and you still don't see!"

In summation, I consider the sort of mathematical proofs these mathematicians have in mind to be perfectly good devices for yielding the sort of knowledge, insights, and reliability described above.

2. BROWN'S "OTHER AVENUES"

One author who has emphasized the "other avenues" mentioned in the above quotation is James Brown, who argues, in *Philosophy of Mathematics*, that there are several ways of acquiring mathematical knowledge distinct from the method of constructing purely logical proofs. For example, he believes (with Hilary Putnam and Mark Steiner) that the kind of inductive reasoning ingeniously employed by Euler to arrive at the result

$$\frac{\pi^2}{6} = 1 + \frac{1}{4} + \frac{1}{9} + \frac{1}{16} + \cdots$$

is more than a mere heuristic method—he is convinced that such reasoning can yield mathematical knowledge. (I shall give more of the details of Euler's reasoning shortly.)[4]

Perhaps a more controversial claim Brown makes is that the use of pictures or diagrams, as much as logical proofs, can engender in us the rational conviction and knowledge of the truth of mathematical propositions, claiming that "they can provide solid evidence, too, evidence which is as rigorous as any traditional verbal/symbolic proof".[5] It should be noted that diagrams and pictures can do more, according to Brown, than merely aid us in following a logical proof: he thinks that "they can play an essential role in proofs".[6] An example of the sort of picture he has in mind is given on page 44 of his book. To see that $x^2 + ax = (x + a/2)^2 - (a/2)^2$, study the diagram in Figure 2 (which Brown calls a "proof").

---

[4] Brown, 1999: 168–70. I give only a brief sketch of Euler's reasoning in this chapter since this example has been discussed by philosophers in great detail in many places: not only by Brown, who gives Putnam's version as well as his commentary on it, but also by Steiner (Steiner, 1975: 102–6). However, I do give (in Section 4) some of the considerations that convinced mathematicians that Euler had indeed hit upon the correct answer.

[5] Brown, 1999: 192. Cf. "Diagrams are considered by the Greeks not as appendages to propositions, but as the core of a proposition" (Netz, 1999: 35). Also: "*Diagramma* is a term often used by Plato—one of the first, among extant authors, to have used it—either as standing for mathematical proofs or as the *de rigueur* accompaniment of mathematics" (Netz, 1999: 35–6).

[6] Brown, 1999: 39. I understand Brown's claim that some diagram or picture plays an "essential role" in a proof as the claim that the proof in question would not be a proof without the diagram or picture; I do not think it is the claim that whatever is proved could not be proved without the diagram or picture.

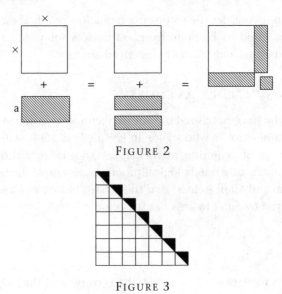

FIGURE 2

FIGURE 3

This diagram needs no commentary. The occurrences of the plus and equality signs in the figure are all that need be noted to see how the reasoning works. (Another example of the sort of diagram that Brown thinks can figure in producing mathematical knowledge—in this case, knowledge of the Pythagorean theorem—is Figure 1 in Section 7 of Chapter 9 above.)[7]

Brown thinks that Platonism, more than any other account of mathematics, is open to the possibility of these "non-standard means" for acquiring mathematical knowledge (Brown, 1999: 14). Indeed, this very openness to the possibility of such "other avenues" to mathematical knowledge is put forward as supporting his overall case for the superiority of the Platonic view of mathematics: Platonism, he claims, provides a better account of mathematics than do its rivals.

The diagram ("picture") in Figure 3 plays a special role in his argument that pictures are "windows to Plato's heaven" (Brown, 1999: 39). The diagram is aimed at showing us that

$$1 + 2 + 3 + \cdots + n = \frac{n^2}{2} + \frac{n}{2}.$$

A question he takes up is: how does the picture "work"? Brown notes that Wittgenstein had written in the *Tractatus* that "For a picture to work there

---

[7] For a deeper grasp of what Brown has in mind, see ch. 3 of Brown, 1999.

must be something in common with what it pictures—'pictorial form'"
(2.161). "The minimal commonality between pictorial form and object is
logical form" (2.18). These passages suggest to Brown that, for Wittgenstein of
the *Tractatus*, there must be a structural similarity between the picture and
what the picture represents. Structural similarity suggests isomorphism
(Brown, 1999: 38). Thus, Brown comes up with a sort of necessary condition
for a picture or diagram to represent a situation: the picture or diagram must
be isomorphic (or at least homomorphic) to the situation represented.[8]
Recent work on visual information provides Brown with support for this
isomorphism condition. He cites Barwise and Etchemendy (1991) and
Hammer (1995) as explicitly adopting such a view.[9]

Reconsider Figure 3. Notice that this picture is a representation for the
case in which $n = 7$. But it is not isomorphic to all the other cases. Brown
writes:

There is no useful homomorphism from our picture to all the natural numbers and
no isomorphism at all. But still the diagram works. It does much more than establish
the formula for $n = 7$; it establishes the result for all numbers. (Brown, 1999: 38)

So how does it work? Evidently, not by representing all the natural numbers.

Here is how Brown explains how the diagram enables us to arrive at our
knowledge of the mathematical truth in question:

[I]t works ... more like an instrument. This, of course, is a realist view of mathematics, but not a realist view of pictures. As telescopes help the unaided eye, so
some diagrams are instruments (rather than representations) which help the unaided
mind's eye. (Brown, 1999: 39)

Brown then puts forward a "bold conjecture": "*Some 'pictures' are not really
pictures, but rather are windows to Plato's heaven*" (Brown, 1999: 39).

Realizing that this "bold conjecture" will seem very implausible to many of
his readers, he offers some analogies in an effort to make the view more
palatable. For example, he notes that some aestheticians analyze David's
"Napoleon" (Napoleon, in a billowing cape, on a white horse) in such a way
that, as a *picture*, it represents Napoleon, but as a *symbol*, it represents leadership, courage, and adventure. These aestheticians maintain, according to

---

[8] A homomorphism is like an isomorphism except the function (or mapping) preserving the
relations need not be a one-to-one function. For a precise definition, see Enderton, 1972: 89–90.

[9] Actually, Hammer gives the isomorphism criterion as a criterion for a representation to be a
"good diagram" (Hammer, 1995: 12).

Brown, that the painting manages to be simultaneously about something concrete and also about something abstract. For Brown, the number-theoretic diagram in Figure 3 does something similar:

It is a *picture* of the special case, $n = 7$, but a *symbol* for every n. Just as we can see courage and adventure depicted in David's painting, so we can see every natural number in the diagram. It's a metaphorical 'seeing' to be sure, but it's a similar sort of perception in each case. If artists can do it, so can mathematicians. (Brown, 1999: 40)

A *Kantian objection*

It is enlightening to examine Brown's response to what he considers a "Kantian objection" to his Platonic explanation:

The claim is this: one sees in the picture the possibility of a reiteration; the diagram can be extended to any number; that's why it works. The objection is anti-Platonistic in that it makes a Kantian point about constructibility. (Brown, 1999: 39)

I am not very clear on just what the Kantian is supposed to be objecting, but Brown has three points to make in response. First, he notes that his account of pictures has two elements: on the one hand, it makes and defends the claim that pictures and diagrams can play essential roles in mathematical proofs; and on the other hand, it offers a Platonic explanation for the fact that a picture or diagram can play such an essential role. (Presumably, the explanation is that such pictures or diagrams are instruments that "aid the mind's eye"—they are "windows to Plato's heaven".) He emphasizes that the Kantian objection is only aimed at the second of these elements and in no way refutes the first claim that pictures can play such a role in proofs.

Second, the Kantian framework, within which the above objection is raised, takes infinity to be a potential infinite. The Platonic framework takes infinity to be an actual (completed) infinite. Since classical mathematics (especially set theory) fits the Platonic view of infinity better than the Kantian one in this respect, the Platonic explanation of how the picture works is favored by mathematics.

Third, he pointedly raises the question: what has the possibility of indefinite reiteration got to do with numbers? The Kantian assumes that we know the fact about numbers *because* we know the fact about the possibility of reiteration, to which Brown replies:

I don't know of any argument for this. One could just as well claim that we see the possibility of iteration *because* we have the prior perception of the number theory result. (Brown, 1999: 40).

Some critical observations

I find it surprising to learn from the above passages that Brown thinks that he has given a "Platonic explanation" for how pictures play an essential role in proofs of mathematical facts. It is reasonable to infer that Brown thinks his Platonic explanations are good explanations, so we should assess the explanatory value of his "explanations".

Notice that Figure 3 can be also used to arrive at the knowledge of the sum of the natural numbers less than or equal to n when it is *expressed in the language of the Constructibility Theory*. Imagine that someone learns to do arithmetic not in the standard way, but rather by learning arithmetic within the framework of the Constructibility Theory. (The reader should reexamine Chapter 7 to see that this is indeed a conceivable state of affairs.) Suppose now that this someone arrives at the formula for the sum of the natural numbers less than or equal to n from studying the above picture and, using this formula, comes to know the sum for the case in which $n = 100$.[10] How can we explain why the picture would work for this situation? We cannot say, with Brown, that the picture would be, for this situation too, a window to Plato's heaven, since the proposition expressing the formula for the arithmetical sum is not a proposition about Platonic entities—it is a formula about the constructibility of open-sentences. The constructibility proposition giving this sum is not a proposition about mathematical entities. Hence the window metaphor does not explain how the diagram is able to engender this modal knowledge.

Should we say that the picture is a window to constructibility? We could say that, but what would it mean? Talk of windows is just metaphorical. What literal content would such an "explanation" have? Brown also tells us that the diagram is an "instrument", but what does that tell us other than that it somehow helps us to grasp certain facts? He also says that it aids "the mind's eye"—a metaphor again. There is also the analogy with David's painting of Napoleon: just as the image of Napoleon is a symbol of leadership, courage, and so on, so also the diagram is a symbol of all the natural numbers. But does this analogy really help? The fact that it is a symbol of all the natural numbers does not explain why the above diagram gives us knowledge that the sum in

---

[10] Boyer relates (Boyer, 1985: 544) the following amusing story about Carl Friedrich Gauss, who at the age of ten was already showing signs of mathematical greatness. It seems that the teacher of the local school in which Gauss was enrolled assigned his students the task of adding all the numbers from one to a hundred, in order to keep them occupied. Almost immediately, Gauss placed his slate on the table, saying "There it is". The teacher was amazed to find the correct answer written on the slate. Evidently, Gauss had used the above formula to obtain the sum.

question is given by the formula '$n^2/2 + n/2$'. 'NN' also is, in many contexts, a symbol of the natural numbers, and that symbol does not give us the knowledge in question. Besides, according to Brown, there are indefinitely many such windows to Plato's heaven. But not just any such window will serve to generate the knowledge of the particular formula for obtaining the sum in question. For example, the diagram given in Chapter 9 to prove the Pythagorean theorem is of no use in finding the sum of the first n terms of the natural numbers sequence. So are we going to say that each of these windows only opens to a particular area of the mathematical universe? Does this metaphorical statement explain why the above diagram works for finding the sum of the initial segment of the natural numbers and not for proving the Pythagorean theorem? *One wants to know how, in particular, the diagram does its work*, and this all the metaphorical explanations Brown supplies fail to do.

An alternative explanation

So let us reexamine the diagram. One cannot just stare at the diagram and expect to just "see" that the formula yields the sum. One has to do a little thinking. To facilitate discussing the diagram, let us order the rows in the triangular figure, starting at the top. In the first row (that is, for $n = 1$), one sees exactly one square (half of which is light and half of which is dark). Thus, for $n = 1$, the number of squares in the nth row is n. Similarly, for $n = 2$, the number of squares in the nth row is n. Similarly, for $n = 3, 4, 5, 6$, and 7, the number of squares in the nth row is n.

Suppose that it is now said:

Yes, I can see that the formula will tell one the number of squares in row n for each of the rows in the diagram (up to, that is, $n = 7$). But how do you know it will work no matter what n is?

Here, let us consider how one could continue expanding the diagram to obtain the above result for $n = 8, 9, 10, \ldots$ To go from the nth case to the $n + 1$st case, one merely needs to add a row to the diagram. In particular, to get the $n + 1$st row, just copy the nth row directly below and then add one more square to the right end of the row (and darken the triangular area that lies to the upper right of the diagonal, as was done for the others). Thus, in order to obtain a diagram of the $n + 1$st case, one will just add a row consisting of exactly $n + 1$ squares. So the number of squares in row $n + 1$ will be $n + 1$. By an intuitive version of mathematical induction, we see that the formula for how many squares that are in row n works for all the natural numbers.

Thus, to calculate the number $(1 + 2 + 3 + \cdots + n)$, one need only to calculate the total number of squares in the diagram which has n rows. But this will be $n^2/2$ (the number of squares in the isosceles triangle whose side is n) plus $n/2$ (the number of squares which the darkened triangles in the diagram make up).

This diagrammatic proof is completely convincing. And it illustrates a lesson we learned in the discussion of the proofs of Fermat's Last Theorem, the Fundamental Theorem of Algebra, and the Pythagorean Theorem in Chapter 9: taking a type of structure (in this case, the type of structure attributed to the natural numbers) and embedding it in another more complex type of structure (in this case, a type containing some of the relations of plane Euclidean geometry) can yield helpful and fruitful mathematical insights.

Brown's response

Let us now consider the three responses Brown made to the Kantian objection to see if they undermine the above account of how the diagram can yield the mathematical knowledge in question, without an appeal to Platonic entities and the metaphor of seeing these entities through a window. The first does not merit any comment on my part, since it cannot be construed as a defense of his Platonic explanation.

The second response attributes to the Kantian framework a view of infinity according to which there are no actual infinities but only potential ones. It is argued that, since contemporary mathematics (especially set theory) fits the alternative view of infinity, that is the view that there are actual completed infinities—a view espoused by Platonism—the Platonic framework fits contemporary mathematics better than the Kantian.

This response of Brown's, however, does no damage to my account, since nothing in my explanation of the workings of the diagram presupposes the doctrine that infinity is only potential and not actual. The doctrine of the potential infinite plays no role either in my explanation or in my general philosophical position.

Turning to Brown's third response, could he claim, as he did for the Kantian case, that there is no good reason to suppose that we know the fact about the sum of the numbers from 1 to n *because* of the reasoning I have described above? I suppose that Brown could say: "One could just as well claim that we see the possibility of expanding the diagram to one with $n + 1$ rows *because* we have the prior perception of the number theory result."

Such a response, however, would be lacking in any real plausibility. Our seeing the possibility of expanding the diagram in the way described in no way depends upon our knowing (or being convinced of the correctness of) the formula for calculating the sum of the first n terms in the natural number sequence. The precise directions that are given for expanding the diagram— just copying the nth row and then adding one more square—does not presuppose the correctness of the formula for calculating the sum. Also, relying upon the correctness of a form of mathematical induction to arrive at the correctness of the formula for the number of squares in row n in no way depends upon knowing already the correctness of the formula we arrived at for that.

So I see no reason at all for accepting Brown's metaphysical explanation for how the diagram works. His explanation that the diagram is a window to Plato's heaven, enabling the "mind's eye" to "see" the correctness of the formula is not, as I see it, a genuine explanation.

My structural account of mathematics is open to alternative routes

Before ending my discussion of the use of diagrams and pictures, it should be emphasized that nothing I have claimed in this work is incompatible with there being the various alternative methods of obtaining the sort of mathematical knowledge discussed above. Inductive reasoning, the use of pictures, diagrams, and even computer calculations to arrive at knowledge of the if-then sort I have been discussing in previous chapters are all consistent with my account: any reasonable way of seeing what would have to be true in a kind of structure would be welcome.[11] I see nothing special about Platonism in that regard.

3. THE SECOND PUZZLE

Let us now consider the second puzzle, from the perspective of the structural view of set theory sketched earlier. We know that there have been and are set theorists who do not believe in the existence of sets and yet continue to do fruitful work in set theory, advancing the field by providing set theorists with new and important theoretical developments. This appears puzzling when compared with what occurs in the empirical sciences, since typically a scientist who does not believe in the existence of the principal sorts of entities studied in a scientific theory does not go on to work with the theory, carrying out new and important experiments that develop and expand the theory to new frontiers.

[11] Of course, my overall position is not compatible with the Platonic explanations that Brown gives of the use of pictures to arrive at mathematical knowledge, since he regards some pictures as "windows to Plato's heaven".

Now, what does a mathematician working in set theory do? Typically, the mathematician proves theorems of set theory.[12] So we should ponder the sort of information a theorem of set theory provides to someone who regards the axioms of set theory in the way emphasized by my structural account of mathematics: that is, as, so to speak, "specifying conditions that any structure of the type being studied must satisfy". If we regard the axioms of set theory as specifying what (so-called) 'sets' there must be in the domain and also how the things in the domain must be related to each other in order that there be a realization of the type in question, then a proof of a theorem of set theory assures us that any realization satisfying these conditions must also satisfy the conditions given by the theorem. To put it another way, a proof of a theorem provides us with further information about what must hold in any realization of the axioms, and it does so regardless of what literal meaning the theorem may have.

As I mentioned earlier, Gödel maintained that our "set-theoretical concepts and theorems describe some well-determined reality, in which Cantor's conjecture [the continuum hypothesis] must either be true or false" (Gödel, 1946b: 263–4). I certainly do not wish to contest Gödel's understanding of what the sentences of set theory truly mean. Nothing in my explanations of these puzzles requires that I provide a definite analysis of the meaning of the theorems of set theory or of any other mathematical theory.

Consider what a proof of a theorem establishes in the case of first-order Peano arithmetic. There are philosophers of mathematics (for example, Mark Steiner) who believe that statements of arithmetic are statements about specific abstract objects.[13] As in the case of set theory, I shall not contradict such philosophers or attempt any type of linguistic analysis of number-theoretic statements. I do not claim to know what '$2 + 2 = 4$' means, when it is asserted by a mathematician, whether she is a mathematical Platonist, a formalist, a logicist, or an intuitionist; it could mean different things to these different people. I certainly do not claim to know what a schoolchild may mean by '$2 + 2 = 4$'. Still, there is a way of understanding the primitive symbols of the

---

[12] Thurston, commenting on Jaffe and Quinn, 1993, writes: "Jaffe and Quinn analyze the motivation to do mathematics in terms of common currency that many mathematicians believe in: credit for theorems." But he goes on to say: "I think that our strong communal emphasis on theorem-credits has a negative effect on mathematical progress" (Thurston, 1994: 172). What Thurston wants to emphasize is "human understanding". He believes that mathematicians would be happier, more productive, and more satisfied if they expressly aimed at such a goal instead of the goal of obtaining "theorem-credits".

[13] See Steiner, 1975, where such a position is defended, especially in ch. 2.

theory *relative to any realization of the axioms*: the quantifiers can be taken to
have as domain the totality of objects in the structure, the individual constants can be taken to refer to specific objects in the structure, and the
operation symbols can be taken to be refer to the specific operations as is
specified by the axioms. In this way, we can regard a proof of a theorem in this
system as assuring us that any realization of the axioms of first-order Peano
arithmetic must also be a realization of the conditions given by the theorem.
Thus, the proof of the Commutative Law of Addition assures us that any
domain of objects that are so related as to satisfy the axioms must also be such
as to satisfy the Commutative Law of Addition when the primitive symbols
are understood in the way described above. In this respect, the situation is no
different in the case of set theory.

So (in summary), how do I explain the fact that there are mathematicians who continue to work in set theory, proving important theorems and
making important breakthroughs, even though they do not believe that
any sets exist; whereas, in the empirical sciences, when a scientist does
not believe in the existence of the theoretical entities postulated in a particular scientific theory, then typically this scientist will not believe in the
theory, work with the theory, perform experiments based upon the theory, or
develop explanations in terms of the theory? We can now see an important
difference in the kind of theories that are being developed in the two contrasting disciplines. In the case of the empirical sciences, the sentences
expressing the laws, principles, and known facts of the science are typically
taken to be literally true, approximately true, or at least true as some sort of
idealization. Furthermore, these laws, principles, and known facts are what
get confirmed, applied, and developed. By contrast, in the case of mathematics, the sentences expressing the theorems of a mathematical system
need not be taken to be true at all in any of the ways described above. The
truth of a theorem ϕ is not crucial, because what is established by a proof of ϕ
is a fact of the form:

Any realization of the axioms (or, alternatively, any structure of the kind
being investigated in the system) must satisfy ϕ.

It is this information (the structural content of ϕ) that is most important in
applications of mathematics. Thus, whether or not one believes in the actual
existence of such postulated entities as sets, numbers, functions, and spaces,
*proofs in mathematics will generate the sort of knowledge that is of genuine interest
to mathematicians and that has been found to be of real value for empirical scientists.* Whether or not one believes in the actual existence of such postulated

entities, there would still be an important point in working in the mathematical area and in continuing to prove theorems in that area.

4. CRITICISMS OF IF-THENISM

The if-thenist (deductivist) position has been criticized by a number of philosophers, most notably by Michael Resnik and Penelope Maddy. Since my view is so similar to if-thenism, it would be natural for some philosophers to assume that my position will be open to these very criticisms. For this reason, it should be enlightening to investigate these criticisms.

The logical language objection

Maddy raises a number of "annoying difficulties" which, she feels, undermine the original plausibility of the if-thenist's position (Maddy, 1990: 25). She first asks: *which logical language is appropriate for the statement of the mathematician's system of axioms?* What is the problem here? What difference does it make which logical language a philosopher uses for the formulation of her mathematical system? Well, the set of sentences that follow from the axioms of the system is certainly dependent upon the choice of the logical language: if one chooses first-order logic, for example, the set of consequences of the axioms will not be identical to the set of consequences of the axioms if one chooses, say, second-order logic. Although Maddy is not explicit about just what problem she is raising, her nagging question suggests to me that she is supposing that, according to the if-thenist, the mathematician first selects some axioms for study and then must choose a logical language in which to formulate the axioms. Maddy seems to be suggesting that the if-thenist will not be in a good position to explain why mathematicians pick any specific logical language.

But why cannot the if-thenist insist that choosing the logical language in which to formulate a mathematical system to develop and study is an integral part of choosing the axiomatic system itself? Suppose, for example, the mathematician starts with a set of axioms expressed in a natural language and which are infected with ambiguities and vagueness. Choosing a formal language may be part of the process of removing ambiguities and vagueness from the axioms. In that case, the first problem will become just a special case of Maddy's second problem (or "annoying difficulty") that supposedly cast doubt on the if-thenist's position.

The axiom system problem

Maddy's second "annoying difficulty" is: *why do mathematicians choose the particular axiom systems they do to study?* Now why is this an annoying difficulty? Maddy seems to have in mind the fact that mathematicians do not just arbitrarily lay down any old set of axioms and study the consequences one gets in such a system. Mathematicians do not discuss and argue about new axioms for set theory as if it were just a matter of setting up a new system of mathematics. Some set theorists, at least sometimes, argue about whether a new axiom should be accepted or be considered to be true. This sort of phenomenon, I believe Maddy is arguing, does not fit the if-thenist's picture of mathematics as concerned only with what follows from what. If the assertions of set theory have the content attributed to them by the if-thenist, then disputes about the acceptability or truth of new axioms should not make sense. But, evidently, they do. So there must be something wrong with if-thenism.

This objection is similar to one that Resnik put forward, with the words:

According to the deductivist, it would be perfectly legitimate for mathematicians to make up axiom sets through some random method and then proceed to investigate their logical properties. But mathematics does not proceed in this way. Not only have the new axiom systems historically developed from issues and problems encountered in the study of previous mathematical theories, but also certain consistent axiom systems are not considered worthy of serious mathematical investigation. (Resnik, 1980: 132)

I do not find Resnik's objection to be convincing. Why is the deductivist committed to the position that it is perfectly legitimate for mathematicians to proceed in the way described in the above quote? Such a position certainly cannot be inferred from the basic doctrine I have attributed to deductivism. After all, the deductivist says what the contents of mathematical claims are— not why we investigate the claims we do. Furthermore, if this is a valid objection to deductivism, then why is it also not a valid objection to Platonism? The Platonist maintains that mathematical theorems are true assertions about abstract mathematical entities and that mathematics is the search for truths about these entities. But that does not imply that mathematicians should find just any axiom system whose assertions are all truths about mathematical entities "worthy of serious mathematical investigation". Consider, for example, the axiomatic theory whose axioms are all conjunctions of the form

$$\phi \& \phi \& \phi \ldots \& \phi,$$

where ϕ is the sentence '$1 = 1$', and the number of conjuncts is the Gödel number of the Social Security number of someone living in the United States. The Platonist would agree that all the assertions of this theory are true, but is she also committed by her philosophical doctrines to maintaining that this theory is "worthy of serious mathematical investigation"? I don't see why. Similarly, I don't see why the deductivist is committed by her doctrines to the absurdity Resnik has in mind.

It is less clear just how a deductivist might reply to Maddy's objection, since it is less clear just what force she thinks an "annoying difficulty" should have. Certainly, Maddy has raised a serious question for the deductivists, but it is not obvious to me that the deductivist could not make up some sort of story about what disputes about the acceptability of some axiom or other come to—this in a way that is consistent with her basic philosophical stance about the asser- tions of mathematics. In any case, what is of concern to me here is whether or not these considerations raise a difficulty for my account of mathematics.

As an objection to my account, the above considerations have no force at all. It is clear that nothing in my account implies that mathematicians should only be concerned about what follows from what. According to my view of the subject, mathematicians are, and should be, concerned about kinds of structures, about how best to characterize certain kinds of structures, about what axioms should be taken to be realized by certain kinds of structures, about how to formalize mathematical reasoning about a certain kind of structure, and about which kinds of structures are most fruitful, mathemati- cally and scientifically—to give just a few examples of what is perfectly con- sistent with my view.

Problems with number theory

Another of the objections Maddy raises to if-thenism proceeds from the question: which premises (or axioms) are to be presupposed in cases like number theory, where assumptions are usually left implicit (Maddy, 1990: 25)? What problem about specifying the axioms of number theory does Maddy have in mind? Aren't there axiomatizations of this branch of mathematics? After all, there are many axiomatizations of the natural number system. Well, suppose that one picked some specific formalization of number theory in, say the version of Peano arithmetic given in Chapter 7. Then according to the if-thenist, what the number theorist attempts to establish are the consequences of the axioms of that theory. The "truths" of number theory are just those sentences that assert that some sentence of that theory is a consequence of the axioms.

One trouble with this picture is this: mathematicians frequently go outside the system of natural numbers to prove truths of number theory. Let us return again to the example of Fermat's Last Theorem. Recall that Wiles's proof involved proving the Taniyama-Shimura conjecture, which required making use of some very sophisticated theorems of analysis.[14] Are we to say that Fermat's Last Theorem is not a truth of number theory? That it is really a truth of analysis? Intuitively, it would seem that Fermat's Last Theorem is just a truth of mathematics—mathematicians don't seem to distinguish the system in which a theorem is proved in the way the if-thenist suggests. Indeed, the if-thenist's account suggests that mathematicians in general, and Wiles in particular, should have been concerned with various if-then statements of the form 'if such and such axioms are true, then Fermat's Last Theorem is true', when in fact, mathematicians act as if they are concerned with getting at the truth of Fermat's Last Theorem, using whatever system or mathematical framework seems to be most useful for establishing the truth value of the sentence.

But this is not a problem for my structural account, since the above facts are all compatible with my view that sees the mathematician as theorizing about kinds of structures. My view is certainly not an espousal of the if-thenist's doctrine that the mathematician only wants to establish statements of the form 'if axioms A are true, then ϕ is true'. On the contrary, I view the researchers as having answered a question about a type of structure (the natural numbers) which is embedded in a larger kind of structure (classical analysis) and they use features of the larger kind of structure to answer questions about the embedded kind of structure.

Resnik's Euler objection

Resnik has raised an objection to deductivism that proceeds from the fact that "throughout the history of mathematics mathematicians have been convinced completely of results by means of arguments that fall short of deductive proof" (Resnik, 1980: 131). Before considering the sort of examples from the history of mathematics that support this claim, let us ask: why is this an objection to deductivism? Presumably, Resnik is arguing as follows. These mathematicians were willing to assert some mathematical proposition ϕ, even though they would not have been willing to assert the deductivist's 'ϕ follows deductively from axioms A' (perhaps because there was no widely accepted set of axioms for the area of mathematics being researched). This is strong evidence that what the mathematician is asserting (in asserting ϕ) is not what the deductivist maintains the mathematician is asserting.

---

[14] See Singh, 1997 for a detailed discussion of these matters.

Let us now examine the kind of example from the history of mathematics that Resnik has in mind. Bernoulli set the following problem: find the sum of the sequence (of reciprocals of squares)

$$1 + \frac{1}{4} + \frac{1}{9} + \frac{1}{16} + \frac{1}{25} + \cdots$$

By very clever analogical reasoning, Euler came up with the solution $\pi^2/6$. The reasoning (being based, as it was, on mere analogical reasoning) was not something that could be considered a deductive proof of the solution from some set of axioms. However, he then checked his conclusion both by calculating the value of the series to twenty places and also by carrying out a decimal expansion of $\pi^2/6$ to twenty places. What he found by means of these calculations was that they coincided. Then he applied his method of reasoning to the sum of reciprocals of the fourth powers and came up with the result that the sum $= \pi^4/90$. Again he tested this conclusion by calculating the two sides of the equation arrived at, and he again got agreement to many places. Finally, he applied his method to another similar sequence and came up with a result that had been proven to be correct. All of this strongly suggested not only that the general analogical method he used was reliable, but also that his $\pi^2/6$ answer to the first problem was correct—so much so that he became convinced that his answer was correct. Here's how Steiner sums up Euler's case:

Euler had not proved his result: [his reasoning] rested on a method whose principles are very subtle and which Euler did not know. But when we narrow our focus from the method to its product,

$$\frac{\pi^2}{6} = 1 + \frac{1}{4} + \frac{1}{9} + \cdots$$

and ask ourselves to what extent *it* was confirmed for Euler, we must admit that Euler had a right to be confident in his discovery beyond any doubt. ... 20 decimal places of agreement between $\pi^2/6$ and the sum of the reciprocals of the squares is conclusive evidence that agreement holds throughout. (Steiner, 1975: 106–7)

Granted that Resnik has raised a serious objection to deductivism, the question we need to consider here is whether he has raised a serious objection to my account of mathematics. It should be clear to the attentive reader that he has not. As I have already emphasized (in Section 2 above), my account is certainly compatible with there being ways of obtaining the kind of structural information crucial in mathematics that are alternatives to the principal (deductive) ways of obtaining this information. Indeed, nothing in my account precludes there being such alternative ways.

What about mathematical theories that are not axiomatized?

Maddy and Resnik have sketched another serious problem with if-thenism that is closely related to the preceding. Putnam seems to assume (or at least require) that mathematics must be axiomatized, since his account of the assertions of pure mathematics seems to be appropriate only for the case of assertions from an axiomatized theory. Thus, both Maddy and Resnik have, in effect, asked: what were mathematicians establishing prior to the twentieth century, before mathematics could be regarded as a fully axiomatized theory? What were they establishing or justifying when they put forward proofs of some mathematical theorem, say of analysis, when analysis was not axiomatized?[15]

Can we raise this very same objection to my view? Unlike Putnam, I have not claimed that pure mathematics makes any such assertions as the one attributed to pure mathematics by Putnam. That is, I have not claimed that every theorem of mathematics is an assertion of the if-then form presented by Putnam. Remember, I make no claims about the literal meaning of any sentence of a mathematics.

Have I required that mathematics must always be axiomatized in order that a proof of a theorem establish a truth of the sort that I have described? Not at all. *There is no reason, according to my account, why a mathematician may not be reasoning about a type of structure that has not been axiomatized.* Take the case of number theory before any axiomatization of it had been given. Through an analysis of mathematical practice, Dedekind was able to come up with a characterization of the kind of structure studied by number theorists that is widely regarded as satisfactory.[16] But nothing in my account requires that,

---

[15] Resnik has emphasized the sort of example that Polya made famous in which mathematicians are convinced of the acceptability of certain mathematical statements, even though no genuine proof is given: "In the cases in question, however, either the mathematicians were unable to cite any axioms, or else the axioms they did cite were insufficient to generate the result" (Resnik, 1980: 132).

[16] I am not suggesting, however, that every feature of Dedekind's characterization was discovered through analysis. What is taken to be a primitive relation of the natural number structure and what is taken to be defined is, to a large extent, a matter of taste or mathematical aesthetics. However, Dedekind did succeed in characterizing a "set theoretical structure which has the essential properties of the natural numbers" (Gillies, 1982: 60–1). Also, it should be noted that Dedekind is widely regarded as having formulated the axioms of number theory known as "Peano's axioms", which (it is thought) Peano borrowed and made famous, so that the axioms are now, unjustly, known as "Peano's axioms". Donald Gillies takes issue with this common belief, writing: "I think it is correct to speak of Peano's axioms rather than Dedekind's axioms; for Dedekind was not trying to axiomatize arithmetic, but rather to define arithmetical notions in terms of logical ones" (66). Unlike the case of Peano's formulation, Gillies argues, there is no natural route from Dedekind's formulation to first-order Peano arithmetic (68).

prior to the publication of Dedekind's system, number theorists had in mind such an explicit characterization. Or consider Newton's work in analysis. Now, I do not claim to know what was going on in Newton's head when he developed the calculus, but from my vantage point, I can regard what he did in a way that is perfectly compatible with the general view I have put forward. Newton seems to have thought that he was reasoning about physical space when he developed the calculus. But he represented this space as having a mathematical structure of the sort that is familiar today. So in a way, he can be regarded as theorizing about a kind of structure—a kind of structure that includes a substructure of the sort physical space was represented to have. The theorems of pure mathematics we are able to abstract from Newton's work can be regarded as having the sort of structural content I attributed to the theorems of set theory. A proof of theorem ϕ found in Newton's writings can be regarded as establishing that any structure of the sort he had in mind will be such as to satisfy the conditions given by the theorem (that is, as establishing that the sentence expressing the structural content of ϕ is indeed true).

Well, some will want to know, what sort of structure did Newton have in mind exactly? Here, we are asking for a precise characterization of the three-dimensional continuum and the available functions of ordered triples of real numbers Newton had in mind—and it is quite possible that this sort of structure simply cannot be specified in any precise way. But that is not surprising. Like so much about the thoughts and theories of bygone ages, there is a certain amount of vagueness, imprecision, and uncertainty that we just have to live with. And it would be a serious mistake to attribute to these thinkers more precision and clarity than they could possibly have possessed.

The justification for accepting rules of inference

Another objection that Resnik has raised to if-thenism (or deductivism) concerns the notion of truth in mathematics. If, as Nidditch has claimed (Nidditch, 1960: 287), to say that some mathematical proposition ϕ is true is merely to say that ϕ is a theorem of some mathematical system S, then the traditional notion of truth in mathematics (mathematical truth) seems to disappear. Resnik responds with the words:

[In] eschewing traditional mathematical truth, deductivism may undercut the epistemology for logic itself ... The point is this: part of the usual justification for accepting a rule of deductive inference is that following it invariably leads from premises that are true ... to conclusions which also are true. But since deductivism provides for no concept of mathematical truth beyond logical truth, it can be sure only of the correctness of the logical rules in their nonmathematical applications. ... the Brouwer-Hilbert doubts about the use of logic in applications to abstract infinite structures—even those only hypothetically existing—remain unanswered. (Resnik, 1980: 135–6)

If Resnik were to advance this objection against my view, I would, first of all, question its premise that, in logic, "the usual justification for accepting a rule of deductive inference is that following it invariably leads from premises that are true ... to conclusions which also are true". As I see it, the usual justification, in mathematical logic, for accepting a rule of deductive inference is that following it invariably leads from premises that are *true in a structure* to conclusions which are also *true in that structure*. In support of this contention, I can cite what is done in practically any mathematical logic text covering the semantics of first-order logic, higher-order logic, or even modal logic. Accordingly, I see no reason why I cannot give such justifications.

Secondly, I certainly do not eschew the traditional notion of truth and, in this way, undercut the epistemology for logic: the notion of truth is an essential element of the Constructibility Theory.[17]

Frege's problem

One other objection that Maddy brings up in her assessment of if-thenism should be considered. This objection is characterized as "a version of Frege's problem":[18] How can the fact that one mathematical statement follows from another be correctly and usefully used in our scientific investigations of the physical world? This is a question that, according to Maddy, completely undermined if-thenism for Putnam and Russell (Maddy, 1990: 25).

The comparable question for my structural account is: how can any fact of the form

any realization of the axioms (or, alternatively, any structure of the kind being investigated in the system) would have to satisfy——

be correctly and usefully used in our scientific investigations of the physical world? Does this question undermine my view of mathematics? This is, essentially, the question that the previous chapter was directed at answering. It can hardly be said that I have nothing to say in response to this version of Frege's problem.

---

[17] It should be emphasized, however, that I do not assume that truth applies to all the sentences of English or to all declarative sentences of English or to some large, unspecified body of sentences of natural languages. Truth is required only for a restricted body of modal sentences.

[18] Frege's problem is expressed as: "what makes these meaningless strings of symbols useful in application?" (Maddy, 1990: 24).

The problem of satisfiability of mathematical axioms

Resnik sees a problem for the deductivist in knowing whether a mathematically complex set of axioms is satisfiable. "My concern", he tells us, "is with the application of sophisticated mathematical theories, such as the theory of real numbers or analysis" (Resnik, 1980: 134). To determine if the axioms of such theories are satisfiable, one would have to be convinced (he thinks) that complex infinite structures exist. This is a difficulty for the deductivist, since he believes that no such structure can be found among the objects of the physical world. Evidently, the only recourse open to the deductivist is to appeal to abstract objects, which would undercut one of the main motivations for being a deductivist.

It should be obvious to the reader that this objection raises no problems for my account, since it was already shown in Chapter 7 that the theory of real numbers and analysis can be shown to be satisfiable, using the Constructibility Theory, without presupposing the existence of mathematical objects.

5. THE MAIN STUMBLING BLOCK OF THE ELIMINATIVE PROGRAM

I shall now take up Charles Parson's problem of "avoiding vacuity in the conditionals used to interpret mathematical statements"—a problem that Stewart Shapiro called the main stumbling block of the eliminative program. Let Φ be a statement of number theory. This must be understood, according to Shapiro's version of eliminative structuralism, to be some such statement as:

$[\Phi]^*$: For every system S, if S exemplifies the natural number structure, then $\Phi[S]$,

where $\Phi[S]$ is the sentence one obtains from Φ by restricting the variables in Φ to the objects in S and interpreting the non-logical constants in Φ in terms of the objects in S (and the phrase 'exemplifies the natural number structure' is expressed in a way that does not presuppose the existence of structures). It is argued by Shapiro that, if the eliminative structuralist's background ontology is finite, she will be forced to hold that there is no system that exemplifies the natural number structure, in which case, both $[\Phi]^*$ and $[-\Phi]^*$ would have to be considered to be true. In this way, Shapiro arrives at the conclusion that the eliminativist whose background ontology is finite would be committed to the truth of even the negations of theorems of mathematics. Not surprisingly, Shapiro declares that, as an account of the content of the sentences of mathematics, the analysis that eliminative structuralism delivers is unacceptable.

Do the above considerations, with minor revisions, pose a major stumbling block for my own eliminative account of mathematics? Let us consider a first-order theory, say **PA**, for specificity. Suppose that θ is a sentence of **PA** expressing what we would ordinarily call the negation of some theorem of **PA**, say '$2 + 2 \neq 4$'. Is my account committed to the position that θ would have to be considered to be true? Not at all. As was shown in Chapter 8, it is possible to construct realizations of **PA**, so the reasoning never gets off the ground. Besides, there is no need for me to maintain that θ, literally construed, is true, since I have not committed myself to any analysis of the literal meaning of any sentence of **PA**. But I have put forward a view about what we can infer about θ from a proof of it. Am I then committed, by the above considerations, to the truth of the structural content of θ? Am I committed to maintaining the truth of:

Every realization of **PA** would have to be a realization of θ?

Obviously not. First of all, I would be committed to the above absurdity only if there were a proof of θ; but we have been given no reason at all for thinking that there is a proof of θ. Secondly, since it is possible to construct realizations of **PA**, we have good reasons for thinking that no proof of θ is possible. Thirdly, the structural content of θ is modal in nature and not a material conditional. So again, my account is in no way committed to the sort of absurdity attributed by Shapiro to the eliminativist.

11

Field's Account of Mathematics and Metalogic

My anti-realist position throughout this work has been that the theorems of mathematics, literally and Platonically construed, are not true. Such a position may suggest to some that I am a (mathematical) "fictionalist"—where a fictionalist is one who holds that the assertions of mathematics express, for the most part, "untruths", as do the sentences of works of fictions.[1] Now the best-known fictionalist is Hartry Field, so it may be enlightening to the reader to learn just how my view of mathematics differs from Field's. As I shall emphasize below, our views on mathematics are, in several respects, quite similar.

In the course of discussing the question of how mathematics is applied to the physical world, Field informs us that he has arrived at a surprising result: "that to explain even very complex applications of mathematics to the physical world ... it is not necessary to assume that the mathematics that is applied is true" (Field, 1980: p. vii). From this quotation, one might conclude that my account of how mathematics is applied is close to that of Field's, especially given that both accounts are classified as "nominalistic". Even what he goes on to say about these applications is close to what I have been saying: "it is necessary to assume little more than that mathematics is consistent" (p. vii).

Of course, what Field has in mind on this score is conservatism (as was explained in Chapter 5)—something very different from the structural ideas I had in mind when I made similar claims. Since Field believes that the only non-question-begging argument for maintaining that mathematics is a body of truths "rests ultimately on the applicability of mathematics to the physical world", he feels he can conclude from his "surprising result" that "*no* part of mathematics is true" and that "the problem of accounting for the knowledge of mathematical truths vanishes" (Field, 1980: viii).

---

[1] Colyvan defines 'fictionalism' this way: "A fictionalist about mathematics believes that mathematical statements are, by and large, false" (Colyvan, 2001: 4–5).

Given that Field is convinced that the problem of accounting for the knowledge of mathematical truths vanishes, it is obvious that he thinks that *all* applications of mathematics to the physical world can be explained in terms of the supposed conservatism of mathematics. If only some cases of applications of mathematics could be so explained, the problem of accounting for the knowledge of mathematical truths could not be said to "vanish". Indeed, that Field thinks *all applications of mathematics can be explained by citing its conservatism alone* is explicitly expressed when he declares that a proponent of his view "is going to have to maintain that the utility of mathematics in *all* applications is accountable for in terms of conservatism alone" (Field, 1989c: 62, italics mine).

In this chapter, I shall indicate how my account of mathematics differs from Field's and why I do not accept many of Field's doctrines. In particular, I shall make clear why I have not made any use of Field's conservation principles (described earlier in Chapter 5) anywhere in my work. Not only have I given a quite different account of how mathematics is applied in the empirical sciences, my view of the very nature of mathematics itself differs considerably from Field's. I shall first explain why my account of mathematics cannot be accurately characterized as a form of "fictionalism".

1. WHY I SHOULD NOT BE CALLED A "FICTIONALIST"

A fictionalist holds that the assertions of mathematics are, for the most part, untrue assertions. Now, despite appearances to the contrary, I have not maintained that the theorems of mathematics are not true. What I have claimed is that these theorems, if literally and *Platonically construed*, are not true. However, I have been careful to avoid taking a stand on the question of how the theorems of mathematics should be construed—what these theorems actually mean. In particular, I have not maintained that the theorems of mathematics are correctly understood to be Platonic in content. As I have repeatedly emphasized, my structural view does not depend upon any specific "reading" or "interpretation" of these theorems. Thus, my position cannot be correctly characterized as "fictionalist" (as the term was characterized above).

Field, on the other hand, definitely commits himself to a Platonic understanding of the theorems of mathematics: for Field, the theorems of set theory are correctly interpreted to be assertions about real objects that exist in the (non-physical) world. Under his reading of the theorems of set theory, they imply or presuppose the existence of things that don't in fact exist. That is why he believes that these theorems are false.

I shall now detail why I have not been convinced by Field's attempt to justify his use of the conservation theorems.

2. FIELD'S METALOGICAL THEOREMS

Most readers of *Science Without Numbers* (Field, 1980) will be struck by Field's heavy reliance, in his theorizing about mathematics, on theorems of mathematical logic that he classifies as "metalogical", where a theorem or theory is said to be "metalogical" if, roughly, it is about logical systems or about logical notions such as logical consequence or derivability. Thus, the completeness and soundness theorems of first-order logic, as well as Gödel's incompleteness theorems, are metalogical theorems upon which Field bases many of his conclusions.[2] Other metalogical theorems that Field makes use of, in that book, are his "representation theorem" and "uniqueness theorem", to which he appeals in justifying his claim that his "intrinsic" version of Newtonian space-time has all the nominalistic consequences of the Platonic "extrinsic" version.[3] All of the above-mentioned metalogical theorems are proven using mathematics, so this raises doubts about how Field, a nominalist, can reasonably believe these theorems. I will now investigate Field's attempt to allay such doubts, and for this reason I will provide an analysis of Field's views on metalogic.

Other metalogical theorems that Field crucially relies upon, in developing his philosophy of mathematics, are his conservation principles. Let us consider Field's justifications of the various conservation principles he formulates, as well as of his grounds for accepting certain other metalogical theorems.

Field's proof of his conservation principles

Field gives basically two proofs of his conservation principles in his book. One is model-theoretic, made within the framework of a standard set theory which asserts the existence of inaccessible cardinals, yielding a principle that says, roughly, that ZFU* is semantically conservative over nominalistic theories. The other is proof-theoretic, presupposing enough function theory and arithmetic to generate standard proof theory, yielding a principle that says, roughly, that ZFU* is syntactically conservative over nominalistic theories.

---

[2] For example, he writes: "The justification for the shift from semantic to syntactic notions is of course the Gödel completeness theorem for first-order logic" (Field, 1980: 115 n. 30).

[3] See Field, 1980: ch. 6, where he states these theorems and applies them in justifying his claim to be producing a "nominalistic treatment of Newtonian space-time".

How can Field use mathematics to prove his conservation theorems?

What is striking about these proofs is that the mathematics used to prove these central theorems of Field's philosophy is supposed by Field to be false. How can Field rely upon theories that he holds to be false? By what reasonable principles of reasoning can Field claim to know these theorems when his supposed knowledge of them is based upon axioms and postulates that he holds to be false? Here's what he writes:

It may be thought that there was something wrong about using platonistic methods of proof in an argument for nominalism. But there is really little difficulty here: if I am successful in proving *platonistically* that abstract entities are not needed for ordinary inferences about the physical world or for science, then anyone who wants to *argue* for platonism will be unable to rely on the Quinean argument that the existence of abstract entities is an indispensable assumption. The upshot then ... is that platonism is left in unstable position: It entails its own unjustifiability. (Field, 1980: 6)

If these proofs were only put forward as, one might say, "reductio ad absurdum arguments" against the Platonist's position, then the above reply would make some sense. But it seems clear that Field *believes* the various metalogical theorems and principles he cites in his book: he certainly writes as if he believes, for example, that the conservation principles he cites are true. And he bases many of his theoretical claims and explanations that are key to his nominalism on these (and other) metalogical principles. Furthermore, it is clear that he expects his readers to be convinced that these principles and metalogical theorems he claims to prove are true. So the question arises, what grounds does Field have for his belief in his conservation principles and the other metalogical theorems he cites: in particular, what grounds does Field have that are not based upon the truth of mathematics? Taken as a response to this last question, the above passage makes no sense at all.

Field is certainly aware of the weakness of completely resting his justification of the conservation principles on his "reductio" type of justification. He tells us that "some story has to be told about how the nominalist is justified in appealing to [the Platonic proofs of conservatism] outside the context of a *reductio*" (Field, 1980: 110). Saying that the story would be a long one, he then gives the "essential idea" as follows:

[W]e've seen that the nominalist has various initial quasi-inductive arguments which support the conclusions that it is safe to use mathematics in certain contexts; if he then *using mathematics in one of those contexts* can prove that it is safe to use mathematics in those contexts, this can raise the support of the initial conclusion quite substantially. (Field, 1980: 110)

These quasi-inductive arguments are, I shall argue later, of dubious value. I also have doubts that the above rough sketch can be rigorously developed to yield a genuine nominalistic justification for belief in his conservation principle. In any case, he has never attempted in print to fill out this very rough sketch—he has certainly never published the kind of proof described in the quotation—so these ideas cannot now be regarded as an acceptable justification of his conservation principle.

Field seems to be aware of at least some of the defects in his position vis-à-vis his acceptance of metalogical theorems, since, subsequent to publishing *Science Without Numbers*, he has written several articles which supply at least a sketch of what he hopes will be the kind of justification that would satisfy the nominalist's needs.

In *Realism, Mathematics, and Modality* (Field, 1989*d*), he gives an answer of sorts to the question of whether he believes the various metalogical theorems he cites in his various works in the philosophy of mathematics. There, he takes up the question of whether or not the claims of metalogic need to be true to be good. The "hard-headed view" attempts to explain the utility of metalogic without attributing truth to its assertions. Field investigates this option and finds that he cannot give a satisfactory defense of it, so he decides to adopt "an attitude of belief toward our normal metalogical claims", while attempting to "account for these metalogical claims without introducing mathematical entities". This "soft" stand is thought to require an expansion of his logic to include the modal operator 'it is logically possible that'.[4]

3. FIELD'S JUSTIFICATION FOR ACCEPTING STANDARD METALOGICAL RESULTS

In *Realism, Mathematics, and Modality*, Field attempts to account for our metalogical beliefs by first taking the modal operators 'it is logically necessary that' and 'it is logically possible that' (using '$\Box$' and '$\Diamond$' to symbolize these

---

[4] Field, 1989*c*: 75. In a postscript written for the publication of this article in a collection of his papers, he writes: "The 'soft' stand on metalogic now seems to me clearly right on the crucial point: we need to utilize a notion of logical possibility in dealing with metalogical notion" (Field, 1989*c*: 78).

operators) as primitive notions of his nominalistic theory,[5] and then adopting the following principles:

(MPT#) If □ (NBG → there is a model for 'A') then ◇A

(MS#) If □ (NBG → there is a proof of '−A' in F) then −◇A

(ME#) If □ (NBG → there is no model for 'A') then −◇A

(MC#) If □ (NBG → there is no proof of '−A' in F) then ◇A

where NBG is the conjunction of the axioms of Neumann–Bernays—Gödel set theory,[6] and F is assumed to be some suitable formal system.[7] These principles are supposedly Field's nominalistic versions of the following Platonic principles:

(MPT) If there is a model for 'A', then ◇A

(MS) If there is a proof of '−A' in F, then −◇A

(ME) If there is no model for 'A', then −◇A

(MC) If there is no proof of '−A' in F, then ◇A

It is then argued that the nominalist has as much reason for believing the hatched principles as the Platonist has for believing the Platonic, unhatched versions.[8] Field concludes that his kind of nominalist "can use the hatched schemata in pretty much the same way the platonist used the unhatched ones: to find out that A is, or is not, logically consistent, he claims that it suffices to derive a model-theoretic or proof-theoretic statement from standard mathematics" (Field, 1989a: 108).

To support this conclusion, he constructs (in Field, 1992) a "nominalistic" proof of a version of his conservation principle (expressed this time in terms of his modal operators)—a proof that mirrors one of his Platonic proofs of conservation but that is supposed to be "nominalistic" insofar as it attempts to avoid the use of mathematics by making use of the hatched principles.

---

[5] It is assumed by Field that the proper modal logic for these operators is a version of S5.

[6] Actually, Field says that NBG is Neumann–Bernays–Gödel set theory (Field, 1989a: 88). However, he later makes it clear that, in the context in question, he intends 'NBG' to stand for the conjunctions of the axioms of the set theory (see Field, 1992: 112–14).

[7] See Field, 1989a: 102 for his description of F. The four principles are given on 108. I have written the four principles in exactly the way Field did except for the fact that I have used the arrow '→' where Field used the "hook" symbol for the material conditional.

[8] See Field, 1989a: 108–10, where this is argued.

Field's deflationism

Before assessing Field's use of these principles, I should like to clarify an aspect of his theory of metalogic that it is easy to overlook. Field makes it clear that the only modal operators he is willing to countenance in his theory of metalogic are 'it is logically possible that' and 'it is logically necessary that'. What lies behind his restricted view of what modal operators are permissible in his theory is his "logicism". He believes that what differentiates a person with lots of mathematical knowledge from a person with little is—apart from empirical knowledge of such things as what mathematicians accept and what they use as axioms—*purely logical knowledge*: they know such things as that certain mathematical claims follow from certain other mathematical claims and that certain bodies of mathematical claims are logically consistent (Field, 1989a: 85). This form of logicism Field calls "deflationism" (it deflates the body of mathematical knowledge we have to merely logical knowledge).

Now the deflationist can make sense of a certain kind of modal knowledge: knowledge of possibility. However, he tells us, "the modal knowledge which deflationism allows is knowledge of purely logical possibility—deflationism does not allow knowledge of mathematical possibility in an interesting sense" (Field, 1989a: 85 n. 7). What Field means by 'it is logically possible that' can be illustrated by some examples. The sentence

$$(\exists x)(x \text{ is a bachelor } \& \ x \text{ is married})$$

is regarded by Field as logically consistent. Thus, he thinks that the sentence

$$\Diamond(\exists x)(x \text{ is a bachelor } \& \ x \text{ is married})$$

is true when the operator '$\Diamond$' is understood to be his "logically possible that" operator.[9] This inference makes it clear that his modal operator '$\Diamond$' is restricted to a very narrow notion of 'logically possible' and does not correspond to the "broadly logical" operators modal logicians frequently make use of.[10]

---

[9] See Field, 1989a: 87, where such examples are discussed in detail.

[10] In Field, 1992, he emphasizes the restricted sense of his modal operators by using a modal logic that limits logical truths to sentences "true by logical form alone" (114–15). There, he tells us, "This logical form principle partly captures the idea that the modalities in question are purely logical".

My doubts about deflationism

I do not find deflationism at all plausible. I do not believe that what distinguishes a child who knows a significant amount of algebra from the child who
knows very little algebra is purely logical knowledge of the sort described by
Field. The child in an elementary school class on algebra and learning the
binomial theorem does not learn that the theorem is a logical consequence of
some set of axioms. First of all, I doubt that there are many children who even
know what a *logical consequence* (in Field's sense of that term) is; and secondly,
I doubt that many children can give any reasonable set of algebraic axioms of
which the theorems being learned, such as the binomial theorem, are logical
consequences. I also doubt that what distinguishes what Wiles came to know
about numbers, in discovering his proof of Fermat's Last Theorem, from what
the average American knows about numbers is that Wiles learned that the
Fermat sentence is a logical consequence of some set of axioms, whereas the
average American is ignorant of such a logical fact.[11]

It can be seen that Field's deflationism is basically a form of if-thenism, and
many of the objections to if-thenism discussed in the previous chapter apply
equally well to this radical form of logicism. Here, the reader may find it
enlightening to reconsider the examples discussed in the previous chapter of
mathematicians coming to know theorems of analysis before any axiomatizations of analysis had been formulated.

How can Field make use of the consequence relation?

One objection that was raised to Field's reliance upon the various metatheorems listed earlier in this chapter was that these theorems are expressed
in terms of the notion of consequence—a model-theoretic notion that is
defined using set theory. How can Field, a nominalist, accept and make use of
such a notion?[12]

It can now be seen how Field responds to this objection. In his nominalistic
versions of the metatheorems, occurrences of the Platonist's

$$\phi \text{ is a consequence of } \theta$$

---

[11] Vineberg, commenting on the above point, has noted (in a personal communication) that
"most mathematicians show very little interest in logic ... However, if mathematical knowledge
were fundamentally knowledge of logical consequence, one would expect mathematicians to pay
close attention to logical form, to study formal logic in great detail, and that logic would be an
essential part of the curriculum, yet this is not the case at all".

[12] See, for example, Chihara, 1990: 162.

are replaced by occurrences of his deflationist version:

$$\Box(\theta \rightarrow \phi).$$

Similarly, the Platonist's

$$\theta \text{ is satisfiable}$$

is replaced by the deflationist's

$$\Diamond\theta.$$

Thus, he suggests (Field, 1989a: 104) that the nominalist can understand Platonic model theory (with its four principles described above) as a sort of device to find out about logical necessity and logical possibility.

What do the hatched principles say?

Before examining Field's use of the above four hatched principles in his metalogical investigations, I suggest that we examine more closely just what the principles say. Let us consider the antecedent of (MS#), that is, '$\Box$(NBG $\rightarrow$ there is a proof of '$-A$' in F)'. Since NBG is a conjunction of sentences of first-order set theory, there is no problem about determining what the antecedent of the conditional inside the scope of the modal operator is. But what about the consequent of that conditional? There are two possibilities: it might be the very English "sentence-form" written there;[13] or it might be the *formal mathematical formula* of NBG that "expresses" what that English sentence-form says, via Gödel numbering and the use of formulas of NBG for primitive recursive relations that *represent*[14] relationships that are ordinarily expressed in English. The basic point I wish to make can be made whichever of the above two possibilities Field may have in mind. However, I am inclined to think that the latter reading is probably the correct one.

Here are my reasons. First, at one point, Field describes what occurs as the consequent of one of his four principles. He tells us that 'there is an F-derivation of x' is a sigma-1 formula (Field, 1989a: 122). Since the term 'sigma-1 formula' has been defined in the first instance for formalized mathematical formulas, this suggests that he is thinking of what occurs

---

[13] I use the expression 'sentence-form' rather than 'sentence' to indicate that the letter 'A' occurring in the consequent is functioning as a schematic letter standing in place of a sentence.

[14] I am using the term 'represent' in the technical sense used by mathematical logicians when they develop the "arithmetization of syntax" in proving the Incompleteness Theorems. See, for details, Enderton, 1972: sect. 3.4.

within the scope of the operator as a formula of NBG. After all, the English 'there is an F-derivation of x' is strictly speaking an open-sentence, not a formula. But we do call strings of primitive symbols of NBG 'well-formed formulas'. So we do have some reason for thinking that Field is thinking of the consequent in question as the well-formed formula of NBG that "expresses" what the English open-sentence does.

Second, reconsider his statement above: "to find out that A is, or is not, logically consistent, it suffices to derive a model-theoretic or proof-theoretic statement from standard mathematics" (Field, 1989*a*: 108). Two of the four hatched principles are supposed to allow us to find out about the logical consistency (or inconsistency) of A from the *derivation* of a proof-theoretic statement from standard mathematics (NBG). Now a derivation, in logic, is the formal analogue of what passes for proof in ordinary unformalized mathematics (recall the discussion in the previous chapter). But 'derivation' does not mean what 'proof' does. One can have a derivation of a formula in some formal system when one does not have a proof (in the ordinary sense of that term). For example, one can have a derivation of ϕ from ϕ, but such a derivation would not ordinarily be considered some kind of proof of ϕ (what would a proof of ϕ from ϕ be but a case of begging the question?). Also, the existence of a derivation is the existence of a sequence of formulas constructed in accordance with specific syntactical rules. Thus, the sentences in a derivation need not even be meaningful: one can have a derivation consisting of sentences that are not interpreted, and such a derivation would not ordinarily be called a proof. Since Field talks of *deriving* a proof-theoretic statement from NBG, this too suggests that Field is thinking of the proof-theoretic statement as something that can be *derived* from NBG, that is, as a sentence of NBG.

Third, by taking the consequent of what occurs within the scope of the modal operator to be a formula of NBG, the whole conditional can be seen to be a formula of NBG. That would make more sense than having the sort of linguistic monstrosity implied by the alternative reading, according to which the antecedent is a formula of NBG and the consequent is a sentence-form of English.

Fourth, in the original version of "Is Mathematical Knowledge Just Logical Knowledge?", Field wrote of a proof of a similar conditional that "the entire proof can be done in...number theory" (Field, 1984: 40). Since the entire proof of such a conditional could not be carried out in number theory if the consequent were a sentence-form of English, we have additional evidence that Field is taking the consequent to be a formula of NBG.

I shall thus begin my analysis of Field's hatched principles with what I consider the more plausible reading, according to which the sentence-form occurring within the scope of the modal operator in each of the four principles is regarded as "standing for" (that is, standing in place of) a formula of NBG that "expresses" what the sentence-form "says". With this in mind, let us reconsider (MC#):

If □(NBG → there is no proof of '–A' in F) then ◇A

Can Field use the hatched principles to find out what is logically consistent?

Suppose that the sentence ϕ of NBG, which expresses via standard Gödelian techniques "there is no proof of '–A' in F" (for some specific sentence A), is actually derived in NBG. (I assume here, for the sake of argument, that the deflationist can actually construct a derivation of ϕ in NBG.) Why should Field be allowed to conclude that ◇A? Remember, ϕ is simply a sentence of set theory which, taken literally, is only about sets and classes: such phrases as 'there is', 'proof of', and 'in F' are not part of the vocabulary of NBG. Let us grant, also for the sake of argument, that if Field knows that there is no proof of '–A' in F, then he can conclude ◇A. What I am questioning is his right to infer, from □(NBG → ϕ), that it is logically necessary that, if the axioms of NBG are true, then there is no proof of '–A' in F.

Since ϕ is an enormously complicated and lengthy sentence of NBG, to be able to infer, from the truth of ϕ, that there is no proof of '–A' in F, one needs to make use of a mapping that takes us from natural numbers (or more specifically, certain sets that, by means of various definitions, function as the natural numbers in NBG) to symbols and sequences of symbols. One also needs to rely upon a significant number of mathematical and metamathematical results encompassing primitive recursive functions, the representation of primitive recursive functions in PA or a similar formal system, in addition to standard theorems of number theory (such as the Chinese Remainder Theorem[15] or equally recondite theorems). But Field does not believe in the existence of sets or natural numbers, and he does not believe that the various theorems of number theory or function theory needed to carry out this reasoning are true. So it is questionable that he is in a position to infer the consequent of his hatched principle from merely knowing the antecedent. The lengthy metalogical reasoning that he presents (in Field, 1989a) still leaves us with the sort of substantial gap with which we started: why is he allowed to use Platonic mathematics and metamathematics in

---

[15] For a statement and proof of this theorem, as well as a discussion of how it is used to prove various representation theorems in logic, see Enderton, 1972: sect. 3.7.

drawing inferences about metalogic? In short, Field has not shown that the hatched principles are ones that the nominalist (deflationist) is justified in believing.

I assumed above, for the sake of argument, that the nominalist could somehow literally construct a derivation of φ within NBG. In fact, no nominalist could (practically speaking) produce such a derivation. Indeed, it is questionable that any nominalist could even write down φ: it would be too lengthy and too complicated to be written. What Field may have thought is that the nominalist could, by means of some sort of metalogical proof that '−A' is not derivable in NBG (perhaps by constructing a deriva-tion of 'A' in NBG), infer that φ is derivable in NBG. However, to make such an inference, he would have to rely upon the use of the sort of meta-mathematical results that Gödel established when he proved his famous theorems. But here again, we come upon the need to rely upon math-ematics and metamathematics to arrive at the conclusion that φ is derivable in NBG. In short, Field again needs to base his reasoning on Platonic the-ories to obtain the result he envisages—something he had hoped to avoid with his new theory of metalogic.

Suppose now that Field has in mind the alternative (less plausible) reading of what occurs within the modal operator. Let us consider (MC#) for specificity, and suppose that A is a specific sentence. The consequent of the conditional occurring inside the scope of the necessity operator of (MC#) is now supposed to be an English sentence. The problem this time is in proving this English sentence from the axioms of NBG. It should be obvious that there is absolutely no way that one can construct a purely logical proof of such a sentence from the axioms of NBG alone; one will need much more than this. To generate such a proof using Gödel numbering and number-theoretic representations, one would need auxiliary axioms about the meaning of the words and symbols occurring in the consequent, such as 'there is', 'proof in F', and '−A', as well as axioms about the crucial relationships between the sets talked about in NBG and about such things as proofs, derivational system F, and '−A'.

To see the hopelessness of trying to prove the consequent from the axiom of NBG using only pure logic, notice that if the sentence

$$(\exists x)(x \text{ is a bachelor } \& \ x \text{ is married})$$

is logically consistent (as Field has affirmed), then it is obvious that

NBG & it is not the case that there is no proof of '−A' in F

is also logically consistent. Hence,

$\diamond$(NBG & it is not the case that there is no proof of '-A' in F)

is true. From this, we can infer (using modal logic) that

$$\diamond-\text{(NBG} \rightarrow \text{there is no proof of '-A' in F),}$$

and hence that

$$-\square\text{(NBG} \rightarrow \text{there is no proof of '-A' in F).}$$

In short, (MC#) would be useless for learning about the logical consistency of A, since the antecedent would be false no matter what A is. I thus conclude that Field has failed to justify his acceptance and use of the various metalogical theorems upon which he bases many of his theoretical conclusions about the nature of mathematics in *Science Without Number*.

I suspect that what led Field astray in his metalogical investigations was the ambiguity (noted above) in each of the principles. On one reading, the antecedent of the principle can (in principle, but not in practice) be obtained, but then Field is not justified in inferring the consequent of the principle from the antecedent. Upon the other reading, the antecedent is always false, no matter what A is. However, if one does not clearly distinguish between these two possible readings, one can easily be led to think that the antecedent can be obtained (using the mathematical reading and forgetting that one needs mathematics to even infer that the antecedent can be obtained), and one can also think that the consequent can be inferred from the antecedent (using the alternative reading), and hence that the principle can be validly used by the nominalist to find out that A is, or is not, logically consistent.[16]

Regarding Field's "nominalistic proof of the conservativeness of set theory" (Field, 1992), since the proof is based upon the above deflationist principles, I do not see how one can accept his thesis that he has constructed a satisfactory nominalistically acceptable proof of conservativeness.

4. FIELD'S OTHER JUSTIFICATIONS OF HIS CONSERVATION PRINCIPLE

Field does not rest his case for the conservation principle solely on his mathematical justifications. He also provides other kinds of justifications for

[16] The above analysis of Field's justification of his hatched principles is a more elaborate and detailed version of an analysis that I gave in the postscript to the Appendix of Chihara, 1990. My earlier objection to Field's justification has received no notice in the literature, and I am inclined to think that this lack of attention is at least partially due to the fact that it was too brief and lacking in the sort of detail that is memorable. I hope the present version will capture the attention of those who think that Field's version of nominalism is the only reasonable option to Platonism.

believing the principle. One is directed at justifying the following variant of the principles given earlier:

> Principle C: Let A be any nominalistically statable assertion, and let N be any body of such assertions. Let S be any mathematical theory. Then, A* is not a consequence of N + S unless it is a consequence of N alone,

where M is a non-logical predicate interpreted to mean 'is a mathematical entity' and A* is the sentence that results from restricting all the quantifiers occurring in A with the formula '$-M\alpha$' (for appropriate variable α).

Principle C is taken by Field to be equivalent to:

> Principle C″: Let A be any nominalistically statable assertion. Then A* is not a consequence of S unless it is logically true.

Now, Field thinks it is obvious that Principle C″ is true, the reason being that, since mathematics is widely regarded as *true in all possible worlds* and *a priori true*, "it is hard to see how any knowledgeable person could regard our mathematical theories in these ways if those theories implied results about concrete entities alone that were not logically true" (Field, 1980: 12).

He then argues that

the failure of Principle C would show that mathematics couldn't be 'true in all possible worlds' and 'a priori true'. The fact that so many people think it does have these characteristics seems like some evidence that it does indeed satisfy . . . Principle C. (Field, 1980: 13)

In assessing this reasoning, it should first be noted that Principle C is stated in a way that makes it applicable to mathematics in general (and not restricted to formal mathematical theories of first-order logic): N is to be "any mathematical theory" (and is not required to be ZFU). The widely held beliefs that are supposed to provide us with evidence of the truth of his conservation principles, Principle C and Principle C″, are not beliefs about ZFU but about "mathematics": it is mathematics (and not merely ZFU) that is supposed, by a large number of people, to be both true in all possible worlds and also a priori true.

Now I shall give reasons shortly for questioning the truth of Principle C— and this, without questioning Field's premise that it is widely believed that mathematics is true in all possible worlds and is a priori true. So the above premise will be judged not to provide much justification for belief in Principle C. Furthermore, even without considering my grounds for questioning the truth of Principle C, I fail to see how Field himself can regard the above justification as providing very strong support for his belief in Principle C. The set of people who believe that mathematical theories are true in all possible worlds and are a priori true is a proper subset of the set of people who believe that mathematical theories are true. But he is convinced that the members of the larger set are just plain wrong: he is sure that mathematical theories are false. So the beliefs that are supposed to provide the evidence for Principle C are all regarded by Field to be false. I thus find myself wondering how Field can think that the cited *false beliefs* that many people are supposed to have provide him with strong support for his belief in Principle C. If Field is convinced that all those people who believe that mathematical theories are true—a truly vast totality—are all wrong, then he should also be convinced that vast totalities of people can be convinced that mathematics has some fundamental property, and be wrong. So why should we infer mathematics has some fundamental property (that implies its conservatism) from the mere fact that many people believe that mathematics has this fundamental property? Huge numbers of people believe that God exists. Does Field think that this provides strong evidence that God exists?

Field argues that if Principle C were false, then mathematics would not be true in all possible worlds and a priori true. The fact that many people believe that mathematics is true in all possible worlds and is a priori true is supposed to constitute significant evidence for the truth of Principle C. But why should the fact that many people have these (admittedly false) beliefs about the nature of mathematics provide us with strong evidence for the hypothesis that Principle C is true? What if many people believed directly that Principle C were true. Would Field consider that supposed fact as strong evidence that Principle C is true? I don't see how he plausibly could.

In a footnote, Field takes a quite different tack regarding the use of his conservation principle in justifying his belief that his version of Newtonian physics has the same nominalistic consequences as the Platonic version. He suggests that "the nominalist need not ultimately rely on such Platonistic proofs of the adequacy of his systems":

[The nominalist] could simply spin out deductions from nominalistic axiom systems like the ones suggested later in the monograph. In this sense, the reliance on platonistic proofs could be regarded as a temporary expedient. (Field, 1980: 107–8 n. 5)

But this suggestion does not really answer the objection. It is clear that Field believes that his conservation principles are true. The question is: what rational grounds does he have for this belief? True, he could proceed as he suggested in the footnote cited above, and this might conceivably convince him, if he were able to spin out a sufficiently large class of deductions from his nominalistic axiom system, that standard Platonic Newtonian physics was a conservative extension of his nominalistic version of Newtonian physics, and that ZFU is syntactically conservative over his nominalistic first-order version of Newtonian physics. Of course, if he were so convinced, we now know that he would be wrong to so conclude.[17] Such an unfortunate result indicates the pitfalls of proceeding in the way being suggested by Field. Justifying belief in logical principles by spinning out a lot of deductions can lead one to believe a lot of false principles. No doubt, one could justify the belief in Frege's Axiom V in just such a way. In any case, it is clear that neither he nor anyone else has yet carried out any such tedious task. So we have not been given an acceptable justification for his belief in the conservation principles.

5. FIELD'S ARGUMENTS THAT GOOD MATHEMATICS IS CONSERVATIVE

At one point, Field asserts that *"Good* mathematics is conservative" (Field, 1980: 19). Why should we believe this claim? Here's how he argues:

[I]t would be extremely surprising if it were to be discovered that standard mathematics implied that there are at least 10^6 non-mathematical objects in the universe, or that the Paris Commune was defeated; and were such a discovery to be made, all but the most unregenerative rationalist would take this as showing that standard mathematics needed revision. *Good* mathematics *is* conservative; a discovery that accepted mathematics isn't conservative would be a discovery that it isn't good. (Field, 1980: 13)

Now the persuasive force of Field's argument is generated by the particular examples he has chosen. Admittedly, it is hard to see how any good mathematical theory could logically imply the particular statements he gives. After all, standard mathematics makes no mention of non-mathematical objects or the Paris Commune, so it is hard to see how it could possibly imply that there are 10^6 non-mathematical objects or that the Paris Commune was defeated, unless it was inconsistent. Field is, in effect, arguing that any mathematics that was not conservative would not be good, and he tries to convince us of this position by getting us to imagine our discovering that standard mathematics was not only non-conservative, but non-conservative in an extreme

---

[17] See Chihara, 1990: 157–8.

way. We are inclined to agree that if mathematics were non-conservative in the way imagined, then mathematics would not be good. But might not mathematics be non-conservative in a less extreme way?

Consider the following argument:

> An illegal immigrant, who is both deeply committed and well trained to commit acts of terrorism, is a danger to this country.

> Therefore, *every* illegal immigrant is a danger to this country.

Any reasonable person would see the weakness of the inference. Similarly, a mathematics that was non-conservative in the way Field has described would undoubtedly not be good. But it in no way follows that *any* mathematical system that was not conservative would not be good.

Must good mathematics be conservative? Or equivalently, must every non-conservative mathematics be not good? My position is: it all depends upon what is meant by 'conservative'. The term can reasonably be understood in such a way that some non-conservative mathematical theories can indeed be good. Thus, to carry out this investigation, we first need to clarify what it means to say that a mathematical theory is conservative over nominalistic theories. Field gives many versions of his conservation principle, and he provides us with many explications of what he means by 'conservative'. But there is still much that needs to be clarified.

What is a reasonable version of the conservation principle?

Field states the precise versions of his conservation principle in terms of axiomatized first-order theories. But his writings are filled with versions of the principle that are not restricted only to formalized mathematical theories. (Some of these "unrestricted versions" will be given below.) As I pointed out at the beginning of this chapter, Field thinks that *all applications of mathematics can be explained by citing its conservatism alone*. Furthermore, time and time again he asserts that "good" mathematics (and not just "good" formalized versions of mathematics) is conservative over theories about the physical world and over nominalistic theories. For example:

> Given *any* particular application of mathematics, then, it is natural to ask whether the utility of mathematics in that application is due to its conservatism or to its truth. It is clear that an anti-realist of the sort we have been considering (one who believes that a mathematical theory that is sufficiently rich to be interesting needn't be true to be good, that it need only be conservative) is going to have to maintain that the utility of mathematics in *all* applications is accountable for in terms of conservativeness alone. (Field, 1989c: 62, italics mine)

Notice that Field speaks, here, of "any particular application of mathematics" and not some application of a restricted version of mathematics (such as ZFU), and he also claims that the utility of mathematics "in *all* applications" (and not just applications of mathematics to certain formalized versions of physics which have been purged of all references to mathematical entities and all modal terms) is to be explained in terms of conservativeness alone.

What is a realistic version of the conservation principle?

This takes me to the problem of trying to state a reasonable version of the conservation principle that treats both mathematics and science in a realistic manner. Here's how Field stated his conservation principle early in *Science Without Numbers*:

[M]athematics is ... useful in enabling us to draw nominalistically-statable conclusions from nominalistically-statable premises; *but here, unlike in the case of physics, the conclusions we arrive at by these means are not genuinely new, they are already derivable in a more long-winded fashion from the premises, without recourse to mathematical entities.* (Field, 1980: 10)

Notice that, in this statement of the principle, there is no restriction of the mathematics and the scientific theory (or assertion) mentioned in the principle to first-order theories.

In an article he published later, in response to certain objections raised by Shapiro, he revised his characterization of conservation, so that the principle is thereafter given in terms of the semantical notion of consequence instead of the syntactical notion of derivation:

What I should have said is that mathematics is useful because it is often easier to see that a nominalistic claim *follows from* a nominalistic theory plus mathematics than to see that it *follows from* the nominalistic theory alone. (Field, 1989b: 127, italics mine)

In line with this revision, here is how Field states his conservation principle in a later article:

(C) A mathematical theory M is conservative if and only if for any assertion A about the physical world and any body N of such assertions, A doesn't follow from N + M unless it follows from N alone. (Field, 1989c: 58)

The need to clarify the conservation principle

In order to give an accurate assessment of the plausibility of this conservation principle, it is important to see where the statement of the principle is unclear or vague. The notion of an assertion following from another, I take it, is to be understood as *logically following from* in accordance with his deflationism and theory of metalogic. The truly unclear terms used here are "mathematical theory" and "assertion about the physical world". Field avoided having to use these unclear terms in his first-order statements of the principle by using a specific first-order theory (namely ZFU) as his "mathematical theory" and by requiring that the vocabulary of his nominalistic theory be disjoint from the vocabulary of ZFU. But in any realistic statement of the principle, we must face the difficulty of trying to clarify such unclear or vague notions.

What is a mathematical theory?

Why not specify that a mathematical theory is any theory that recognized historians of mathematics both acknowledge to be a mathematical theory and also treat as a mathematical theory in their histories of mathematics? By that criterion, Euclid's geometry is a mathematical theory *par excellence*. Descartes's analytic geometry is a mathematical theory, as is Newton's theory of the differential and integral calculus. But Euclid's geometry is a theory about (physical) space and hence its assertions can be regarded as being "about the physical world". It is easy to see how, when Euclid's geometry is added to a body P of assertions about the physical world, an assertion could follow that does not follow from P itself. (One can take as P some statements expressed in the vocabulary of the geometry itself.) In short, if Euclid's geometry is a mathematical theory, then one could make a serious case for the proposition that mathematics is not always conservative.

Newton's calculus is also a theory about (physical) space. The mathematical notions of his calculus are so intertwined with his physics of motion and time that it can be regarded as partially a physical theory.[18] It would be reasonable to regard such a mathematical theory as being about the physical world. Then, for this case too, an assertion that is about the physical world can be found to follow from the combined theory consisting of his physics and his mathematical theory but not to follow from his physics alone. Thus, a strong case can be made that Newton's version of the calculus is not conservative over his physics.

Suppose that Field replied that the cases I cite are examples of mathematical theories that are not "good". I do not wish to insist that these theories are good. Certainly, they are not the sort of mathematical theory that con-temporary mathematicians use and develop. Still, Euclid's geometry was

---

[18] See Chihara, 1990: 167–8 for more details and references.

widely regarded, for thousands of years, as a paradigm of what a good mathematical theory should be.[19] If Euclid's geometry is not a good mathematical theory, what criteria are we to use to determine what is and what is not a good mathematical theory? And what justifies using those criteria? Obviously, we need some sort of criterion of "goodness" for mathematical theories that is not just an ad hoc stipulation that rules out counter examples to the conservation thesis.

It might be suggested that Field could defend his principle by requiring that the mathematical theory be a completely "pure" mathematical theory— where a pure mathematical theory is not *in any way about the physical world*. Let us call this suggested requirement the "pure mathematics requirement". Now what would it mean to say that a theory is not "in any way about the physical world"? Even Frege's Logicist version of arithmetic can be considered to be, in a way, "about the physical world".[20] The statement '2 $\neq$ 1' is about an extension (the number 1) under which falls the concept *is a moon of the earth*—a concept that can reasonably be taken to be "about the physical world". Or consider ZFU (Field's favored mathematical theory). This set theory talks about physical objects (the urelements), and hence is in some way about the physical world. Evidently, the "pure mathematics requirement" implies that even ZFU would not be classified as a completely "pure" mathematical theory. Not surprisingly, Field himself in effect rejects the "pure mathematics requirement".[21]

What counts as an assertion about the physical world?

I now turn to the other vague notion that Field uses to express his general conservation principle: what is to count as an "assertion about the physical world". Take a typical statement of science in which mathematical terms occur:

[VCR] The velocity of a chemical reaction is given by the formula:

$$dx/dt = k(s - x)$$

---

[19] Thus, Kline writes: "the best mathematicians, scientists, and philosophers before about 1800 regarded [the *Elements*] as the ideal of rigorous proof." He then goes on to quote the number theorist Henry John Stephen Smith as proclaiming: "The methods of Euclid are, by almost universal consent, unexceptional in point of rigour" (Kline, 1982: 103).

[20] It might be replied that Frege's system of arithmetic is inconsistent and hence is not good. But one can specify a version of Frege's system of arithmetic that is arguably consistent. See in this regard Boolos, 1987.   [21] See Field, 1989c: 55–6.

where s is the original amount of substance per unit volume, x is the amount transformed in time t, and $s - x$ is the amount remaining unchanged at the end of time t.

Surely, such a statement would be "about the physical world" if anything is. Yet, it is easy to specify another statement which clearly is about the physical world—The amount x of the substance transformed in time t is equal to $s(1 - e^{-kt})$—which logically follows from the combined theory consisting of [VCR] and the (differential and integral) calculus but which does not *logically follow* from [VCR] alone.[22] Thus, the calculus is not, according to this analysis, conservative. But is not the calculus a good mathematical theory?

This time, it might be suggested that, for an assertion to be "about the physical world", it would have to be nominalistic and "*purely* about the physical world", presupposing no mathematical entities and making no reference to any mathematical entities such as numbers or derivatives. Thus, it could be argued that the statement [VCR] is not "about the physical world". Similarly, it could be argued against the previous objection based upon taking Euclid's geometry to be a "good mathematical theory", that the sentences of Euclid's geometry are not "about the physical world".

It is questionable, however, that such a response obviates the geometry example, because (as was argued in Chapter 1) the statements of Euclid's geometry do not presuppose the existence of geometric objects but are statements about the possibility of constructing geometric objects. One will still have a good case for asserting that the statements of Euclid's geometry are about the physical world.

Besides, the requirement that no statement about the physical world can presuppose the existence of any mathematical objects seems implausible, since it would imply (at least according to standard realistic interpretations of such sentences involving mathematical terms that Field generally gives) that such simple statements as "There are five apples on the table" and "The fifth set was won by Sampras" are not "about the physical world". More importantly, the suggested revision wouldn't undercut the force of the above [VCR] example, since one can construct a statement, using the Constructibility Theory in place of standard mathematical theory, expressing what [VCR] does, but in which no reference is made to any mathematical entity, so that the conclusion that the calculus is not conservative would still be forthcoming.

---

[22] The reader should keep in mind the precise sense of 'logically follow' that is relevant here: ϕ logically follows from (or is a consequence of) θ iff $\Box$ $(\theta \rightarrow \phi)$. Remember: 'John is unmarried' does not logically follow from 'John is a bachelor' in Field's sense of the term 'logically follow'.

What the above investigations make clear is that one can reasonably understand the crucial terms of the statement of the principle in such a way that mathematics turns out not to be conservative over bodies of statements about the physical world after all.

A reconsideration of Principle C

Let us now reconsider Principle C to see if it, too, is open to refutation by counterexample. The main difference between Principle C and (C) is that the former maintains that mathematics is conservative over bodies of *nominalistically statable assertions*, whereas the latter holds that mathematics is conservative over bodies of assertions *about the physical world*. Thus, the focus this time is on nominalistically statable assertions.

What is a "nominalistically statable assertion"? I would think that a "nominalistically statable assertion" would be any assertion that does not presuppose, imply, or require the existence of any mathematical or other "abstract entities". Thus, the following should count as a "nominalistically statable assertion":

[NSA] If $(\exists x)(\exists y)(y \in x)$, then there are infinitely many objects,

where '$\in$' is specified to express the membership relation of set theory.[23] Statement [NSA] does not presuppose, imply, or require the existence of either mathematical objects or other abstract entities, so it is surely nominalistically acceptable. But the theory that results from adding to [NSA] the axioms of a standard version of finite set theory yields the statement that there are infinitely many objects—something that does not logically follow from [NSA] alone. Thus, many versions of finite set theory are not conservative over bodies of nominalistically statable assertions.

How Field might respond

Here's one way Field might respond to the various examples I have given of mathematical theories that are not conservative over bodies of assertions about physical reality or over nominalistic theories: he might *restrict* his conservation principle to only first-order theories of the sort described earlier in Chapter 5 and, by suitably restricting the sort of theories the mathematical and the nominalistic bodies of assertions are to be, avoid the above objections. Thus, he might attempt to advance and to justify only the highly

---

[23] I assume here, for the sake of argument, that '$\in$' has a definite meaning and that such sentences as '$(\exists x)(\exists y)(y \in x)$' have definite truth values.

restricted version of the conservation principle such as the formalized versions sketched above in Chapter 5, and in this way obviate the sort of objections to the truth of the principle that I have raised. I shall take up such a response in the next section.

6. A COMPARISON BETWEEN TWO VIEWS OF MATHEMATICS

Mark Balaguer recently compared my constructibility version of mathematics quite unfavorably with a view of mathematics that he calls "fictionalism". There are two versions of fictionalism that he discusses in detail: Field's version and his own version. In this chapter, I shall regard Balaguer's comparison as pertaining only to Field's version. In Appendix B, I shall regard Balaguer's comparison as pertaining to his own brand of fictionalism.

A comparison to Field's fictionalism

My constructibility theory is thought by Balaguer to be inferior to Field's fictionalist version, on the grounds that my view is "non-standard" and seems to "fly in the face of mathematical practice" (Balaguer, 1998: 103). In his book, Balaguer goes on to claim that my view (indeed, *every* non-fictionalist nominalistic version of mathematics) "has no advantage over fictionalism" and, for that reason, is "inferior" to Field's view (103). The idea that my view has no advantage over fictionalism is not justified by Balaguer; it is just asserted. Perhaps he believes that this remarkable claim is self-evident. Perhaps it is based upon another unjustified claim he makes, that "there is nothing in mathematical practice that runs counter to [Field's] fictionalism" (103). Both unjustified claims strike me as blatantly false.

In assessing Balaguer's claims, we need to keep in mind two distinct goals that the developers of these rival nominalistic versions of mathematics may have had in mind: (1) they may have been aiming at providing a response to what I called in Chapter 5 "Quine's challenge to the nominalist"; or (2) they may have been attempting to provide an analysis of (actual) mathematical practice and (actual) applications of mathematics in science—an analysis that does not presuppose the existence of mathematical objects. (Such a goal is similar to that of what Balaguer calls the "hermeneutical project": (Balaguer, 1998: 3).)

If we are concerned to assess Balaguer's claims with respect to goal (1), then the lack of agreement or fit of my Constructibility Theory with actual mathematical practice is irrelevant. There is nothing in the challenge that requires the nominalistic alternatives being developed to fit actual mathematical practice (nor did Quine have in mind any such requirement). For this reason, I emphasized several times in my earlier book (Chihara, 1990) that I was in no way concerned to develop a nominalistic system of mathematics that mirrored or provided an analysis of actual mathematical practice. Besides, insofar as we are concerned with responses to Quine's challenge, there is no reason why one would have to pick a *single* response as "the best"—that there are many nominalistic alternatives to Platonic versions of mathematics undermines much of the force of Quine's argument for Platonism.

Suppose, this time, that we wish to assess Balaguer's claims with respect to goal (2). Then my competitor to Field's fictionalism is not the Constructibility Theory but rather the structural account of mathematics of the present work.[24] In what follows, I shall take Field's fictionalism to be aimed at providing an analysis of (actual) mathematical practice and (actual) applications of mathematics in science and compare the view to my structural account.

I would like to begin by assessing Balaguer's claim that no non-fictionalist view of mathematics has any advantages over Field's account. It should be clear from the previous section that Field's view of mathematics suffers from serious and fundamental problems—problems that do not plague my account. I have argued that Field has not been able to provide a convincing justification of the basic principles of his account, namely, the conservation principles. I have even given grounds for concluding that these principles are false, at least when they are given various plausible readings. Thus, I conclude that my structuralist account has significant advantages over Field's fictionalism.

Now I left open one possible way of rebutting my arguments that these principles are false: Field could restrict his principles to only first-order theories such as ZFU and the nominalistic version of Newtonian physics sketched by Field. So there is some value in assessing this way of responding to my arguments. But before continuing this line of investigation, it should be noted that restricting his principles to only first-order theories does not

---

[24] Because I have taken pains, throughout this work, not to presuppose any specific analysis of the meaning of any mathematical theorem, it might be thought that I have not been concerned with putting forward an account of actual mathematical practices and theories. But such a thought would be mistaken. What I have been concerned to show is that our actual mathematical and scientific practices are such that we do not have to know the actual meaning of any mathematical theorem in order to apply the theorem in the structural way described throughout this work. Thus, the Platonist takes the theorem of number theory 'There are infinitely many primes' to be an assertion about abstract entities that is true. Field also understands that theorem to be an assertion about abstract entities, but he takes it to be false. I take no stand on what that theorem means.

get him off the hook with respect to one problem, namely the problem of justifying his use of Platonic mathematics to prove his conservation principles. He still needs to give a satisfactory explanation of why a nominalist is allowed to use classical mathematics in that way. But here I want to return to the theme I emphasized in the Introduction to this book: one of the principal goals of the philosophy of mathematics is to produce a coherent overall general account of the nature of mathematics (where by 'mathematics' I mean the actual mathematics practiced and developed by current mathematicians)—one that is consistent not only with our present-day theoretical and scientific views about the world and also our place in the world as organisms with sense organs of the sort characterized by our best scientific theories, but also with what we know about how our mastery of mathematics is acquired and tested. To achieve such a goal, we need to analyze and understand as best we can the nature of *present-day mathematics and science*—not some logician's unrealistic idealization of mathematics and science.

As I indicated earlier, much of what Field says about mathematics seems to be directed at accomplishing just this goal. Thus, he writes:

The form that my own realism takes is very straightforward: it involves no claim that mathematical statements mean anything other than what they appear to mean; instead, it simply claims that there are no mathematical entities, and hence that mathematical statements which assert that there are such entities are not true ... Obviously I am not proposing that we replace the usual mathematics by a new mathematics that rejects all existential assertions ... I am proposing that mathematics 'does not have to be true to be good' ... Instead, I proposed as my alternative to truth a slight strengthening of consistency, which I called *conservativeness*. To say that a theory is conservative is to say that it is consistent with every internally consistent theory that is 'purely about the physical world' in that it involves no reference to mathematical objects. (Field, 1989d: 239–40)

Again, he writes:

[I]n explaining the application of mathematics to the physical world we never need assume that the mathematics is true, we need only assume that it is strongly consistent (i.e., conservative). (Field, 1989a: 97)

Notice that, in these quotations, Field places no restriction on the kind of mathematics being discussed. He says "we *never* need assume". And recall the quotations given at the beginning of this chapter. Field was quoted as claiming "*no* part of mathematics is true" and that "that to explain even very complex applications of mathematics to the physical world ... it is not necessary to assume that the mathematics that is applied is true". These are not qualified statements about particular formalized versions of mathematical theories. The clear implication of these quotations is that all applications of mathematics can be accounted for in terms of conservation.

By restricting his principles to first-order theories of the sort described earlier, he may be able to avoid having to respond to my arguments that his conservation principles are false, but he would then have to similarly restrict the very general claims about mathematics that are sprinkled throughout his writings on the topic. Such restrictions would seriously diminish the attractiveness of his account of mathematics. (Surely, one reason Field's instrumentalist view of mathematics has received so much attention from philosophers of mathematics is that he seemed to provide a very simple overall account of the utility of mathematics for science.) Moreover, he would have to abandon any claim to have dealt with the central goal of the philosophy of mathematics discussed above, namely, that of producing the sort of coherent overall general account of the nature of mathematics that is consistent with our present-day theoretical and scientific theories about the world.

Let us now compare this hypothesized first-order restricted version of Field's instrumentalistic account of mathematics with the structural account of this work. We see that, contrary to Balaguer's remarkable claim that "there is nothing in mathematical practice that runs counter to [Field's] fictionalism", it would be Field's version that is "non-standard" and that "flies in the face of mathematical practice", the reason being that real mathematics—the mathematics taught, applied, and researched in practically all the mathematics departments of all the universities in the world—is not a formalized theory developed from the axioms of ZFU or any such first-order set theory. Also, as was emphasized in the previous chapter, very, very few, if any, of the mathematical proofs produced by practicing mathematicians and published in mathematics journals are first-order derivations. Furthermore, genuine scientific theories (such as quantum mechanics, molecular genetics, and astrophysics) are not axiomatized first-order theories either. So far as I know, no one has been able to formalize and axiomatize such theories in first-order logic—certainly no one has as yet published an adequate axiomatized first-order version of any of the substantial theories listed.[25] And indeed, why

---

[25] Cf. Suppes's comment: "Theories ... like quantum mechanics, classical thermodynamics, or a modern quantitative version of learning theory, need to use not only general ideas of set theory but also many results concerning the real numbers. Formalization of such theories in first-order logic is utterly impractical" (Suppes, 1967: 58).

should we accept Field's belief that "there is an alternative formulation of science that does not require the use of any part of mathematics that refers to or quantifies over abstract entities" (Field, 1980: 2)? Field seems to believe that our actual scientific theories are all axiomatized.[26] But such a belief is false, and may be based upon the highly questionable Quinean doctrine that scientific theories ought to be formalized and axiomatized in first-order logic. In any case, nothing Field has done in his book provides us with any good reason for believing that any such reformulation of science is forthcoming. Notice that the nominalistic version of Newtonian gravitational theory Field sketches in *Science Without Numbers* is not even claimed to be logically equivalent to standard Newtonian gravitational theory: it is supposed to be a nominalistic theory that can be used as a kind of replacement for the more standard Platonic versions. So there is no question that Field's physics is non-standard.

Besides, there are theoretical reasons why no purely extensional versions of physics (which first-order versions are supposed to be) are both true and accurate representations of the theories physicists have actually adopted.[27] Quine's idea that scientific theories *ought to be* formalized in the extensional language of first-order logic strikes me as far-fetched and unsupported by strong rational arguments. However, since I have already expressed my criticisms of Quine's "language of science" thesis elsewhere, I shall not repeat them here.[28]

Field's realistic thesis about the semantics of set theory

In my discussion of the van Inwagen puzzle, I contrasted two very different ways of understanding the theorems of set theory. There was the Gödelian realistic view, according to which the theorems are taken to be straightforward assertions about the real world: the statement that there is a null set, for example, is understood to be the assertion of the existence in the real world of a set that has no members. I also described an alternative structural way of understanding these theorems, according to which the assertion that there is a null set is taken to be an assertion about all structures of a certain sort: in all such structures, there will be a position (or point) that will be such that no other position (or point) will be in the membership relation to it. Both Peter van Inwagen and David Lewis viewed the theorems of set theory in a realistic

---

[26] Thus, he writes: "one can always reaxiomatize scientific theories so that there is no reference to or quantification over mathematical entities in the reaxiomatization" (Field, 1980: p. viii).

[27] See Chihara, 1990: 10–12. For a more detailed discussion of this point, see Joseph, 1980.

[28] See Chihara, 1990: 10–12.

manner, resulting in the puzzle about how we could have obtained a true understanding of either the membership relationship or the intrinsic nature of sets.

Now what Field makes clear in his presentation of his theory of metalogic is that he also thinks of the axioms of set theory (NBG) as assertions about real objects that are supposed to exist in the real universe. This can be seen by his discussion of how he thinks we should apply his metalogical principles. He tells us that we need to have knowledge (or at least good reason to believe) that the conjunction of the axioms of NBG is logically possibly true. On the other hand, he asserts that we do not know (or have good reason to believe) that this conjunction is true, "since [the axioms] assert the existence of mathematical entities" (Field, 1989a : 90). Clearly, then, he must be thinking of the axioms as existential assertions about the real world.[29]

But it is not just that Field thinks of the axioms in this way—*the axioms must be taken to be assertions of that sort* if such principles as (MTP#) are to make the kind of sense Field attributes to them. For example, Field thinks that the deflationist can infer, from □(NBG → there is no proof of '−A' in F), that it is logically necessary that, if the axioms of NBG are true, then there is no proof of '−A' in F—which highlights the realistic sense he attributes to the conjunction of the axioms of NBG. As further evidence of Field's Gödelian way of understanding the axioms of set theory, I cite Field's statement that the consistency of set theory can be expressed:

$$\Diamond (S_1 \ \& \ S_2 \ \cdots \ \& \ S_n),$$

where the S_i are the axioms of the theory, "to include the assertion that there is a set of all urelements, and the separation and replacement schemas are to be understood as allowing the distinctively physical vocabulary of [the formal language] to occur in their instances" (Field, 1992: 112–13).

Some reasons for doubting Field's realistic thesis

It should be emphasized that the realistic semantical interpretation of statements of mathematics that I am questioning seems especially questionable when one is concerned with the statements of the mathematical theories practicing mathematicians employ in their day-to-day work. These are theories built on things that mathematicians are taught in elementary schools,

---

[29] Actually, his writings are filled with passages that make it abundantly clear that he regards the sentences of set theory, as did Gödel, to be about the real world. See, for example, the discussion of set theory in Field, 1989c: 55–8.

secondary schools, and universities. Early in their education, students learn to make existential statements of mathematics. Given any rational number, they are taught, there is a rational number that is larger. They are also told that there are infinitely many prime numbers, that there is a null set, that for any set S, there is a set whose members consist of all the subsets of S, and so on. Now do students learn that such statements of mathematics are statements about real objects in the real world? If so, how do they learn such things? Is this what they are taught? And by whom?

It is ironic that Field, a nominalist, not only takes the semantics of set theory in such a realistic way, but bases his metalogic on this kind of realistic semantical analysis. Surely, there are advantages in keeping an open mind about such matters and in not committing oneself to any such questionable semantical thesis—at least in the absence of a thorough linguistic investigation into its plausibility. That there are disadvantages can be seen by reexamining one of the puzzles with which I began this work.

Recall that, in Chapter 3, according to my analysis, Lewis and van Inwagen got into their puzzle about set theory because they had adopted the realistic reading of the theorems of set theory. Thus, by adopting Field's realistic semantical analysis of the statements of set theory, a fictionalist would also be saddled with the task of explaining not only how we can be said to have a genuine understanding of the relation of membership and of the nature of sets, but also why we use the "structural" criteria we do to test a student's grasp of set theory. Here again, there are advantages to a structural understanding of the semantics of set theory, since it yields a nice explanation of the puzzle as well as an explanation of the appropriateness of use of the criteria of understanding we in fact use.

7. THE FUNDAMENTAL THEOREM REVISITED

Balaguer's claim that my view of mathematics "has no advantage over fictionalism" will now be examined in the light of the Fundamental Theorem of the calculus, which is heavily used in practically all engineering and scientific work that make use of the calculus. Now how does Field explain the utility of mathematics in science? By appealing, *in every instance*, to the conservatism of mathematics. The focus of this section, the Fundamental Theorem, will shed unfavorable light on Field's view that *all* cases of applications of mathematics should be explained in terms of conservatism.

Early researchers in the development of the calculus certainly made use of the Fundamental Theorem in applying the integral calculus to empirical problems of science—this despite the fact that they could not have delimited and articulated in any precise way the mathematical theory they were employing (one cannot, even now, accurately express, as first-order axiomatized theories, the various theories these researchers were using). Furthermore, some early developers of the calculus seemed to utilize principles and concepts that were muddled if not downright inconsistent.[30] But conservatism is a logical feature of theories—not individual sentences. (The sentences of mathematics that express the Fundamental Theorem are supposed by Field to be false.) Thus, if the mathematical theory being used by these scholars either is inconsistent or cannot be definitely specified, claims that it is conservative can neither be established nor justified, and a fortiori one cannot explain the utility of the specific mathematical applications targeted in terms of conservatism. Besides, in many of these early cases, it is impossible to separate all the sentences used by the theorizer into two mutually exclusive types: sentences about the physical world and sentences of pure mathematics.[31] So there are additional obstacles to proving that the mathematical theories they were using were conservative over their physical theories. The problems inherent in the Fieldian approach to explaining the utility of the calculus during this period can be seen from the following description of the development of the calculus:

In the eighteenth century the problems considered to be most important were those which could be treated without paying attention to the foundations of the calculus. No strict line was drawn between the calculus and its applications. Many of the results obtained in the calculus had immediate physical applications; this circumstance made attention to rigor less vital, since a test for the truth of the conclusions already existed—an empirical test. (Grabiner, 1981: 16–17)

It is hard to see how Field's explanation of the usefulness of mathematics for such cases can even get off the ground.

[30] For an indication of the muddled state of mathematics (especially, the calculus) in the seventeenth and eighteenth centuries, see Kline, 1982: ch. 6 (entitled: "The Illogical Development: The Morass of Analysis"). Of course, the confusing and confused descriptions of the fundamental concepts of the calculus were well documented by such critics as Bishop Berkeley (Berkeley, 1956), and as Judith Grabiner notes, "Berkeley's attack on the calculus pointed out real deficiencies" (Grabiner, 1981: 27). Grabiner also argues that the criticisms leveled against the limit concept employed by Newton and his British colleagues by Berkeley "were almost impossible to answer in eighteenth-century terms" (Grabiner, 1981: 33).

[31] See Chihara, 1990: 167–9, for details on why I believe that Newton could not have separated all the theoretical sentences of his work into the two mutually exclusive types described above.

Surely, there is something wrong-headed (and hopeless) about attempting to explain the applications of the Fundamental Theorem, during this highly active period of development, in terms of the conservativism of the various mathematical theories being developed. Would it not be more reasonable to suppose that the reason that the Fundamental Theorem was so useful in science and engineering at that time was that it provided the researchers with *valuable information*?

Consider the following use of the Fundamental Theorem in drawing conclusions in pure mathematics. Let:

(1) $\quad \sum(k) = 1 + 2 + 3 + \dots + k$

where k is a natural number. One can construct a mathematical proof of the theorem that:

(2) $\quad \sum(k) = \int_0^k x\,dx + k/2$

Now let $G(x)$ be the function $x^2/2$. Since $G(x)$ is an indefinite integral of x, by applying the Fundamental Theorem to (2), we obtain:

(3) $\quad \sum(k) = k^2/2 + k/2$

Now recall the "theorem" discussed in the previous chapter that was justified by appeal to Figure 3. That "theorem" was basically (3). In other words, the above result obtained by using the Fundamental Theorem is confirmed by the earlier result we arrived at in our reasoning about the rational numbers. The Fundamental Theorem is a theorem about the continuum, and hence involves a more complex structure than that of the rational numbers. The fact that we obtain results, using the Fundamental Theorem, that are consistent with, and that are confirmed by, results we obtain in reasoning about the rational numbers or even the natural numbers strongly suggests that the theorem provides us with genuine information about structures.

Of course, a Fieldian might reply that such consistency results are only a consequence of our working with theories that happen to be consistent with each other. But such a reply would not be very convincing. After all, these results could very well have all been arrived at by the early workers who developed the calculus, and there is no good reason for thinking that these researchers were working with a consistent collection of theories. As I have been emphasizing in this chapter, there are grounds for thinking that their mathematical theories were fundamentally confused and incoherent. So there are reasons for being skeptical of such a Fieldian explanation.

The mathematical theories which the early researchers in the calculus used may have been fundamentally inconsistent and muddled, but such important individual theorems as the Fundamental Theorem were simply too fruitful and too strongly confirmed in the above way to be regarded as just false statements that happen to be theorems in consistent collections of mathematical theories. For even as the calculus was being significantly revised, through conceptual clarification and arithmetization, the Fundamental Theorem was never seriously regarded as dispensable. All of this suggests that the usefulness of such theorems should not be explained in the Fieldian way in terms of the conservatism of the successive mathematical theories of which these theorems are theorems, but rather in terms of the usefulness of the *information* they provide. Such a suggestion is in line with my own contrasting view that theorems have a kind of content that provides scientists and engineers with valuable information that can aid them in their theorizing both in mathematics and in science.

Appendix A.   Some Doubts About Hellman's Views

Hellman's account of mathematics runs into trouble when statements of applied mathematics are given roughly the same form as statements of pure mathematics. Thus, a simple cardinality statement, such as

There are more spiders than apes

(and a definite finite number of each)

is given a modal-structuralist translation of the form:

$$\Box Q(\Omega(X, f) \rightarrow - - -),$$

where 'Q' is to be replaced by two universal quantifiers with variables 'X' and 'f' respectively and '$- - -$' expresses a condition which results in a statement that says (in ordinary language) that if X and f were any ω-sequence, then there would be a 1–1 correspondence between the class of spiders and ….

Unfortunately, the standard S5 necessity operator raises difficulties here, since it implies that all possible worlds (to use the possible worlds semantics of necessity) are relevant to the evaluation of the sentence that follows the operator—even worlds in which there are fewer spiders than there are in the actual world or even worlds in which there are more apes than there are in the actual world. Thus, he is forced to adopt a "non-interference proviso": "We must", he says, "*stipulate* from the outset that the only possibilities [possible worlds] we entertain in employing the '$\Box$' are such as to leave the actual world entirely intact."[1] In other words, the only possible worlds considered to be in the range of the modal operator '$\Box$' are those in which all those facts of the actual world remain as facts. This implies that if there are apes in the actual world, then the "non-interference proviso" requires that all the possible worlds that are to be entertained in employing the '$\Box$' will be worlds in which there are apes. It follows that, when such a modal operator is

---

[1] Hellman, 1989: 99. For an explanation of why Hellman needed such a proviso, see Resnik, 1997: 70–2.

being employed, if

There are apes

is true, then

□ (There are apes)

must also be true. This shows that the resulting system incorporates a non-standard necessity operator and a non-standard modal logic.[2] (What does □φ mean in this system? Certainly, nothing very close to what 'It is necessary that φ' is ordinarily thought to mean.) Consequently, the modal structuralist's applications of mathematics will be, at best, a very inelegant matter: one system of modal logic is to be used for the sentences of pure mathematics, and another system of modal logic is to be used for the sentences of applied mathematics. How this is all supposed to mesh together when dealing with complicated reasoning in a highly theoretical part of physics is never made clear nor, evidently, worked out.

But there are other difficulties. Hellman's "modal structuralism" ran into difficulties when he attempted to construct, within the proposed framework, modal-structuralist versions of contemporary scientific theories that can be regarded as nominalistic. Thus, Hellman felt forced to admit, at one time, that "there is a strong case that modern physical theories—especially General Relativity and Quantum Mechanics—require (the possibility of) mathematical structures so rich that even the chances of a "modal nominalism" in any reasonable sense are dim" (Hellman, 1989: 95).

In his more recent work, however, Hellman has been more optimistic about his goal of reproducing a nominalistic version of modern physical theories within his modal-structuralist framework. He now feels that he has a way of going beyond the limitations of second-order real analysis imposed on him by the framework he had adopted in his earlier book (Hellman, 1989). In outline, Hellman makes use of the devices of plural quantification and mereology, and he then uses the work of John Burgess, Alan Hazen, and David Lewis to obtain relations within this framework.[3] This allows him to expand his system to third-order real analysis, within which he can obtain the applications of general relativity and quantum mechanics, which gave his earlier theory difficulties.[4]

---

[2] This despite Hellman's claim that the logic of his system is that of Cocchiarella, 1975.

[3] See Lewis, 1991: Appendix on Pairing.

[4] For details, see Hellman, 1996: sect. 2, as well as Hellman, 1999: sect. 2. In an email message to me, Hellman has expressed regret that, in his more recent writings, he had not explicitly expressed his withdrawal of his earlier admission of serious problems for his nominalistic program.

Two questions about Hellman's new mathematical system now arise: first, is it truly nominalistic? And second, are the revised assumptions of the system acceptable? The first question arises because Hellman's nominalistic thesis depends essentially upon some claims put forward by George Boolos. Boolos has provided us with an effective method for translating the sentences of second-order logic into "English sentences" containing plural quantification, where "English" is taken to be standard English, supplemented with various subscripted pronouns, the subscripts functioning as indices of cross refer-ence.[5] Boolos claims that:

[N]either the use of plurals nor the employment of second-order logic commits us to the existence of extra items beyond those to which we are already committed. We need not construe second-order quantifiers as ranging over anything other than the objects over which our first-order quantifiers range. (Boolos, 1984: 449)

Thus, Hellman's nominalistic thesis rests upon Boolos's thesis regarding second-order logic and plural quantification, and one may have doubts about these claims. Hellman does not give any genuinely independent arguments to support his acceptance of the plural quantification thesis, relying primarily on Boolos's argumentation.[6] Still, there is room for serious debate about the validity of these arguments. Parsons, for example, has questioned both the effectiveness of Boolos's argumentation and the correctness of the conclu-sions he drew, as has Resnik.[7]

I have worries of a somewhat different nature. Initially, my worries con-cerned the reasonableness of the assumptions needed to carry out the Burgess–Hazen–Lewis constructions. In what follows, I shall focus on the Burgess method, although it can readily be seen that my worries carry over to the other two methods. Burgess's method of constructing relations requires the hypothesis that there are infinitely many "atoms" and no "atomless gunk".[8]

---

[5] See Boolos, 1984: 444 for details. [6] See, for example, Hellman, 1996: 112.

[7] See Parsons, 1996: 296–300 for Parsons's argument that a structuralist (of the sort advocated by Hellman) who uses second-order logic in the way recommended by Boolos "will not be able to avoid ontological commitments more uncomfortable on balance than that to mathematical objects, either to Fregean concepts or to multiplicities that are not unities". See also Resnik, 1988b for additional considerations that are brought to bear against the Boolos thesis. The reader may also wish to consider the arguments put forward by Lewis in partial support of Boolos in Lewis, 1991: 65–71. More recently, Agustin Rayo and Stephen Yablo have produced a clever and cautiously argued article supporting various aspects of the Boolos position. The article (Rayo and Yablo, 2001) certainly adds a stimulating new element to the controversy.

[8] Lewis, 1991: 124. Hellman does not seem to take account of the second part of the assumption that requires there to be no gunk: after outlining the improvements in his account that he has achieved by means of the Boolos thesis and the Burgess–Hazen–Lewis results, he claims

Now, an "atom" is defined to be an "individual" with no proper parts (parts that are not identical to the individual). An "atomless gunk" ('gunk' for short) is defined to be an individual possessing no parts that are atoms.

So how plausible is Burgess's hypothesis? To be in a position to answer this question, we need to become much clearer about "individuals" and the relation of part to whole. We shouldn't assume that the notion of part being used is just the ordinary notion that we are all familiar with. After all, if I brought a watch to a jeweler and asked him to tell me how many parts it had, the answer he would give would not be at all acceptable to Lewis and Burgess.

To see what Lewis has in mind when he speaks of parts, let us ask: if there were such things as classes, what things would be the parts of the class of all cats? One might think that classes do not have any parts. Or one might argue that it makes no clear sense to speak of the parts of a class. But these are not the responses to the question that Lewis would have given. Lewis believes that the question has a clear and definite answer: the parts of the class are simply the non-empty subclasses of the class, that is, all the various non-empty classes of cats. And Lewis does not believe he is *stipulating* that this be the case—he thinks he is merely stating what is the case.[9]

Clearly, one needs to be given additional information before Burgess's hypothesis about "atoms" can be understood. Let us begin with a few definitions and then go on to mereology.

Definition: Something is a *fusion* (or *mereological sum*) of some things iff:

(1) it has all of these things as parts,

and

(2) it has no part that itself has no part in common with any of those things.

Thus, we are told by Lewis that the fusion of all cats is that "large, scattered chunk of cat-stuff which is composed of all the cats there are, and nothing else" (Lewis, 1991: 1).

Definition: Classes are things that have members.

It follows, of course, that there is no such things as a memberless class. By definition, a class has members.

that "the strength of full, classical third-order number theory ... is attained within a nominalist modal-structural system without postulating more than a countable infinity of atoms" (Hellman, 1996: 112).

[9] Thus, he writes: "Mereology does apply to classes. For classes do have parts: their subclasses. Maybe they have other parts as well; that remains to be seen" (Lewis, 1991: 3).

Definition: Individuals are things that are members but do not themselves have members. (Lewis, 1991: 4)

Here are Lewis's basic axioms of mereology:

Transitivity: If x is part of some part of y, then x is part of y.

Unrestricted Composition: Whenever there exist some things, then there exists a fusion of those things.

Uniqueness of Composition: It never happens that the same things have two different fusions. (Lewis, 1991: 74)

From the above definitions and axioms, it should be clear to the reader that, according to the above framework, there are some mighty strange "individuals": there is, according to Unrestricted Composition, an individual consisting of my right thumb, the Eiffel Tower, the planet Mars, and all the citizens of China. (I am assuming, for the sake of argument here, that my thumb, the Eiffel Tower, Mars, and citizens are "individuals".) Evidently, this axiom implies that there exists an individual that is scattered all across the entire universe. The above axioms, which are supposed to be a priori truths, indicate just how metaphysically liberal the devotees of mereology are: they are quite willing to believe in the existence of all sorts of things that ordinary folk do not. It is thus noteworthy that Hellman's nominalistic ontology, insofar as it includes the ontology of mereology, has the quality of ontological lushness generally associated with Platonism—not nominalism.

Let us now consider Burgess's hypothesis of the existence of infinitely many "atoms". It seems to me that, before asking if there are infinitely many such individuals, we should first determine if there are any at all? Do atoms exist? If one believes, as Field does, that there are space-time points, then perhaps a space-time point is quite a good candidate to put forward as an atom. Since such an entity is supposed to be extensionless, one can certainly argue that such a thing would have no proper parts. And since Field argues that space-time points are nominalistically acceptable, Hellman can appeal to Field's works as providing him with plausible support for his contention that there actually are atoms.[10] Of course, if we accept the hypothesis that space-time points are atoms, then we can hardly deny that there are infinitely many atoms. However, it should be mentioned that Field's thesis that space-time points are both genuine entities and nominalistically acceptable can be, and

---

[10] Hellman does express some reservations about Field's thesis that space-time points are nominalistically acceptable, in Hellman, 1996: 113.

has been, disputed. One might, for example, theorize that space-time points are classes, that is, ordered 4-tuples of real numbers reduced via the well-known constructions to sets. Specific writers who have disputed the above thesis are Resnik, who devotes a considerable portion of one paper to a refutation of it (Resnik, 1985), and David Malament, who has written:

Suppose it is agreed that space-time points are "entities that exist in their own right." Still, philosophers with nominalist scruples might well be uncomfortable with them. They certainly are not concrete physical objects in any straight-forward sense. They do not have a mass-energy content (unlike, for example, the Klein-Gordon field itself). They do not suffer change. It is not even clear in what sense they exist *in* space and time.[11]

We now need to consider the second part of Burgess's hypothesis: the hypothesis that there is no gunk. How do we know that there is no individual in reality possessing no part that is an atom? Should we ask the physicist to determine for us if there exists any gunk? I doubt that many physicists would take seriously any such request. Well, does Burgess, Hazen, or Lewis provide any metaphysical or logical arguments to support the hypothesis that there is no gunk? None that I can find. Has Hellman provided us with grounds for accepting such an assumption? Not at all. So far as I can see, then, we have been given no good reason for accepting the assumption.

Now Hellman has a response to this worry. Consider again the modal-structuralist interpretation of arithmetic. Recall that, if ϕ is a sentence of arithmetic, then the modal-structuralist interpretation of the sentence has the following form:

$$\Box Q(A \to \phi),$$

where Q is a string of universal quantifiers, A "expresses" the conjunction of the axioms of second-order arithmetic, and the necessity operator is given a possible-worlds semantical analysis. Now the Burgess constructions are to be regarded as carried out inside the scope of the modal operator, so that these constructions can be regarded as modalized in the following way (utilizing for perspicuity the possible-worlds manner of indicating the semantics of the situation):

In every world w in which there are infinitely many atoms and no gunk, $Q(A \to \phi)$ holds in w.

In other words, the Burgess hypothesis functions as a sort of condition on the worlds in which $Q(A \to \phi)$ holds. Thus, Hellman argues that his use of the

---

[11] Malament, 1982: 532; see also 532 n. 11.

Burgess construction does not require the hypothesis in question to be true of the actual world—it only requires the Burgess construction to be carried out in all of the hypothetical worlds in which the above conditions on atoms and gunk hold. For the nominalist to be able to apply the modal-structuralist version of mathematics to actual physical systems, she need only assume the compossibility of whatever gunk there actually exists with "enough atoms" to code the gunky individuals so that, by means of the Burgess construction, one could get the effect of ordered n-tuples and hence functions and relations.

Hellman's response shows how we do not have to believe that the Burgess hypothesis about atoms and gunk is true in the actual world, but it still requires us to believe that the Burgess hypothesis is a genuine hypothesis about a possible world, that is, it requires us to believe that the hypothesis expresses a bone fide condition on worlds so that, in any particular world, the hypothesis is either true or false. I have some doubts about that. My doubts stem, in part, from the suspicion that such terms as 'thing', 'class', 'part', 'individual', 'fusion', 'atom', and 'gunk' are simply not sufficiently definite in meaning to give the sentences expressing the supposed "hypotheses" a determinate truth value in all "possible worlds".

I look out of my window and see a spectacular rainbow. I ask myself, "Is that a thing?" How can I tell? We have been given no criterion for determining. It is hard to see how science can answer such a question in the absence of anything more definite than what we are given by Lewis and his collaborators. But even if we could determine whether a rainbow is a thing, we would still need to determine whether it is an individual or a class. It might be thought that this should be easy to answer, since all we need to do is to determine whether the rainbow has members. But how do I determine that? Remember the Lewis–van Inwagen questions: is membership an *intrinsic* or *extrinsic relation*? If the latter, is it *internal* or *external*? Again, is this something that the empirical sciences can determine for us? Questions of this type abound when attempting to classify the rainbow as an atom or gunk.

The problems multiply when we turn to "possible worlds", since the above questions arise with a vengeance when an alien "universe" is to be sorted into things, individuals, classes, atoms, and gunk. What happens in this "universe" may be so very different from what occurs in our universe—not only may the phenomena in this "universe" differ radically from what is in ours, but the laws of nature, space, and even time may be completely different. This raises a real question in my mind as to whether Hellman's suggested nominalization of the modal-structuralist interpretations of mathematics will work in the way he has outlined.

Appendix B.   Balaguer's Fictionalism

In this appendix, I shall investigate Mark Balaguer's claims about fictionalism, when the term 'fictionalism' is taken to stand for his own version of fictionalism. Return to his extravagant claim (discussed in Chapter 11) that all non-fictionalist nominalistic accounts of mathematics have "no advantage over fictionalism" and, for that reason, are "inferior" to it. As I noted earlier, Balaguer does not really justify such a remarkable claim—he only baldly proclaims it.

Well, is it true? Balaguer's fictionalist account of mathematics is strikingly thin, unspecific, and, about most aspects of mathematics, vague. Thus, *many* non-fictionalist accounts of mathematics have significant advantages over Balaguer's fictionalism: many such accounts will be more robust, more finely detailed, more specific, and definitely clearer than Balaguer's. Another disadvantage of Balaguer's fictionalism is that it is saddled with existence assertions about, and reference to, types of entities that are even more ontologically dubious than the mathematical entities that fictionalism is supposed to eschew. A yet more serious disadvantage of Balaguer's account, when compared with many of its rivals, is the highly questionable central principle upon which it is based, and this is compounded by the seriously flawed argumentation Balaguer adduces in its support.

Before explaining why I make the above charges, I shall first outline the strategy Balaguer gives his fictionalist for avoiding any ontological commitment to the existence of mathematical objects. Roughly, his fictionalist argues that all of what empirical science "says" about the world can be divided into two kinds: (a) there are the nominalistic things it says about the physical world; and (b) there are the Platonistic things it says about the mathematical world. Then the fictionalist can claim that science is only concerned with getting right the nominalistic things it says about the physical world and it can disregard what it says about the mathematical world.

One can see a similarity that this strategy has to the constructive empiricist strategy advocated by Bas van Fraassen. Van Fraassen argues in *The Scientific Image* (van Fraassen, 1980) that the aim of science is only to produce *empirically adequate* theories, that is, theories that are correct about what is observable.[1] Science does not aim to produce theories that are correct about what is not observable. Thus, he argues, there is no reason for the constructive empiricist to commit himself to the existence of such unobservables as photons and quarks. Similarly, by Balaguer's strategy, there is no reason for the fictionalist to commit himself to such Platonic entities as numbers and sets.

Balaguer's fictionalism can be characterized as "nominalism without toil": with the above strategy, it can avoid the irksome toil of producing a detailed nominalistic version of mathematics. This is quite an advantage. I can imagine the ghost of Russell saying: yes, it has the advantages of theft over honest toil.

Let us now try to get a more accurate statement of Balaguer's reasoning. We first need to state just what principles Balaguer attributes to science on which to base the above strategy. These are the principal ones:

(NC) Empirical science has a purely nominalistic content that expresses its "complete picture" of the physical world.

(COH) It is coherent and sensible to maintain that the nominalistic content of empirical science is true and that the Platonistic content of empirical science is fictional.[2]

Notice that there is much in the very statements of these principles that is murky. What is "empirical science"? Is it a set of propositions? Is it a set of beliefs? Is it an active practice (with Kuhnian paradigms) in which living humans participate? Is it a set of theories? Is it the set of all theories that any living scientist now accepts? Is it the set of all theories that "most" scientists now accept? Does it contain particular assertions by specific scientists about specific pieces of equipment, machinery, or instrumentation? Does it contain beliefs or propositions about particular experiments conducted by means of specific pieces of equipment, machinery, and/or instrumentation? Does it contain assertions or belief about scientific principles and practices? Does it contain modal assertions or laws? All of the above?

What about "content"? What is that? We can gather from (NC) that content is something that can express a "picture". We can also infer from (COH) that a content can be true and that a content can be fictional. Contents can also "say" things—say things about facts. For example, we are

---

[1] This is the work's central thesis. See, for example, van Fraassen, 1980: 3.

[2] Balaguer, 1998: 131. The title of the first of these principles '(NC)' undoubtedly stands for 'Nominalistic Content'; the second probably stands for 'Coherence'.

told that the nominalistic content of empirical science "says that" facts of a certain sort (that is, "purely nominalistic facts") obtain (Balaguer, 1998: 134). Indeed, Balaguer implies that one can specify the content of an assertion of science by specifying the set of facts that is said by the assertion or science to obtain (Balaguer, 1998: 134).

But what are "facts"? I do not think we can assume that what Balaguer means by 'fact' is just what any English speaker means by that term. For one thing, facts, for Balaguer, either obtain or do not obtain. Thus, if John is in his office, we can infer that *the fact that John is in his office* obtains and we can also conclude that *the fact that John is not in his office* does not obtain, but we cannot conclude that *the fact that John is not in his office* does not exist. (The fact exists—it just does not obtain.)

Now Balaguer believes that there are a huge variety of facts: there are, he tells us, logical facts, physical facts, mathematical facts, nominalistic facts, Platonistic facts, mixed facts, pure facts, bottom-level facts, facts that super-vene on other facts, and so on. But he never informs his readers just what principle(s) of fact existence he is using. Nor does he ever articulate any cri-terion of fact identity that he accepts. In other words, he makes reference to, and quantifies over, a type of abstract entity without ever setting out, even in a rough way, the basic principles of the theory of these entities which he is presupposing. It seems surprising that, in articulating his strategy for the fictionalist, he is willing to rely upon the existence of a type of abstract entity whose fundamental features are so obscure, when the mathematical entities his fictionalist eschews are, by contrast, so precisely known. I, for one, believe that there are much stronger reasons for believing in mathematical objects than there are for believing in the "facts" that Balaguer places his confidence in.

Why should we accept the fictionalist's principles, (NC) and (COH)? Since (NC) is used to justify (COH), (NC) is the more fundamental principle. So let us concentrate on it. Principle (NC) tells us that empirical science has a nominalistic content that is "complete" in the sense that the picture of the physical world that this content expresses is not lacking any facts that are expressed about the physical world by the science. What about the enormous number of mathematically expressed facts about the physical world that empirical science contains? Balaguer is convinced that the full content of empirical science can be divided into two sorts: one type ("the nominalistic content") that only talks about purely nominalistic facts and another type ("the Platonistic content") that only talks about purely Platonistic facts, where the two sorts of facts are "independent" of each other—"independent" in the sense that fact *A* is independent of fact *B* iff *A*'s holding does not *depend upon B*'s holding and also *B*'s holding does not depend upon *A*'s holding. Thus, if *A* is independent of *B*, then *B*'s not holding tells us nothing about whether or not *A* holds. The full content of empirical science, it is argued, is captured by the two contents: the purely nominalistic content and the purely Platonistic content. Furthermore, what empirical science says about the physical world is fully captured by its nominalistic content.

Balaguer attempts to justify (NC) by first showing that a much simpler empirical assertion such as

(**) The number of ants in the United States is more than 157 times greater than the number of fleas in China

has a purely nominalistic content that captures its "complete picture" of the physical world.[3] He reasons as follows:

(1) Numbers (and other mathematical objects) aren't causally relevant to ants or fleas.

(2) It follows that, if (**) is true, then it is true in virtue of nominalistic facts about both the ants in the United States and the fleas in China and also Platonistic facts about mathematical objects such as cardinal numbers, and these two different sorts of facts are independent of one another.[4]

(3) The nominalistic facts mentioned in (2) constitute the purely nominalistic content of (**).

(4) Since the purely Platonistic content of (**) concerns only the mathematical world, the purely nominalistic content of (**) must constitute the "complete picture" of the physical world expressed by (**).

(5) Thus, (**) has a purely nominalistic content that constitutes the "complete picture" of the physical world expressed by (**).

He then generalizes from what he believes he has shown for specific sentences (such as (**)) to what he is convinced he can show for the whole of empirical science, proclaiming: "It should be clear that this argument can be applied to all of empirical science" (Balaguer, 1998: 134).

I doubt very much that one can make such a huge jump to the whole of science as easily as Balaguer seems to think. Imagine trying to express, in a nominalistically acceptable way, every scientific assertion about the physical world that the quantum physicist makes. Can the contents of all such

---

[3] The specific example Balaguer uses is slightly different.   [4] See Balaguer, 1998: 129.

assertions be divided in the way required? What about modal or subjunctive conditional assertions involving both physical objects and mathematical objects that the physicist may make? Can they all be rendered in the way described by (NC)? That such tasks could be carried out, even in principle, boggles the mind. But there is no need for me to argue that point, since I believe Balaguer's argument, even for the specific sentence (**), is seriously flawed.

Before investigating the logic of this argument, let us ask: how does one determine what is, and what is not, a purely nominalistic fact? Consider a specific example: there being more ants in the United States than there are fleas in China. Is that fact a purely nominalistic fact? One might suppose that it is, on the grounds that no mathematical object is *explicitly* mentioned in the linguistic expression of that fact. (It would not take much ingenuity to develop difficulties for Balaguer's position if one used what is explicitly mentioned to determine what is a purely nominalistic fact.) On the other hand, according to standard Platonic analyses of such linguistic expressions, mathematical entities (such as one-one correspondences) are being implicitly mentioned or presupposed, so it can be argued that the fact is not purely nominalistic. In that case, whether or not a fact is purely nominalistic would depend upon what analysis of the linguistic expression of the fact is correct. It can be seen that a worrisome unclarity infects a fundamental concept of the above reasoning.

If one examines the steps in this argument, one in particular stands out as highly questionable. Why can one infer (2) from (1)? It certainly does not follow by any generally accepted rule of logical inference. Yet, Balaguer seems to think that (2) obviously follows from (1).[5] This is surprising, given that he nowhere gives us a general method for specifying or expressing the nominalistic content of (**). Could he come up with a plausible candidate for the content of (**) that is somehow divisible into two component parts as (NC) requires, so that its purely nominalistic content can be reasonably said to capture the "complete picture" of the physical world expressed by (**)? I very much doubt it.

So why is he so sure that (2) follows from (1)? I believe that he has a sort of "picture" of the situation being described by (**), according to which there are two "worlds" being described: the physical world and the math-ematical world. Premise (1) tells us, he thinks, that the two worlds are

---

[5] Indeed, at one point in his argumentation for (5), he implies that we are "forced" to draw the inference to step (2). See Balaguer, 1998: 133.

utterly separate and *unrelated*, so that whatever is true of one of the worlds would have no implications at all for what is true of the other. In other words, Balaguer seems to think that "not causally relevant" is the same as "not relevant at all".

Why do I believe that what lies behind Balaguer's conviction that (2) follows from (1) is a belief that the things in the two worlds are utterly unrelated to one another? Consider the following quote:

[I]f we assume that (NC) is true, then [the claim that there are no causally efficacious mathematical objects] suggests that (COH) is also true, because it suggests that the truth value of the platonistic content of empirical science is simply irrelevant to the truth value of its nominalistic content. (Balaguer, 1998: 132)

But why should Balaguer think that the "truth value of the platonistic content of empirical science" is irrelevant to the truth value of its nominalistic content? Evidently, because he thinks that causally irrelevant is tantamount to irrelevant.

Consider another quote:

Empirical science *knows*, so to speak, that mathematical objects are causally inert... Thus, it seems that empirical science *predicts* that the behavior of the physical world is not dependent *in any way* upon the existence of mathematical objects. (Balaguer, 1998: 133, my emphasis)

Why does Balaguer think that empirical science predicts that the behavior of the physical world does not depend "in any way" upon the existence of mathematical things? Evidently because he thinks that "does not depend causally" is tantamount to "does not depend in any way".

But those are highly questionable assumptions that should bring to mind the "metaphysical problem of applicability" discussed earlier in Chapter 9. Recall that Steiner's response to that problem was to point to Frege's analysis of mathematics, according to which "mathematical entities relate, not directly to the physical world, but to concepts; and (some) concepts, obviously, apply to physical objects".[6] To recapitulate his response, according to Frege's analysis of arithmetical reasoning, cardinal numbers are extensions under which concepts fall, and some concepts are, essentially, properties that some physical objects have and some physical objects do not have. Thus, even though numbers are not *causally related* to any physical objects, they certainly are *related* to physical objects.

---

[6] The reader may wish to review my discussion of this problem in Chapter 9, Section 1.

Actually, there are many philosophical accounts of mathematical entities according to which such things as functions, sets and numbers are related to physical objects. So it is by no means unreasonable to demand grounds for supposing that "no causal relationships" is tantamount to "no relationships at all".

Of course, Balaguer might attempt to reply to this point by arguing that all accounts of mathematics that postulate relationships between mathematical objects and physical objects, such as Frege's analysis of cardinal number, are fundamentally mistaken; but even if he could produce a credible refutation of such accounts (which I doubt), that would not justify his claim that (2) "follows from" (1), for he would obviously need much more to mount such a refutation than merely the premise that mathematical objects are not causally related to physical objects.

Reconsider (**). Given the Fregean analysis of cardinality or practically any other widely held Platonic analysis of cardinality, (**) clearly expresses a great many complex relationships between biological organisms (ants and fleas) and mathematical objects (cardinal numbers). The content of (**) would have to say that certain facts about these relationships between biological organism and mathematical objects do obtain. In other words, the facts relating biological organisms to mathematical objects would have to be in the content of (**). Now (5) in effect tells us that the content of (**)—the set of facts that are said to obtain by (**)—is the union of two disjoint sets of facts, the purely nominalistic facts that make up the purely nominalistic content of (**) and the purely Platonistic facts that make up the purely Platonistic content of (**). There is no room in the content of (**), according to (5), for another non-empty set of facts that make reference to both biological organisms and mathematical objects, since (5) tells us that the purely nominalistic content constitutes the "complete picture" of the physical world expressed by (**). Thus, if the content of (**) is completely exhausted by the purely nominalistic content and the purely Platonistic content, as is implied by (5), how can the content of (**) contain any facts giving the clearly expressed relationships between the ants, the fleas, and cardinal numbers? Obviously it cannot. The conclusion to draw is that (2) is just false and not a consequence of (**), all of which throws considerable doubt on the inference from (1) to (2), as well as on (NC)—the fundamental principle supporting Balaguer's supposedly superior form of fictionalism. On the basis of his questionable proof of (NC), Balaguer claims to "provide a complete refutation of the Quine–Putnam argument against fictionalism" (Balaguer, 1998: 16). Needless to say, I regard this "complete refutation" as a complete failure.

Bibliography

Adams, Ernest W. 2001. *Surfaces and Superposition*. Stanford, Calif.: CSLI Publications.

Adams, R. M. 1979. Primitive Thisness and Primitive Identity. *Journal of Philosophy* 76: 5–26.

Albert, David. 1992. *Quantum Mechanics and Experience*. Cambridge, Mass.: Harvard University Press.

Artmann, Benno. 1999. *Euclid: The Creation of Mathematics*. New York: Springer-Verlag.

Azzouni, Jody. 1994. *Metaphysical Myths, Mathematical Practice*. Cambridge: Cambridge University Press.

Bach, Craig. 1998. Philosophy and Mathematics: Zermelo's Axiomatization of Set Theory. *Taiwanese Journal for Philosophy and History of Science* 10: 5–31.

Balaguer, Mark. 1998. *Platonism and Anti-Platonism in Mathematics*. Oxford: Oxford University Press.

Barbut, Marc. 1970. On the Meaning of the Word 'Structure' in Mathematics. In M. Lane, ed., *Introduction to Structuralism*, 367–88. New York: Basic Books, Inc.

Barrett, Jeffrey A. 2001. *The Quantum Mechanics of Minds and Worlds*. Oxford: Oxford University Press.

Barwise, Jon and Etchemendy, John. 1991. Visual Information and Valid Reasoning. In W. Zimmerman and S. Cunningham, eds., *Visualization in Teaching and Learning*, 9–24. Washington, D.C.: Mathematics Association of America.

Benacerraf, Paul. 1965. What Numbers Could Not Be. *Philosophical Review* 74: 47–73.

—— 1973. Mathematical Truth. *Journal of Philosophy* 70: 661–79.

Berkeley, Bishop George. 1956. The Analyst. In J. R. Newman, ed., *The World of Mathematics*, 1. 288–93. New York: Simon & Schuster.

Bernays, Paul. 1950. Mathematische Existenz und Widerspruchsfreiheit. In *Études de philosophie des sciences en hommage à Ferdinand Gonseth*, 11–25. Neuchâtel: Éditions du Griffon.

—— 1967. Hilbert, David. In P. Edwards, ed., *The Encyclopedia of Philosophy*, 3: 496–504. New York: Macmillan.

Birkoff, Garrett and MacLane, Saunders. 1953. *A Survey of Modern Algebra*. New York, Macmillan.

Blank, Jiri, Exner, Pavel, and Havlicek, Miloslav. 1994. *Hilbert Space Operators in Quantum Physics*. New York: American Institute of Physics Press.

Boolos, George. 1984. To Be Is to Be a Value of a Variable (or to Be Some Values of Some Variable). *Journal of Philosophy* 81: 430–49.

—— 1987. The Consistency of Frege's Foundations. In J. J. Thompson, ed., *On Being and Saying: Essays for Richard Cartwright*, 3–20. Cambridge, Mass.: MIT Press.

Borowski, E. J. and Borwein, J. M. 1991. *The HarperCollins Dictionary of Mathematics*. New York: HarperCollins.

Boyer, Carl B. 1956. *History of Analytic Geometry: Its Development from the Pyramids to the Heroic Age*. Princeton Junction, N.J.: The Scholar's Bookshelf.

—— 1985. *A History of Mathematics*. Princeton, N.J.: Princeton University Press.

Brouwer, L. E. J. 1967. On the Significance of the Principle of Excluded Middle in Mathematics, Especially in Function Theory. In J. van Heijenoort, ed., *From Frege to Gödel: A Source Book in Mathematical Logic, 1879–1931*, 334–45. Cambridge, Mass.: Harvard University Press.

Brown, James R. 1999. *Philosophy of Mathematics: An Introduction to the World of Proofs and Pictures*. London: Routledge.

Burgess, John. 1983. Why I Am Not a Nominalist. *Notre Dame Journal of Formal Logic* 24: 93–105.

—— 1990. Epistemology and Nominalism. In A. D. Irvine, ed., *Physicalism in Mathematics*, 1–15. Dordrecht: Kluwer Academic Publishers.

—— 1998. Occam's Razor and Scientific Method. In M. Schirn, ed., *The Philosophy of Mathematics Today*, 195–214. Oxford: Oxford University Press.

—— and Rosen, Gideon. 1997. *A Subject with No Object: Strategies for Nominalistic Interpretations of Mathematics*. Oxford: Oxford University Press.

Calinger, Ronald, ed. 1982. *Classics of Mathematics*. Oak Park, Ill.: Moore Publishing Company.

Carnap, Rudolf. 1964. The Logicist Foundations of Mathematics. In P. Benacerraf and H. Putnam, eds., *Philosophy of Mathematics: Selected Readings*, 31–41. Englewood Cliffs, N.J.: Prentice-Hall.

Cartwright, Nancy. 1983. *How the Laws of Physics Lie*. Oxford: Oxford University Press.

Chihara, Charles S. 1973. *Ontology and the Vicious-Circle Principle*. Ithaca, N.Y.: Cornell University Press.

—— 1980. Ramsey's Theory of Types: Suggestions for a Return to Fregean Sources. In D. H. Mellor, ed., *Prospects for Pragmatism: Essays in Memory of F. P. Ramsey* 21–47. Cambridge: Cambridge University Press.

—— 1981. Quine and the Confirmational Paradoxes. In P. French, T. Uehling, and H. Wettstein, eds., *Midwest Studies in Philosophy, 6: The Foundations of Analytic Philosophy*, 425–54. Minneapolis: University of Minnesota Press.

—— 1982. A Gödelian Thesis Regarding Mathematical Objects: Do They Exist? And Can We Perceive Them? *Philosophical Review* 91: 211–27.

—— 1987. Some Problems for Bayesian Confirmation Theory. *British Journal for the Philosophy of Science* 38: 551–60.

—— 1990. *Constructibility and Mathematical Existence.* Oxford: Oxford University Press.

—— 1994. The Howson–Urbach Proofs of Bayesian Principles. In E. Eells and B. Skyrms, eds., *Probability and Conditionals*, 161–78. Cambridge: Cambridge University Press.

—— 1998. *The Worlds of Possibility: Modal Realism and the Semantics of Modal Logic.* Oxford: Oxford University Press.

—— 1999. Frege's and Bolzano's Rationalist Conceptions of Arithmetic. *Revue d'histoire des sciences* 52: 343–61.

Cocchiarella, N. B. 1975. On the Primary and Secondary Semantics of Logical Necessity. *Journal of Philosophical Logic* 4: 13–27.

Coffa, Alberto. 1986. From Geometry to Tolerance: Sources of Conventionalism in Nineteenth-Century Geometry. *From Quarks to Quasars: Philosophical Problems of Modern Physics*, 3–70. Pittsburgh, Pa.: University of Pittsburgh Press.

—— 1991. *The Semantic Tradition from Kant to Carnap: To the Vienna Station.* Cambridge: Cambridge University Press.

Cohen, Paul. 1966. *Set Theory and the Continuum Hypothesis.* New York: W. A. Benjamin.

—— 1971. Comments on the Foundations of Set Theory. In D. Scott, ed., *Axiomatic Set Theory*, 9–15. Providence, R.I.: American Mathematical Society.

Colyvan, Mark. 1999. Contrastive Empiricism and Indispensability. *Erkenntnis* 51: 323–32.

—— 2001. *The Indispensability of Mathematics.* New York: Oxford University Press.

Corry, Leo. 1999. Hilbert and Physics, 1900–1915. In J. Gray, ed., *The Symbolic Universe: Geometry and Physics 1890–1930*, 145–88. Oxford: Oxford University Press.

Dauben, Joseph. 1990. *George Cantor: His Mathematics and Philosophy of the Infinite.* Princeton, N.J.: Princeton University Press.

Demopoulos, William. 1994. Frege, Hilbert, and the Conceptual Structure of Model Theory. *History and Philosophy of Logic* 15: 211–25.

—— ed. 1995. *Frege's Philosophy of Mathematics.* Cambridge, Mass.: Harvard University Press.

Detlefsen, Michael. 1986. *Hilbert's Program: An Essay on Mathematical Instrumentalism.* Dordrecht: D. Reidel.

Dilke, O. A. 1987. *Mathematics and Measurement.* London: British Museum.

Dorrie, Heinrich. 1965. *100 Great Problems of Elementary Mathematics: Their History and Solution.* New York: Dover.

Dummett, Michael. 1964. Truth. In G. Pitcher, ed., *Truth*, 93–111. Englewood Cliffs, N.J.: Prentice-Hall, Inc.

—— 1991. *Frege: Philosophy of Mathematics.* Cambridge, Mass.: Harvard University Press.

—— 1993. What Is Mathematics About? *The Seas of Language*, 429–45. Oxford: Oxford University Press.

—— 1998. Neo-Fregeans: In Bad Company? In M. Schirn, ed., *Philosophy of Mathematics Today*, 369–87. Oxford: Oxford University Press.

Earman, John. 1992. *Bayes or Bust*. Cambridge, Mass.: MIT Press.

Eells, Ellery. 1982. *Rational Decision and Causality*. Cambridge: Cambridge University Press.

Enderton, Herbert. 1972. *A Mathematical Introduction to Logic*. New York: Academic Press.

Feferman, Solomon. 1998a. Deciding the Undecidable: Wrestling with Hilbert's Problems. Feferman 1998c: 3–27.

—— 1998b. Gödel's Life and Work. Feferman 1998c: 127–49.

—— 1998c. *In the Light of Logic*. New York: Oxford University Press.

—— 1998d. Infinity in Mathematics: Is Cantor Necessary? Feferman 1998c: 28–73.

—— 1998e. Introductory Note to Gödel's 1933 Lecture. Feferman 1998c: 165–73.

—— 1998f. Weyl Vindicated: *Das Kontinuum* Seventy Years Later. Feferman 1998c: 249–83.

—— 1998g. Why a Little Bit Goes a Long Way: Logical Foundations of Scientifically Applicable Mathematics. Feferman 1998c: 284–98.

Field, Hartry. 1980. *Science Without Numbers*. Princeton, N.J.: Princeton University Press.

—— 1984. Is Mathematical Knowledge Just Logical Knowledge? *Philosophical Review* 93: 509–52. (Reprinted with revisions as Field, 1989a.)

—— 1985. Comments and Criticisms on Conservativeness and Incompleteness. *Journal of Philosophy* 82: 239–60.

—— 1989a. Is Mathematical Knowledge Just Logical Knowledge? In Field, 1989d: 79–124.

—— 1989b. On Conservativeness and Incompleteness. In Field, 1989d: 125–46.

—— 1989c. Realism and Anti-Realism About Mathematics. In Field, 1989d: 53–78.

—— 1989d. *Realism, Mathematics, and Modality*. Oxford: Basil Blackwell.

—— 1992. A Nominalistic Proof of the Conservativeness of Set Theory. *Journal of Philosophical Logic* 21: 111–23.

Fine, Kit. 1978. Model Theory for Modal Logic. Part I: The De Re/De Dicto Distinction. *Journal of Philosophical Logic* 7: 125–56.

Forbes, Graeme. 1985. *The Metaphysics of Modality*. Oxford: Oxford University Press.

Fowler, A. C. 1997. *Mathematical Models in the Applied Sciences*. Cambridge: Cambridge University Press.

Frege, Gottlob. 1959. *The Foundations of Arithmetic*. Oxford: Basil Blackwell.

—— 1971a. On the Foundations of Geometry. *On the Foundations of Geometry and Formal Theories of Arithmetic*, ed. E.-H. Kluge, 22–37. New Haven, Conn.: Yale University Press.

—— 1971b. On the Foundations of Geometry: A Reply to Korselt. *On the Foundations of Geometry and Formal Theories of Arithmetic*, ed. E.-H. Kluge, 49–112. New Haven, Conn.: Yale University Press.

—— 1979a. Sources of Knowledge of Mathematics and the Mathematical Natural Sciences. *Gottlob Frege: Posthumous Writings*, ed. H. Hermes, F. Kambartel and F. Kaulbach, 267–74. Chicago: University of Chicago Press.

—— 1979b. On Euclidean Geometry. *Gottlob Frege: Postumous Writings*, ed. H. Hermes, F. Kambartel, and F. Kaulbach, 167–9. Chicago: University of Chicago Press.

—— 1980. *Philosophical and Mathematical Correspondence*. Chicago: University of Chicago Press.

Freudenthal, Hans. 1962. The Main Trends in the Foundations of Geometry in the Nineteenth Century. In E. Nagel, P. Suppes, and A. Tarski, eds., *Logic, Methodology and Philosophy of Science: Proceedings of the 1960 International Congress*, 613–21. Stanford, Calif.: Stanford University Press.

Geroch, Robert. 1985. *Mathematical Physics*. Chicago: University of Chicago Press.

Gillies, Donald. 1982. *Frege, Dedekind, and Peano on the Foundations of Arithmetic*. Assen: Van Gorcum.

Girvetz, Harry, Geiger, George, Hantz, Harol, and Morris, Bertram. 1966. *Science, Folklore, and Philosophy*. New York: Harper & Row.

Glymour, Clark. 1980. *Theory and Evidence*. Princeton, N.J.: Princeton University Press.

Gödel, Kurt. 1964a. Russell's Mathematical Logic. In P. Benacerraf and H. Putnam, eds., *Philosophy of Mathematics: Selected Readings*, 211–32. Englewood Cliffs, N.J.: Prentice-Hall.

—— 1964b. What is Cantor's Continuum Problem? In. P. Benacerraf and H. Putnam, eds., *Philosophy of Mathematics: Selected Readings*, 258–73. Englewood Cliffs, N.J.: Prentice-Hall.

Goldman, Randolph Rubens. 2000. Gödel's Ontological Argument. Unpublished Ph.D. dissertation. University of California, Berkeley.

Grabiner, Judith V. 1981. *The Origins of Cauchy's Rigorous Calculus*. Cambridge, Mass.: MIT Press.

Grattan-Guinness, I. 2000. *The Search for Mathematical Roots, 1870–1940*. Princeton, N.J.: Princeton University Press.

Gray, Jeremy. 1999. Geometry: Formalisms and Intuitions. In J. Gray, ed., *The Symbolic Universe: Geometry and Physics 1890–1930*, 58–83. Oxford: Oxford University Press.

Hacking, Ian. 1985. Do We See Through a Microscope? In P. Churchland and C. Hooker, eds., *Images of Science*, 132–52. Chicago: University of Chicago Press.

Halmos, Paul R. 1960. *Naive Set Theory*. Princeton, N.J. Van Nostrand.

Hammer, Eric M. 1995. *Logic and Visual Information*. Stanford, Calif.: CSLI Publications.

Hand, Michael. 1993. Mathematical Structuralism and the Third Man. *Canadian Journal of Philosophy* 23: 179–92.

Harman, Gilbert. 1965. The Inference to the Best Explanation. *Philosophical Review* 74: 88–95.

Hawthorne, James. 1993. Bayesian Induction Is Eliminative Induction. *Philosophical Topics* 21: 99–138.

Heath, Thomas. 1956. *The Thirteen Books of Euclid's Elements*. New York: Dover.

Heck, R., Jr. 1995. Frege's Principle. In J. Hintikka, ed., *From Dedekind to Gödel*, 119–42. Dordrecht: Kluwer Academic Publishers.

—— 1997. The Julius Caesar Objection. In R. Heck, Jr., ed., *Language, Thought, and Logic*, 273–308. Oxford: Oxford University Press.

Hellman, Geoffrey. 1989. *Mathematics Without Numbers*. Oxford: Oxford University Press.

—— 1996. Structuralism Without Structures. *Philosophia Mathematica* 4: 100–23.

—— 1998. Beyond Definitionism—But Not Too Far Beyond. In M. Schirn, ed., *The Philosophy of Mathematics Today*, 215–25. Oxford: Oxford University Press.

—— 1999. Some Ins and Outs of Indispensability: A Modal-Structural Perspective. In A. Cantini, E. Casari and P. Minari, eds., *Logic in Florence, 1995*, 25–40. Dordrecht: Kluwer.

Hempel, Carl. 1964. On the Nature of Mathematical Truth. In P. Benacerraf and H. Putnam, eds., *Philosophy of Mathematics: Selected Readings*, 366–81. Englewood Cliffs, N.J.: Prentice-Hall.

Hesse, Mary. 1974. *The Structure of Scientific Inference*. Berkeley and Los Angeles: University of California Press.

Heyting, Arend. 1956. *Intuitionism: An Introduction*. Amsterdam: North-Holland.

Hilbert, David. 1971. *Foundations of Geometry*. La Salle, Ill.: Open Court.

—— 1980. Letter to Frege 29.12.1899. In *Gottlob Frege: Philosophical and Mathematical Correspondence*, ed. G. Gabriel, H. Hermes, F. Kambartel, C. Thiel and A. Veraart, 38–41. Chicago: University of Chicago Press.

—— 1983. On the Infinite. In P. Benacerraf and H. Putnam, eds., *Philosophy of Mathematics: Selected Readings*, 2nd edn., 183–201. Cambridge: Cambridge University Press.

Hintikka, J. and Hintikka, M. 1989. Towards a General Theory of Individuation and Identification. *The Logic of Epistemology and the Epistemology of Logic: Selected Essays*, 73–95. Dordrecht: Kluwer Academic Publishers.

Israel, Giorgio. 1998. Des *Regulae* à la géométrie. *Revue d'histoire des sciences* 51: 183–236.

Jaffe, Arthur and Quinn, Frank. 1993. "Theoretical Mathematics": Toward a Cultural Synthesis of Mathematics and Theoretical Physics. *Bulletin of the American Mathematics Society* 29: 1–13.

Jarrett, Charles. 1978. The Logical Structure of Spinoza's Ethics, Part 1. *Synthese* 37: 15–65.

Joseph, Geoffrey. 1980. The Many Sciences and the One World. *Journal of Philosophy* 77: 773–91.

Jubien, Michael. 1991. Could This Be Magic? *Philosophical Review* 100: 249–67.

Kaplan, David. 1979. Transworld Heir Lines. In M. J. Loux, ed., *The Possible and the Actual*, 88–109. Ithaca, N.Y.: Cornell University Press.

Kitcher, Philip. 1984. *The Nature of Mathematical Knowledge*. New York: Oxford University Press.

—— 1993. *The Advancement of Science*. New York: Oxford University Press.

—— and Aspray, William. 1988. An Opinionated Introduction. *History and Philosophy of Modern Mathematics* 11: 3–57. Minneapolis: University of Minnesota Press.

Klein, Felix. 1939. *Elementary Mathematics From an Advanced Standpoint: Geometry*. New York: Dover.

Kline, Morris. 1982. *Mathematics: The Loss of Certainty*. New York: Oxford University Press.

Kneale, William and Kneale, Martha. 1962. *The Development of Logic*. Oxford: Oxford University Press.

Knopp, Konrad. 1945. *Theory of Functions*, Part I: *Elements of the General Theory of Analytic Functions*. New York: Dover.

Kogbetliantz, E. G. 1969. *Fundamentals of Mathematics from an Advanced Viewpoint*. New York: Gordon & Breach Science Publishers.

Laubenbacher, Reinhard and Pengelley, David. 1999. *Mathematical Expeditions: Chronicles by the Explorers*. New York: Springer-Verlag.

Laudan, Larry. 1997. How About Bust? Factoring Explanatory Power Back into Theory Evaluation. *Philosophy of Science* 64: 306–16.

Lear, Jonathan. 1977. Sets and Semantics. *Journal of Philosophy* 74: 86–102.

Lewis, David. 1983. New Work for a Theory of Universals. *Australasian Journal of Philosophy* 61: 343–77.

——1991. *Parts of Classes*. Oxford: Basil Blackwell.

Liston, Michael. 2000. Review of Mark Steiner, *The Applicability of Mathematics as a Philosophical Problem*. *Philosophia Mathematica* 8: 190–207.

Maddy, Penelope. 1980. Perception and Mathematical Intuition. *Philosophical Review* 89: 163–96.

——1990. *Realism in Mathematics*. Oxford: Oxford University Press.

——1995. Naturalism and Ontology. *Philosophia Mathematica* 3: 248–70.

——1997. *Naturalism in Mathematics*. Oxford: Oxford University Press.

Maki, Daniel P. and Thompson, Maynard. 1973. *Mathematical Models and Applications*. Englewood Cliffs, N.J.: Prentice-Hall.

Malament, D. 1982. Hartry Field's *Science Without Numbers*. *Journal of Philosophy* 79: 523–34.

Malcolm, John. 1991. *Plato on the Self-predication of Forms: Early and Middle Dialogues*. Oxford: Oxford University Press.

Mancosu, Paolo. 1996. *Philosophy of Mathematics and Mathematical Practice in the Seventeenth Century*. Oxford: Oxford University Press.

Mates, Benson. 1972. *Elementary Logic*. New York: Oxford University Press.

Mendelson, Elliot. 1987. *Introduction to Mathematical Logic*. Monterey, Calif.: Wadsworth.

Mostowski, Andrzej. 1967. Recent Results in Set Theory. In I. Lakatos, ed., *Problems in the Philosophy of Mathematics*, 82–96. Amsterdam: North-Holland.

Mueller, Ian. 1981. *Philosophy of Mathematics and Deductive Structure in Euclid's Elements*. Cambridge, Mass.: MIT Press.

Netz, Reviel. 1999. *The Shaping of Deduction in Greek Mathematics*. Cambridge: Cambridge University Press.

Newman, James R., ed. 1956. *The World of Mathematics*. New York: Simon & Schuster.

Nidditch, P. H. 1960. *Elementary Logic of Science and Mathematics*. London: University Tutorial Press Ltd.

Nomoto, Kazuyuki. 2000. Why, in 1902, Wasn't Frege Prepared to Accept Hume's Principle as the Primitive Law for His Logicist Program? *Annals of the Japan Association for Philosophy of Science* 9: 219–30.

Nye, Mary Jo. 1972. *Molecular Reality*. London: Macdonald.

Padoa, Alessandro. 1967. Logical Introduction to Any Deductive Theory. In J. van Heijenoort, ed., *From Frege to Gödel: A Source Book in Mathematical Logic, 1879–1931*, 118–23. Cambridge: Mass.: Harvard University Press.

Parsons, Charles. 1995a. Platonism and Mathematical Intuition in Kurt Gödel's Thought. *Bulletin of Symbolic Logic* 1: 44–74.

——— 1995b. Structuralism and the Concept of Set. In W. Sinnott-Armstrong, D. Raffman, and N. Asher, eds., *Modality, Morality, and Belief: Essays in Honor of Ruth Barcan Marcus*, 14–92. Cambridge: Cambridge University Press.

——— 1996. The Structuralist View of Mathematical Objects. In W. D. Hart, ed., *The Philosophy of Mathematics*, 272–309. Oxford: Oxford University Press.

Peressini, Anthony. 1997. Troubles with Indispensability: Applying Pure Mathematics in Physical Theory. *Philosophia Mathematica* 5: 210–27.

Plantinga, Alvin. 1974. *The Nature of Necessity*. Oxford: Oxford University Press.

Poincaré, Henri. 1952. *Science and Hypothesis*. New York: Dover.

——— 1953. *Science and Method*. New York: Dover.

Pollard, Stephen. 1990. *Philosophical Introduction to Set Theory*. Notre Dame, Ind.: University of Notre Dame Press.

Putnam, Hilary. 1967. The Thesis that Mathematics Is Logic. In R. Schoenman, ed., *Bertrand Russell: Philosopher of the Century* 273–303. London: George Allen & Unwin.

——— 1971. *Philosophy of Logic*. New York: Harper & Row.

——— 1979. What Is Mathematical Truth? *Mathematics, Matter and Method*, 60–78. Cambridge: Cambridge University Press.

——— 1980. Models and Reality. *Journal of Symbolic Logic* 45: 464–82.

Quine, Willard. 1938. On the Theory of Types. *Journal of Symbolic Logic* 3: 125–39.

——— 1959. *Methods of Logic*. New York: Henry Holt & Company, Inc.

——— 1960. *Word and Object*. New York: John Wiley & Sons.

——— 1961a. On What There Is. *From a Logical Point of View*, 1–19. Cambridge, Mass.: Harvard University Press.

——— 1961b. Two Dogmas of Empiricism. *From a Logical Point of View*, 20–46. Cambridge, Mass.: Harvard University Press.

——— 1963. *Set Theory and Its Logic*. Cambridge, Mass.: Harvard University Press.

——— 1966a. Carnap and Logical Truth. *The Ways of Paradox and Other Essays*, 100–25. New York: Random House.

——— 1966b. Posits and Reality. *The Ways of Paradox and Other Essays*, 233–41. New York: Random House.

——— 1966c. The Scope and Language of Science. *The Ways of Paradox and Other Essays*, 215–32. New York: Random House.

——— 1966d. Whitehead and Modern Logic. *Selected Logic Papers*, 3–36. New York: Random House.

——— 1969a. Epistemology Naturalized. *Ontological Relativity and Other Essays*, 69–90. New York: Columbia University Press.

Quine, Willard. 1969*b*. Existence and Quantification. *Ontological Relativity and Other Essays*, 91–113. New York: Columbia University Press.

Ramsey, Frank. 1931. The Foundations of Mathematics. *The Foundations of Mathsematics and Other Logical Essays*, ed. R. B. Braithwaite, 1–61. London: Routledge & Kegan Paul.

Rayo, Agustin and Yablo, Stephen. 2001. Nominalism Through De-nominalization. *Nous* 35: 74–92.

Reid, Constance. 1970. *Hilbert*. New York: Springer-Verlag.

Resnik, Michael. 1980. *Frege and the Philosophy of Mathematics*. Ithaca, N.Y.: Cornell University Press.

—— 1981. Mathematics as a Science of Patterns: Ontology and Reference. *Nous* 16: 529–50.

—— 1985. How Nominalist Is Hartry Field's Nominalism? *Philosophical Studies* 47: 163–181.

—— 1988*a*. Mathematics from the Structural Point of View. *Revue internationale de philosophie* 42: 400–24.

—— 1988*b*. Second-order Logic Still Wild. *Journal of Philosophy* 85: 75–87.

—— 1995. Scientific vs. Mathematical Realism: The Indispensability Arguments. *Philosophia Mathematica* 3: 166–74.

—— 1997. *Mathematics as a Science of Patterns*. Oxford: Oxford University Press.

—— 1998. Holistic Mathematics. In M. Schirn, ed., *The Philosophy of Mathematics Today*, 227–46. Oxford: Oxford University Press.

Robinson, Abraham. 1965. Formalism 64. In Y. Bar-Hillel, ed., *Logic, Methodology and Philosophy of Science*, 228–46. Amsterdam: North-Holland.

Russell, Bertrand. 1920. *Introduction to Mathematical Philosophy*. London: George Allen & Unwin.

—— 1956. The Philosophy of Logical Atomism. In R. C. Marsh, ed., *Logic and Knowledge*, 177–281. London: George Allen & Unwin.

—— and Whitehead, Alfred North. 1927. *Principia Mathematica*. Cambridge: Cambridge University Press.

Shapiro, Stewart. 1993. Modality and Ontology. *Mind* 102: 455–81.

—— 1997. *Philosophy of Mathematics: Structure and Ontology*. Oxford: Oxford University Press.

—— 1998. Review of J. Burgess and G. Rosen, *A Subject With No Object*. *Notre Dame Journal of Formal Logic* 39: 600–12.

—— 2000. *Thinking About Mathematics*. New York: Oxford University Press.

Sherry, David. 1999. Construction and Reductio Proof. *Kant-Studien* 90: 23–39.

Shoenfield, Joseph R. 1967. *Mathematical Logic*. Reading, Mass.: Addison-Wesley.

Singh, Simon. 1997. *Fermat's Enigma*. New York: Walker & Co.

Sober, Elliott. 1993. Mathematics and Indispensability. *Philosophical Review* 102: 35–57.

—— 1999. Testability. *Proceedings and Addresses of the American Philosophical Association* 73: 47–76.

Sober, Elliott. 2000. Quine's Two Dogmas. *Proceedings of the Aristotelian Society, Supplementary Volume* 74: 237–80.

Steiner, Mark. 1975. *Mathematical Knowledge*. Ithaca, N.Y.: Cornell University Press.

—— 1978. Mathematics, Explanation, and Scientific Knowledge. *Nous* 12: 17–28.

—— 1995. The Applicabilities of Mathematics. *Philosophia Mathematica* 3: 129–56.

—— 1998. *The Applicability of Mathematics as a Philosophical Problem*. Cambridge, Mass.: Harvard University Press.

Stenius, Erik. 1974. Sets: Reflections Prompted by Max Black's Paper, "The Elusiveness of Sets". *Synthese* 27: 161–88.

Stewart, Ian. 2001. *Flatterland: Like Flatland, Only More So*. Cambridge, Mass.: Perseus Publishing.

Suppe, Frederick. 1974. *The Structure of Scientific Theories*. Urbana: University of Illinois Press.

Suppes, Patrick. 1967. What Is a Scientific Theory? In S. Morgenbesser, ed., *Philosophy of Science Today*, 55–67. New York: Basic Books.

Tarski, Alfred. 1969. Truth and Proof. *Scientific American* 220: 63–77.

Thurston, William. 1994. On Proof and Progress in Mathematics. *Bulletin of the American Mathematical Society* 30: 161–77.

van Fraassen, Bas. 1980. *The Scientific Image*. Oxford: Oxford University Press.

van Inwagen, Peter. 1986. Two Concepts of Possible Worlds. In P. French, T. Uehling, and H. Wettstein, eds., *Midwest Studies in Philosophy* 11: *Studies in Essentialism*, 185–213. Minneapolis: University of Minnesota Press.

Vaughan, Herbert and Szabo, Steven. 1971. *A Vector Approach to Euclidean Geometry*. New York: Macmillan.

Vineberg, Susan. 1998. Indispensability Arguments and Scientific Reasoning. *Taiwanese Journal for Philosophy and History of Science* 10: 117–40.

Weyl, Herman. 1970. David Hilbert and His Mathematical Work. In C. Reid, ed., *Hilbert*, 245–83. New York: Springer-Verlag.

Wilder, Raymond. 1952. *Introduction to the Foundations of Mathematics*. New York: John Wiley & Sons.

Wittgenstein, Ludwig. 1953. *Philosophical Investigations*. New York: Macmillan.

—— 1958. *The Blue and Brown Books*. Oxford: Blackwell.

—— 1961. *Tractatus Logico-Philosophicus*. Trans. D. F. Pears and B. F. McGuinness. London: Routledge & Kegan Paul.

Wolff, Peter. 1963. *Breakthroughs in Mathematics*. New York: New American Library.

Zermelo, Ernst. 1967. Investigations in the Foundations of Set Theory I. In J. van Heijenoort, ed., *From Frege to Gödel: A Source Book in Mathematical Logic, 1879–1931*, 199–215. Cambridge, Mass.: Harvard University Press.

Index